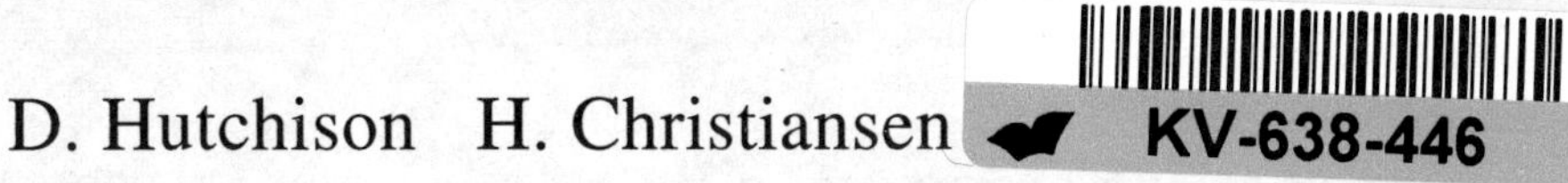

D. Hutchison H. Christiansen
G. Coulson A. Danthine (Eds.)

Teleservices and Multimedia Communications

Second International COST 237 Workshop
Copenhagen, Denmark, November 20-22, 1995
Proceedings

Volume Editors

David Hutchison
Geoff Coulson
Lancaster University, Computing Department
Lancaster LA1 4YR, Lancashire, United Kingdom

Henning Christiansen
Technical University of Denmark, Department ofTelecommunication
DK-2800 Lyngby, Denmark

André Danthine
University of Liège, Research Unit in Networking
Institut Montefiore, B28, B-4000 Liège, Belgium

Cataloging-in-Publication data applied for

Die Deutsche Bibliothek - CIP-Einheitsaufnahme

Teleservices and multimedia communications : Copenhagen, Denmark, November 20 - 22, 1995 ; proceedings / D. Hutchison ... (ed.). - Berlin ; Heidelberg ; New York ; Barcelona ; Budapest ; Hong Kong ; London ; Milan ; Paris ; Santa Clara ; Singapore ; Tokyo : Springer, 1996
(Lecture notes in computer science ; Vol. 1052) (... International COST 237 Workshop ; 2)
ISBN 3-540-61028-6
NE: Hutchison, David [Hrsg.]; European Cooperation in the Field of Scientific and Technical Research; 1. GT; European Cooperation in the Field of Scientific and Technical Research: ... International COST ...

The workshop was sponsored by a grant from the European Community

CR Subject Classification (1991): H.4.3, C.2, I.7.2, B.4.1

ISBN 3-540-61028-6 Springer-Verlag Berlin Heidelberg New York

Printed in Germany

Typesetting: Camera-ready by author
SPIN 10512724 06/3142 – 5 4 3 2 1 0 Printed on acid-free paper

Lecture Notes in Computer Science

Edited by G. Goos, J. Hartmanis and J.

Springer
Berlin
Heidelberg
New York
Barcelona
Budapest
Hong Kong
London
Milan
Paris
Santa Clara
Singapore
Tokyo

Teleservices and Multimedia Communications
Second COST 237 International Workshop
Copenhagen, 20-22 November 1995

Preface

The first COST 237 Workshop was held in Vienna one year ago. The theme of that workshop was **Multimedia Transport and Teleservices**, similar to the title of this COST Action and directly reflecting the research activities of the two COST 237 working groups. The workshop emphasis was on service and architectural aspects of distributed multimedia application support from the transport layer upwards.

During 1995, participation in COST 237 has increased, with more countries joining the Action. We now have a very good coverage, in terms of both countries and research teams within those countries, of the European basic research activity in multimedia communications. COST 237 consists of a network of researchers who collectively have knowledge and expertise of the whole area of multimedia transport and teleservices; most are in university research laboratories, though key industrial players are also directly involved. The participants all have their own national - and international - research activities, and through these they have connections with many other academic and industrial research teams in their own countries and indeed on other continents.

Following the success of the Vienna workshop, the COST 237 Management Committee decided to organise a second international workshop under the title **Teleservices and Multimedia Communications**. The intention was, I believe, twofold. First, we wanted to gather and present the main strands of research that have been developing within the technical scope of COST 237 during the past year; and, second, we strongly wished to maintain the momentum of the action by making sure that we came together as a community a year or so after the first workshop.

The process of putting the workshop together was different from that of the first event, for which we made a traditional call for papers. We decided, for the second workshop, to solicit papers from groups known to be active in our technical areas of interest; this was done via a fairly narrowly targeted call for papers. This resulted in a smaller, but generally higher quality, set of submitted papers than last year. Some 24 papers were successfully solicited in this way. After refereeing, 15 of these were accepted for presentation at the workshop, and assembled in five technical paper sessions as described below.

To start the workshop program, on Monday 20th November, we scheduled a substantial tutorial on the very important topic of MPEG compression techniques and standards which was presented by Heimo Mueller of Joanneum Research in Graz, Austria.

The five technical paper sessions, through Tuesday 21st and Wednesday 22nd November, reflected key areas of research on which the COST 237 community is concentrating its efforts: multipeer communications; teleservices and applications support; broadband network and transport services; quality of service (QoS) assurance; and multimedia support.

Multipeer communications is a development of the earliest COST 237 activity in new transport services, which concentrated originally on peer-to-peer services. Several of the participants have, since then, been working on the more challenging case of communications among multiple entities, including working together on the definition of a multipeer taxonomy within the Enhanced Communications Functions and Facilities (ECFF) project of ISO SC6/WG4. In the first of the five sessions, the three papers presented various aspects of this interesting research area; the third paper was the result of a specific collaboration between two COST 237 participants from different countries, involving an exchange of researchers in order to carry out the work.

Teleservices and applications support was one of the target areas for this Workshop. The topics of the three papers in the second session were: a distributed multimedia system based on the MHEG multimedia/hypermedia interchange standard; performance evaluation of a CSCW application called JVTOS (Joint Viewing Tele-Operation Service); and multimedia teleservices modelling using the OSI Application Layer Structure.

In the third session, the theme was broadband network and transport services, in which the three papers covered: a new concept called the multimedia partial order connection; distributed media scaling support for multimedia communications; and a new transport service over ATM developed in the RACE ACCOPI project.

The fourth session concentrated on QoS assurance: the firtst paper was from Australia, and described a hybrid scheme for application adaptation; the second and third respectively were about admission control for end-to-end distributed bindings, and admission control for deterministic and predictive services.

Session five presented three papers dealing with various multimedia support issues: first, conditional access and copyright protection for image communication; second, the design of a video compression algorithm that is particularly suited to interactive video; and third, a language for representing the requirements of streams for multimedia communications.

The program also contained two invited papers on subjects of strong interest to the COST 237 community, both on Tuesday 21st November: a keynote speech given by Augusto de Albuquerque of DG XIII at the European Commission and entitled “Multimedia in the ACTS Program”; and an invited paper from Giovanni Venuti of CSELT in Torino, Italy entitled “DAVIC Activities and Multimedia Applications”. On the afternoon of Wednesday 22nd November, rounding off the Workshop, there was a presentation and discussion session on "The Internet and Multimedia" led by Frode Greisen of UNI-C in Denmark.

Finally, I wish to record my appreciation of the efforts of many people in bringing about this workshop: to the Steering Committee for the original planning of the event; to my colleagues on the Program Committee, who have organised the acquisition, refereeing and selection of papers; to those authors who responded to our requests and submitted papers, some of which inevitably could not be accepted; and to the local Organizing Committee, particularly Villy Iversen and Henning Christiansen, for taking on the very tough, but all-important, task of hosting the second COST 237 International Workshop.

January 1996 David Hutchison

Program Chairman

David Hutchison	Lancaster University, UK

Steering Committee

André Danthine	University of Liège, Belgium (Chair)
Christophe Diot	INRIA, France
David Hutchison	Lancaster University, UK
Helmut Leopold	Alcatel, Austria
Villy B. Iversen	Technical University of Denmark, Denmark
Theodoros Bozios	INTRACOM, Greece
Manfred Weiss	Alcatel-SEL, Germany (Vice Chair)
Melanie Pralong	Swiss Telecom, Switzerland
Sandor Stefler	PKI, Hungary
Giorgio Ventre	University of Napoli, Italy
Brian O'Donohoe	Telecom Eireann, Ireland

Organising Committee

Villy B. Iversen	Technical University of Denmark, Denmark (Chair)
Henning Christiansen	Technical University of Denmark, Denmark

Contents

Group Support in Multimedia Communications Systems

A. Mauthe, G. Coulson, D. Hutchison and S. Namuye

Computing Department,
Lancaster University
Lancaster LA1 4YR, UK
email: (andreas, geoff, dh, namuye)@comp.lancs.ac.uk

Abstract: Communication among multiple entities is becoming more and more widespread in computing and telecommunications. Although many existing communications protocols and services do offer some limited support for multicast or group communication, the new requirements of multipeer applications make it difficult to find efficient and *comprehensive* solutions. In this paper we discuss the required characteristics of group services and survey the extent of the support provided by today's services and protocols. In addition, a brief outline of standardisation efforts in this area within ISO and ITU is given and selected examples of research projects which deal with different aspects of group communication are presented.

1 Introduction

The use of computers for interpersonal communication among multiple users has provoked a new trend called *group communication*. Although group communication is also used for distributed file systems, distributed data bases, fault tolerant systems, etc., the main requirement for group communication support comes from *interactive multimedia group applications*. In general, the main characteristics of interactive multimedia group applications are heterogeneous end-systems and networks, multiple senders and receivers (which change over time), high data volumes, high data rates and time-dependent data values.

A number of communication protocols and networks (e.g. ethernet, DQDB, IP, XTP) currently offer multicast (1:N) data transmission. Although (1:N) data transmission is the fundamental basis for group communications, it is by no means sufficient for the support of multimedia group applications. New management protocols and service specifications are also required to fully support conferencing and group communication applications. Although such protocols and services are currently under development in an experimental context, it is still too early to evaluate the extent to which these developments actually fulfil real application needs. International standardisation organisations, viz. ISO and ITU, are also addressing these issues and are currently discussing how group communication should be accommodated in new and existing standards. Last, but not least, there are various universities and research institutes working on different aspects of multimedia group communication and its support in the communication sub-system.

The paper is organised in six sections. In section two we briefly discuss the required characteristics of group services. The purpose of this discussion is to provide some criteria according to which group communication protocols and services can be assessed. Section three then presents a survey of existing group communication support provided in today's communication protocols and services. Standardisation efforts within ISO and ITU are outlined in section four. Some selected examples of research projects dealing in particular with communication aspects of multimedia group applications are discussed in section five. A summary and discussion of the results of the survey concludes the paper.

2 Characteristics of Group Services

Different aspects of groups applications, communication requirements and characteristics have been studied by various authors (cf. [7, 18, 34, 43]). However, the nature and extent of support required from the communication system is still controversial. Where and how certain services to support group communication should be provided is also an issue of discussion. There are no general answers to these questions; nor are there any generally recognised criteria by which proposed solutions can be assessed. Group communication is very complex and many aspects new and unique to this area have to be considered. Further, applications using group communication service are also developing rapidly. In the following subsections we outline some characteristics of group services which are demanded by emerging group applications.

2.1 Resource Characteristics

The resource requirements of group applications vary depending on the application, the kind of data, the group size, and the use of optimisation techniques (e.g. multicast, compression), etc. For distributed databases, for instance, a burst of data has to be transferred reliably at random intervals to multiple receivers. Interactive multimedia group applications, on the other hand, require the transmission of discrete and continuous media streams. The following example illustrates the resource requirements of a typical multimedia group application. We assume a car design process with eight participants at four different locations. Interpersonal communication is performed via audio and video. A shared CAD/ CAM and editing program supports the co-operation. An additional VCR component allows joint viewing of videos. Documents in common file format are exchanged among the participants. The document size is rather large since documents contain audio and video as well as text and graphics; however, this data has to be exchanged only once at the beginning of the communication. For stored MPEG video a data rate of 1.5 Mb/s is required. Data traffic arising from the shared CAD/ CAM and editing programs depends on the location of the data. We assume that copies of the documents are stored with each participant, and up-dates have to be transmitted between all participants, i.e. every participant has to send any operation on the shared object to every other participant. For interactive applications usually a data rate of 0.1-1 Mb/s is assumed. We assume an average value of 0.5 Mb/s. Audio and video in conference mode have to be transmitted among all participants, i.e. audio with 64Kb/s and video according to the H.261 standard for H0 ISDN at 0.32 Mb/s. The required data rate for this application with multicast but without any further optimisation is then:

1 * 1.5 Mb/s	= 1.5 Mb/s
8 * 0.5 Mb/s	= 4.0 Mb/s
8 * 0.064 Mb/s	= 0.5 Mb/s
8 * 0.32 Mb/s	= 2.56 Mb/s
	= 8.56 Mb/s

Further optimisation in encoding and a better utilisation of resources is needed to handle these requirements. They could be for instance reduced if certain characteristics of group applications could be exploited. For example the fact that usually not more that one participant talks at the time, or that not all participants operate simultaneously on shared documents could be used to reduce bandwidth requirements.

2.2 Topology Characteristics

Communication between multiple senders and multiple recipients (M:N) describes the most general communication topology, with unicast and broadcast as special cases. Within this general paradigm, a set of basic services characterised by topology can be defined. The *multicast* service is the most fundamental service in this set. In such a service one data copy is transmitted from a single sender to multiple receivers. Two types can be distinguished where multiple senders can transmit data to multiple receivers: first, the data from multiple senders is transmitted over multiple multicast data streams which can be distinguished at the receiver side *but* are managed as one entity; and second all data from the senders is received in one semantically meaningful data stream composed from the data of all senders. We call the former *virtual multipeer* and the latter *multipeer* services. The semantics of multipeer data streams, i.e. how data from different sources is combined, are not entirely clear. In multimedia group communication such a stream can be for instance generated through the use of *mixing filters* [49]. In this case it is not possible to distinguish the individual senders.
The last service type we consider is a *session* where different (virtual) multipeer and multicast connections are jointly established and managed.

2.3 Service Characteristics

Apart from their topology, group services are characterised by conditions according to which success or failure of connection establishment and data transfer can be determined. These specify the number and/ or identity of senders and receivers. We call these conditions *multipeer integrity conditions*. For sessions, multipeer integrity also refers to the number and/ or identity of connections.
Many data streams, especially multimedia data streams, are characterised by *QoS requirements*. However, the specification and provision of QoS in the multipeer case is much more complex than in peer-to-peer or peer-to-multipeer communication. Different senders will send at different rates, and different receivers will have different QoS requirements. In virtual multipeer connections a drop in the provided QoS has to affect all multicast streams equally and not just one. In multipeer connections the throughput might depend on the number of senders and may therefore vary. If mixing filters are employed, the overall throughput required for the mixed data stream is less than the sum of the individual data streams. In sessions it should be possible to trade QoS between different connections and to prioritise their QoS requirements. Thus, QoS support for group communication has to be flexible. QoS is not a one dimensional characteristic of a service, it is rather a service in itself that ensures data transmission with a certain degree of quality from multiple senders to multiple receivers over one or more related connections.

Table 1 summarises the characterisation of group services described above.

Communication Type	***Characteristics***
Multicast	• 1-to-N data delivery • QoS • integrity conditions
Virtual Multipeer	• M-to-N data delivery • M distinguishable data streams at receiver side • joint stream management • joint QoS • multipeer integrity
Multipeer	• M-to-N data delivery • 1 semantically meaningful data stream at receiver side • multipeer QoS • multipeer integrity
Session	• multiple multicast and M-to-N data streams • joint connection management • QoS trading and prioritising • session integrity

Table 1: Multipeer Communication Types

2.4 Other Aspects

In group communications, senders and receivers may join and leave the on-going communication, properties and QoS might change, etc. Thus, services offered to interactive applications must allow for dynamic changes.

There are different policies which can be applied to the communication or the management of groups. This adds additional complexity to services supporting group communication. A clear separation between policies and mechanisms is a good way to deal with this problem.

Performance and time aspects of any operation especially in interactive communication is crucial. These aspects have to be kept in mind when looking at any protocol or service supporting group communication.

3 Group Support in Communication Protocols and Services

3.1 Group Communication Support in Link Layer Protocols

Group and multicast communication is supported by a number of protocols at the data link layer. In LANs, group communication depends mostly on the ability of the underlying network to broadcast messages. This is for instance the case in a ring or bus topology. Message placed onto the network will eventually pass by all connected stations. Hence every station listening with the right address set can pick up a multicast message and pass it to the user. Usually, any station on the ring or bus can send a message to a group without any restriction. Most protocols standardised in IEEE 802.x (viz. token ring, ethernet, DQDB) provide multicast. In order to receive multicast messages on these networks a host has to set the multicast address on the network interface. Ethernet multicast addresses are identified by a 1 in the low-order bit of the high-order byte in the 48-bit Ethernet address. The Token-Ring adapter

allows the setting of only one group address which is identified by setting bit 0 in byte 0 of the address. Traditional ethernet is not suitable for the guaranteed transmission of continuous media data. However further developments like *isochronous ethernet* [42] give QoS guarantees. Token ring is more suitable for the support of continuous media than ethernet but is still far from ideal. DQDB offers asynchronous, synchronous and isochronous services.
In point-to-point networks like ATM the ability to broadcast or multicast messages depends on whether or not copying functions in the switches are available. Different mechanisms to route multicast and broadcast messages in ATM networks have been proposed, e.g. centralised repetition in a copy network, copying before the switch by a serial copy function and distributed repetition in the switch fabric. The QoS provided by ATM is considered sufficient for multimedia applications.

3.2 Group Communication in the Network

3.2.1 Routing Schemes

To support group communication in the network messages have to be copied and routed to all receivers. To provide multicast in WANs existing routing algorithms have to be extended. Commonly, group membership protocols are used to enable routers to learn about members on their sub-network. There are different methods to propagate routing information among multicast gateways. MOSPF for example is an extension of the link state unicast protocol OSPF (Open Shortest Path First) [9]. Changes in group membership are first detected by a router directly attached to the sub-network containing the joining/ leaving participant. The change is then broadcast by this router to every other router in the same routing domain. The state of the domain topology is kept in the (unicast) link-state protocol at every router. A multicast message is forwarded according to a shortest path tree which is determined using group membership and topology information [11].
Inter-Domain Policy Routing (IDPR) also uses link state routing information but, in contrast to MOSPF, source specified packet forwarding is used [36]. Another technique is used by the Distance-Vector Multicast Routing Protocol (DVMRP) which uses variants of reverse path routing to erect a multicast tree. In this scheme a router forwards the first packet sent to a group to all outgoing links. A router on a sub-network without any members sends a prune message towards the source of the packet. This message prunes the branches where no group members reside, hence a shortest path tree that has only leaves containing members of the group is erected. To include new potential members, pruned branches "grow back" and are pruned again if there are still no members. Special mechanisms to prevent packets from looping are employed [11]. Yet another way to establish multicast trees is the Core Based Tree (CBT) scheme. In CBTs there is one router defined as the core from which all branches emanate. Receivers that wish to participate have to find the tree or the core and attach themselves to it [2]. Protocol Independent Multicasting (PIM) is, like CBT, based on the receiver initiated philosophy. Two different type of multicast trees, shared and source-specific multicast trees, can coexist [11, 36].

3.2.2 Internet Multicast Schemes

The Internet Protocol (IP) provides a scheme for multipoint data delivery. IP multicast groups are dynamic, i.e. a host can join or leave a group at any time. Further, a host can be a member of several groups. Groups in IP are addressed using class D addresses which are especially reserved to address multicast groups. IP class D addresses can be easily mapped onto link layer group addresses like ethernet multicast addresses. Some IP multicast group addresses are permanently assigned by the Internet authority to groups which always exist (even if they currently have no members). Other groups are transient, i.e. they are created when needed and discarded when the count of members reaches zero. If a multicast group spans multiple networks, group membership information is communicated by the Internet Group Management Protocol (IGMP). A host that wishes to join a group issues an IGMP request [9].

MBONE is a virtual network that provides multicasting facilities on top of the Internet [13]. It is composed of a number of multicast capable networks. There is one host in each of these networks that runs a multicast routing demon called *mrouted*. The routing algorithm commonly used is DVMRP. Different mrouted demons are connected through so-called tunnels. A multicast packet sent out by a client is picked up by the local mrouted demon, and it chooses the tunnels into which the packet has to be sent out according to its routing table. The mrouted on the other end of the tunnel receives the packet, examines its routing table and forwards it to any client in its sub-net that has subscribed to the group and if necessary to other tunnels. Figure 5 shows a schematic MBONE topology. Some pruning for MBONE is implemented. MBONE also provides a wide variety of video, audio and whiteboard applications and tools such as vat, nevot, nv, wb, etc. [13].

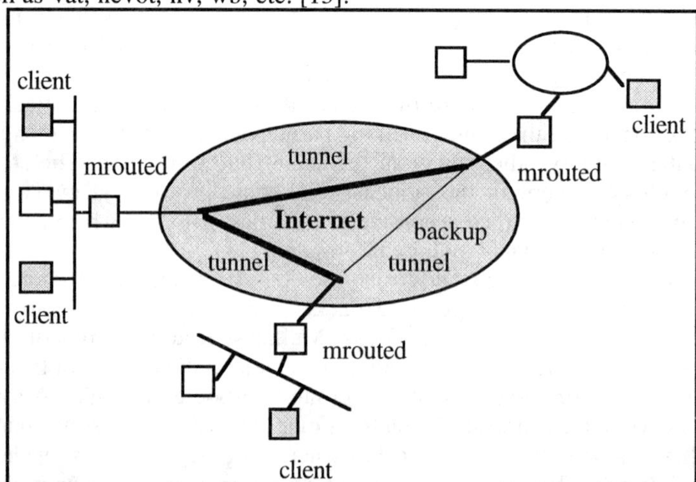

Fig 1: MBONE Topology

ST-II is a stream oriented protocol, designed to provide end-to-end guarantees across the Internet. The offered service is connection oriented; in the multicast case a multi-destination connection is established along a directed multicast distribution tree. During the establishment phase the necessary network resources are reserved. This has to be agreed to a common QoS among all involved entities, i.e. the entire stream has the same QoS. New receivers have to establish a connection with the source. ST-II also allows several streams to be clustered into *stream groups* [12, 46].

RSVP is a resource reservation set-up protocol, designed as a companion protocol to IP [50]. RSVP messages do not carry any application data; in fact, RSVP just controls the packet transmission of IP. Multicast in RSVP is built around existing and future multicast routing protocols. A host first has to send an IGMP message to join a multicast group. Multiple senders can send to the group. To reserve resources a receiver determines its QoS and initiates a *reservation* request. The required resources

along the sub-net(s) from the receiver to the sender are reserved by intermediate RSVP agents. Reservation messages are propagated towards the sender until an existing distribution tree with sufficient resources allocated for the sender group is found. The reservation model has two components, a resource allocation component where the amount of resource is specified and a packet filter component which selects which packets can use the resource. With a *wildcard* filter no source specific filter is required. When a source specific filter is required the receiver might specify a fixed set of senders (*fixed-filter*). With a *dynamic-filter* this set may change [35, 50].
At present the IP protocol architecture is undergoing a number of major changes. All proposals for new or modified protocols are subsumed under IP next generation (*IPng*) or IP version 6 (*IPv6*) [23]. One of the biggest changes is the step from 32 bit addresses to 128 bit addresses. New IP addresses are different in structure, especially as far as multicast addressing is concerned. A multicast address is identified by 1 in the first 8 bits of the address followed by the 4 bit `flgs` field, the 4 bit `scop` field and the 112 bit `group ID`. Group identifiers can be either permanently assigned or transient. The scope of the group is limited, i.e. transient group identifiers are only valid within a given scope. Permanent group identifiers are globally meaningful but their scope can also be bounded. The scope of an address can be intra-node, intra-link, intra-side, intra-organisation, intra-community and global. The 112 bits `group ID` identifies either a permanent or transient group identifier within a given scope [22]. *Cluster addresses* are used to reach the nearest of a set of so called boundary routers of a cluster of nodes which are identified by a common prefix. This allows one to chose the provider which should carry the data [23]. In IPv6 the Internet Group Membership Protocol (IGMP) has been absorbed into the Internet Control Message Protocol (ICMP). There are two classes of ICMP messages, error messages and informal messages. Three informal messages to exchange group information are defined, Group Membership Query, Group Membership Report and Group Membership Termination [10]. In IPv6 a 28 bit Flow Label field in the header is introduced to support QoS. Traffic is handled by the IP router according to the traffic class in the 4 bit traffic class sub-field of the Flow Label [21]. This is particularly useful for data flows such as continuous media streams with QoS specifications.

3.3 Group Support in End-To-End Protocols

3.3.1 Internet Protocols

At the transport layer a group consists of sending and receiving processes. Most traditional transport protocols like OSI-TP4 and traditional TCP do not support any multipeer data communication. New protocols and protocol enhancements do support multicast at the transport layer to some extent. In fact, some TCP implementations already support multicast data delivery on top of IP multicast. A proposal to support *reliable* multicasting on top of IP through a so called Single Connection Emulator (SCE) can be found in [45]. The SCE resides above IP and can be used by TCP. It mimics a single destination network layer interface to TCP. IP multicast is used to provide the necessary multicast functionality.
The Real-Time Transport Protocol (RTP) offers multipeer data transfer (if supported by the underlying network) for data with real-time characteristics. It was designed for the Internet and resides on top of protocols such as UDP, TCP, OSI TP1 or TP4 and ST-II [39]. For multicast data transmission a multicast address is obtained by the initiator which is distributed out-of-band to the participants. RTP packets are received at the port of the underlying transport service, and encapsulated in packets of the

underlying transport system, e.g. if transmitted over UDP, the RTP header and data are contained in a UDP packet. The RTP header itself contains timing information and a sequence number which enables the receiver to reconstruct the timing seen by the source. RTP does not provide mechanisms to ensure timely delivery or to provide QoS guarantees. It also does not guarantee delivery or prevent out-of-order delivery. RTP is accompanied by the Real-Time Control Protocol (RTCP). This provides information about participants in on-going sessions. Using RTCP, each participant sends local information periodically to all other participants as options within RTP packets. RTCP does necessarily have to be used to communicate over RTP [40].

3.3.2 XTP

XTP (Xpress Transfer Protocol) was defined as a protocol comprising network and transport layer functions. A connection-oriented one-to-many multicast, if supported by the underlying network, was proposed. [37, 47]. Four techniques to improve XTP's group communication support were suggested: the *bucket algorithm*, *slotting*, *damping* and *cloning*. The bucket algorithm was proposed as control mechanism to determine how and when to up-date status information. Control information processing is timer based, and 1-reliability can be achieved. Slotting and damping are mechanisms to reduce the number of control packets. Slotting forces a receiver to wait a random time span before generating a control packet. In case of an error a receiver sends control messages to the sender and all other group members. Other receivers damp their own control messages when they contain a larger value than the one of the received control packet. Error recovery is done by go-back-n. With *cloning* a multicast transmitter can group several slave contexts which depend on a single master context. This is a way to achieve concentration where multiple senders can send to the same multicast address.
Originally no group management was defined in XTP 3.x [37]. The new version, XTP 4.0, is defined solely as a transport protocol. Multicast is now an integral part of the protocol and not simply an addendum. XTP multicast provides a simplex data flow from one transmitter to an arbitrary number of receivers. Control procedures defined for unicast apply to multicast as well. Group management to determine membership in the receiver group is also defined. This allows the support of different reliability semantics[1], i.e. k, full reliability, and to determine a sub-set of receivers who have to receive a message. In general, XTP 4.0 supports go-back-n or selective retransmission as its error recovery mechanism. In the reliable multicast case *all* data reported missing by any receiver has to be re-transmitted to the group with either error recovery scheme. Slotting is still used to distribute the number of control messages over time. As in the unicast case, synchronisation handshakes are used for information status up-dates [48]. No data is transferred during a synchronisation handshake.

3.3.3 ITU-T T.120 Series

The Multilayer Protocol (MLP), also known as the T.120 series, recently developed within the International Telecommunication Union - Telecommunication Standardisation Sector (ITU-TSS), is designed to provide support for multimedia applications and services in a heterogeneous environment [3, 8]. The MLP can be

---

[1] Note, reliability or group reliability in XTP is comparable to the concept of multipeer integrity defined in section 2.3.

used with various different types of networks, viz. Public Switched Telephone Networks (PSTN), Integrated Services Digital Networks (ISDN), Circuit Switched Digital Networks (CSDN), Public Switched Digital Networks (PSDN) and Broadband-ISDN (B-ISDN). Further, support for other networks like LANs, TCP/ IP, IPx on LANs, and ATM are under consideration or being developed [6]. Multipoint service is provided on top of reliable point-to-point links in a network independent manner. Groups of point-to-point transport connections are mapped together and build a Multipoint Communication Service (MCS) domain. Each domain represents effectively an independent "conference" or group activity. Within a domain different logical channels can provide (1:1), (1:N), (N:1) data delivery.

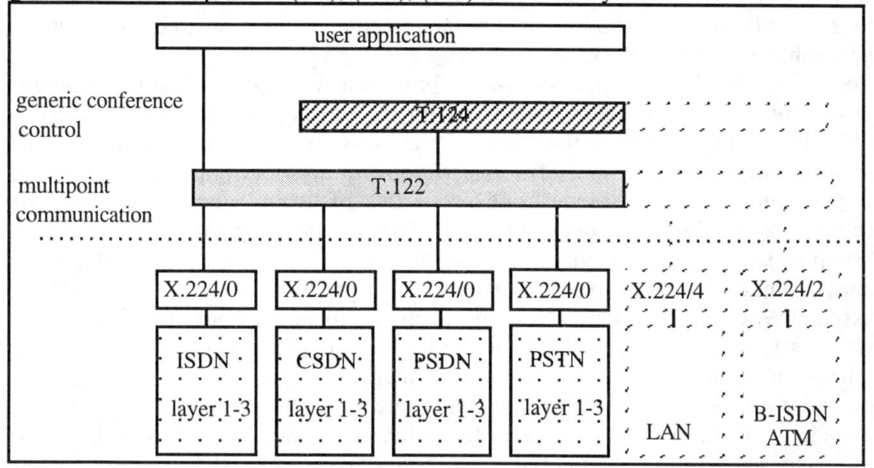

Fig 2: T.120 Communication Structure

The Generic Conference Control (GCC) (T.124) located above MCS provides services for managing multipeer conferences including conference establishment and termination, joining ongoing conferences, managing changes during sessions, conference database management, application database management, bandwidth control and remote actuation. Real-time and non real-time aspects are co-ordinated by the GCC. Apart from these protocols four application internetworking protocols are suggested: An Audio-Visual Control standard (T.AVC), a Binary File Transfer protocol (T.MBFT), a protocol to support shared whiteboarding and exchange of still images (T.SI) and a facsimile service. Figure 2 shows the T.120 communication structure.

4 Group Communication Standardisation

The two main standardisation bodies ISO (International Organisation for Standardisation) and ITU (International Telecommunication Union) have also addressed issues of group communication. The two groups have differing opinions on some issues. However, both groups try to overcome these differences and to produce compatible standards. This effort resulted in a joint meeting in Geneva at which it was decided to merge the existing proposals. The result of this meeting is reported in [26, 30]. However, no new documents or new versions of any of the discussed documents were presented. Therefore we only consider the latest version of the discussed documents and not the meeting reports.

4.1 Group Communication Standardisation in ISO

Group communication issues are within the remit of various committees and working groups in ISO. Transport issues are discussed within SC6/ WG4. This working group was up to now also the most active one as far as group communication is concerned. Work items which address the problem of group communication include the Enhanced Communications Transport Service (ECTS) and Enhanced Communications Transport Protocol (ECTP) specification and the definition of a Multipeer Taxonomy. The Multipeer Taxonomy was developed within the Enhanced Communication Functions and Facilities for OSI Lower Layers (ECFF) project of SC6/ WG4. It is currently available in its fourth draft version [25]. This version was passed on to SC21 as a liaison contribution, because architectural work of this kind is regarded as being in the province of SC21.

The aim of the Multipeer Taxonomy is to provide a layer independent terminology for group communication and a framework for the development of multipeer services [25]. The main concepts defined in this document are the concepts of *group*, *group membership*, *population characteristic*, *group association* and *conversation*. Population characteristics describe properties of group membership like *static* (population of the group does not change) vs. *dynamic* (population may change) group and *definite* (members can be determined) vs. *indefinite* (membership can not entirely be determined) group. The concept of group association is defined as an association established between group members for the purpose of data exchange. Group associations are qualified by a set of characteristics and conditions; particularly interesting is the definition of active-group integrity (AGI) which specifies conditions on the number and/ or identity of communicating participants. Related to group association is the concept of conversation which is defined as the basic component of a group association. Four types of conversations exist: simplex multicast conversations (1:N), duplex multicast conversations ((1:N) followed by (N:1)), unicast conversations (1:1) and N-plex multicast conversation (N:N) [25].

The taxonomy also includes a discussion of the general topology of group associations, i.e. the kinds of integrity applying to the whole group association, elasticity conditions on the communication (such as static vs. dynamic, closed vs. open communication), etc. A service model is also introduced. Finally, issues like membership control, ordering relationships, QoS, naming and addressing and group management are addressed.

The first draft of the ECTS was tabled at the SC6/ WG4 meeting in Beppu, March 1995 [24]. It is an attempt to specify a transport service providing support for group communication. Most concepts defined in the Multipeer Taxonomy can be found in the model of ECTS. A transport group conversation (N:N) is the generic type of a transport conversation, all other types are degenerates. Transport associations are associations among a group for the purpose of transferring data. They may contain one or more transport conversations. Naming and addressing in this model is more complex than in traditional transport services since the new concept of transport conversation endpoint is required. In the current version conversations terminate on a TSAP of a transport association. AGI conditions are specified on transport association and are not negotiable. QoS specification and negotiation is described in detail. QoS is negotiated during the *bind* of a potential participant and not during the join of an association. Issues like QoS relationships between different conversations and prioritising of QoS requirements are not addressed. Sequence primitives and time sequence diagrams for the service are specified.

4.2 Group Communication Standardisation in ITU

ITU-T SG7 Study Group 7 (International Telecommunication Union - Telecommunication Standardisation Sector) recently introduced three documents concerned with group communication. The Multipeer Communication Architecture (MPCA), also known as X.multi [27], a framework document, X.tms [28] which is concerned with multicast transport service issues, and X.nms [29] which deals with network layer multicast. X.tms and X.nms are almost identical documents, using the same structure and wording. We therefore only discuss X.tms.

The aim of X.multi is to provide a common basis for the co-ordination of standards developments for the purpose of multipeer communications [27]. It defines concepts and terms which should be used for conformance and compliance reasons by all standards which provide a multipeer service. Multipeer is defined as communication among more than two participants (M:N). Messages are transmitted either by multicast or broadcast. The only other topologies recognised are multicast as a single service invocation and broadcast. An (N)-group association is defined as a co-operative relationship among (N)-entity-invocations, whereas an (N)-group-connection is an association request by an (N+1)-entity for the transfer of data among a group of (N+1)-entities. There is no restriction on the topology. The general topology consisting of all active group members who are able to send and receive information at the same time, i.e. the general topology is (N:N). The terms *group*, *population characteristic* (static vs. dynamic, known vs. unknown), *communication discipline* (send only/ receive only, send/ receive), *dialogue control*, and *reliability characteristics* are also defined. Further, X.multi contains a section on naming and addressing and on ordering. Active-group integrity (AGI) is specified as conditions which have to be met in order to continue with the data transfer. The document also contains a section where the seven OSI layers and their capabilities in the multipeer case are described.

X.tms defines the notion of group and its meaning in the context of communication. Further, definitions for group connections and ordering semantics are given. The multicast transport service is characterised by the following features: binding of transport multicast (M:N) group connections, dynamic join and leave according to integrity and QoS conditions, transferring of data in multicast to all participants, receiver flow control, dynamic leave and unbinding. A model of multicast transport connections as an (N:N) service is introduced. QoS characteristics including *population* (the way multicast groups are established and maintained), *reliability* (degree of error control) and *transmission characteristics* (e.g. throughput flow control, delay) are defined. No further specification of how QoS negotiation is performed is defined. In contrast to ECTS the QoS negotiation mainly takes place during the join of a transport connection. Sequences of multicast primitives at a TSAP are outlined. Further, the different phases of a group connection from the allocation phase (including bind, unbind, join and leave), to the data transfer phase are briefly discussed.

5 Group Communication Research

In recent years the research community has become more and more interested in the area of group communication. The problem of multimedia group communication is tackled by various research groups at different universities and research institutes. The context in which the research is carried out differs from project to project, depending on overall project goals and objectives of the organisation. Since group communication itself is a multifarious problem there are various different ways to address issues related to group communication. A full discussion of the research currently carried out in this field beyond the scope of this paper. We are aware of several other groups currently working on different aspects of group communication (e.g. [5, 20, 33, 41]). In this section we introduce some of the work relevant for multimedia group communication concentrating on work dealing with resource reservation aspects and end-to-end communication services.

Research in the ***Tenet Group*** focuses on the design and development of real-time communication services and network support for continuous media applications. A suite of protocols which provide QoS guarantees for real-time packet switching networks has been designed and implemented. Earlier work in the area of group communication in the Tenet Group was concerned with *channel groups* and *half-duplex real-time channels*. Channel groups are abstractions by which the user can describe relationships between channels to the network [16]. These relationships would for instance allow resource sharing among related channels, could be considered during routing, indicate establishment relationships, etc.. Multipoint-to-multipoint connections can be emulated using channel groups. A further concept introduced in this context is that of *target set*. A *target set* is a set of tuples indicating the address and QoS specification of every member in the set. Senders establish multicast channels to a target set. A (M:N) semantic can be achieved if multiple senders address the same target set. [14].

The concept of half-duplex real-time channel for multiparty interactive multimedia applications was first discussed in [44]. This abstraction exploits the fact that in applications such as lectures or seminars usually only one participant is allowed to speak at a time. Hence it is sufficient to reserve resources for only one communication channel, if the network can offer guarantees in both directions of the same multicast connection. Application level mechanisms like floor passing ensure that there is only one sender at a time.

More recent work concentrates on *resource sharing*. With resource sharing, related connections can share resources in a controlled manner. In this case a network client specifies how traffic from related connections are multiplexed. The client states the maximum aggregated resource requirements for the group and actually indicates to the network when to use the group rather than individual specifications. During admission of a new participant the group resource allocation is used. There is no admission test needed to admit additional members after the sharing threshold is reached [15]. *Multicast real-time channels* are now part of the Tenet Suite 2. The design of Suite 2 is connection oriented, most of the changes for multiparty communication affect channel establishment. At the moment channels are established from the source to all destinations. During the communication multicast packets have to be copied in the intermediate nodes; the above described resource sharing mechanism requires a modification of the scheduling mechanism [4].

The ***Multipeer Broadband Transport Service*** is a transport protocol developed within the European RACE research programme. It was especially designed for distributed multimedia systems and is considering the special needs of group communication applications [19]. The two main problems addressed in this context are multipeer data transmission and relations between different data types (e.g. audio and video) transmitted over parallel transport connections. The provision of a variable multipoint transmission topology, QoS provision and negotiation, delivery semantics, voting and notification are the main design criteria of the protocol. Apart from the usual simplex and duplex unicast transport connections, different types of multipoint transport connection (MP-TC) are introduced. The basic element of any multipoint transmission topology is a *connection*. Different, related connections are grouped together into *Transport Calls* (TL) which are jointly addressed at a *Transport Call End Point* (TLEP). Figure 6 shows how the different concepts are related. The transport protocol is fully specified including a list of service primitives, time sequence diagrams and state transition diagrams. Apart from the transport protocol a group management service to manage user groups, their members and properties are defined.

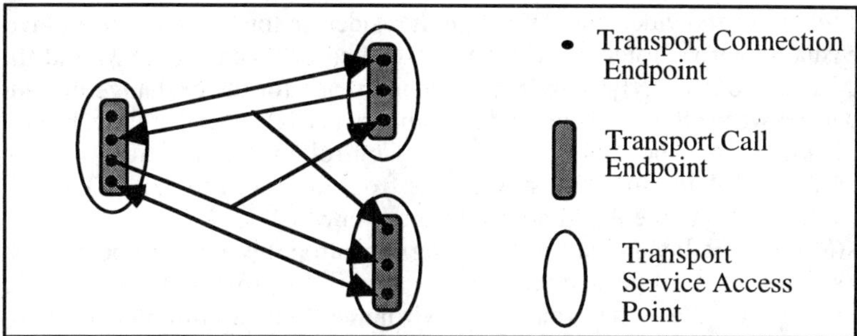

Fig. 3: Multipoint Transport Topology

A ***Multi-user Communication Service*** supporting multimedia applications is proposed in [18]. This service offers basic and general facilities to support communication among multiple participants. A *call model* is introduced in which a call is a dynamic association between service users and service provider. Different functions and service elements to operate on calls are defined. Apart from data exchange, other interactions such as synchronisation between streams are supported. Service elements to establish a call, to add users, to add media, to remove users and to release a call are provided. Further service elements to attach and detach users to *media* are also specified. Communication can only take place among users who are attached to a common medium. A generic service element supports negotiation to change the state of a call. Users can send, receive, and send and receive data. One user is the owner of a call and is involved in all negotiations. Additional users have to be invited to join an on-going call. All service elements apart from detach media, remove media, remove user and release call are confirmed. As soon as the service provider determines that a service element will be successful a confirm is sent to the initiator and all involved users who have already responded. Notification primitives are sent to all the users affected by state changes but who are not involved in the decision process. This is an extension of the standard confirmed and unconfirmed service elements in the OSI-RM. The service was defined independent of any implementation. It merely defines required capabilities of the communication system to support multimedia group applications.

The BERKOM (Berliner Kommunikationssystem) project is a joint Broadband ISDN (B-ISDN) trial project between various different computer manufacturers, research institutions and universities. One of the three working areas in the second phase of the project was the ***Multimedia Collaboration Service*** (MMC) which supports joint working in a distributed environment, i.e. it enables the sharing of applications and provides audio-visual conferencing tools. The different components of MMC are a Conference Manager (CM) to administer and run conferences, Conference Interface Agents (CIAs) which serve as user interfaces, a Conference Directory (CD) to store information about user groups and conferences, an Application Sharing Component

(ASC) for joint viewing and accessing of applications, an Audiovisual Manager (AVM) and the Audiovisual Component (AVC) [1]. The AVM offers a service to share audio and video data; it enables the establishment of multiple audio/video streams between all conference participants and between multimedia applications and all conference participants. All relations in the audio-visual communication system are controlled by the AVM [32]. The AVC manages all endpoints of data streams in the end systems. Communication streams are multicast streams (1:N). There is exactly one AVC per end system. The Audio Visual Data Exchange Protocol (AVXP) is used for audio and video data exchange. It resides on top of the transport layer. The Audiovisual Control Protocol (AVCP) is executed between the AVM and the CM. The CM opens one (N:N) group in every conference for the exchange of audio and video data between all participants and an additional (1:N) group for each shared audio-visual application. The Source and Sink Control Protocol (SSCP) is executed between the AVM and the AVCs. Requests from the AVM to the AVCs and event reports from AVCs to the AVM are exchange over SSCP [32].
MICE (Multimedia Integrated Conferencing for Europe) is a project between various partners from six European countries. Its aim is to provide a service to enable internetworking in a heterogeneous environment via multimedia conferencing technology [38]. The MICE technology allows conferencing between participants located in conference rooms which can accommodate up to ten people, and at workstations. The proposed services are multi-way shared workspace, multi-way video, multi-party voice, and multiplexing and conference management. Audio and video data are treated independently because of the higher delay sensitivity and smaller bandwidth requirements of audio. A core element in the MICE architecture is the Conference Multiplexing and Management Centre (CMMC). Its main function is to allow communication between users at different networks with different facilities, e.g. ISDN, EBONE, EMPB. Further, it provides a relay between different coding standards, multiplexes audio and video, reserves, controls and manages resources, ensures multicast of shared workspace data, etc. [17, 31]. A conference control system based on the CAR conference control system offers functions to create, delete, join and leave a conference, floor control and control of conference video channels inter alia. Practical experience with MICE could be gathered during various conferences and a series of seminars for researchers and students. Seminars were multicast using MBONE.

6 Conclusion

Various services and protocols provide multicast and group communication support to some extent. Active research in this area addresses a multitude of problems. National and international standardisation bodies address the problem of group communication as well and try to produce commonly acceptable standards. Because of the complexity of the problem no straight forward solutions can be found. It is even difficult to find common measures according to which the proposed solutions could be assessed. Some criteria and aspects especially relevant for interactive multimedia group applications were discussed in this paper. In particular, resource requirements, communication topology, multipeer integrity and QoS as well as efficiency and performance issues must be considered.
At the lower layers of the communication architecture support for multicast is offered by many link layer protocols. Unfortunately, the QoS support offered by the most common protocols (viz. ethernet and token ring) is insufficient for multimedia

communication. In the network different mechanisms to support multicast and group communication exist. Various protocols proposed for the Internet support group communication at different levels, e.g. RSVP, ST-II and RTP. In the new version of IP, IPv6, multicast and multimedia data flows are supported to an even larger extent. However, IP was originally not designed for multimedia data communication and many IP networks can still not cope with the bandwidth requirements of digital audio and video. MBONE is frequently used to transmit audio/ video conferences or other special events. Many applications running on top of MBONE are actually dissemination services to a large user group, but little interaction takes place. The quality of the transmitted data is often low and does not satisfy every user.

Multimedia group applications have very high resource requirements which might even exceed the capabilities of emerging high-speed networks. However, there is also a big so far mainly unutilised potential to save resources through mechanisms like efficient multicast and resource sharing. Proposals for mixing filters in RSVP are a step towards exploiting the characteristics of group communication for a better resource utilisation. The work of the Tenet Group on resource sharing should also be mentioned in this context.

One of the biggest problems faced in the design of communication protocols and services is where and how to deal with the additional complexity placed onto the communication system by group applications. For instance, where and how can it be best determined if data transmission is successful or not; or, where and how can multiple connections be managed most efficiently. At the moment all this is very often left to the application. Transport protocols such as XTP 4.0 offer only (1:N) multicast. The Multilayer Protocol of ITU-TSS provides support for conference-like group communication. It resides on top of peer-to-peer connections and was especially developed for a telecommunications environment. It is therefore very difficult to use in an open, heterogeneous environment. The Multipeer Broadband Service and the Multi-user Communication Service are both specified to support group communication over multiple (1:N) channels. They reduce complexity for the application by dealing with multiple connection management at the transport layer. In turn, they are themselves rather complex. Moreover, they are defined in a very abstract way, thus their practical relevance is limited. BERKOM's MMC and the MICE project deal with a whole range of services necessary for multimedia conferencing. It can be observed that all the above services were developed with specific project requirements in mind and the extent to which they can be regarded as general solutions is unclear.

The current discussion about group communication standardisation reflects the diversity of opinions in this area. Whereas ITU and ANSI support with X.tms and X.multi a solution which is deliberately simple, ISO proposes a more comprehensive solution to the problem of group communication. However, there are a number of points which are still controversial or not at all clear. Among those is the detailed specification of an (N:N) service, QoS issues and addressing. Further work and discussion is required resolve the contested questions and to show the feasibility of the proposals.

Many problems in multimedia group communications are still unsolved. Further work is needed, for example in routing, resource management, and in QoS assurance. In multimedia group communication the problem of scarce resources and of their optimal and efficient use is particularly critical. And last, but not least, the ideal solution to deal with the additional complexity placed onto the systems architecture by interactive multimedia group applications has still to be found.

Acknowledgements

For research support the authors are indebted to the Engineering and Physical Sciences Research Council of Great Britain. The work on this project is funded by the EPSRC project GR/J47804 on Multimedia Group Communications Services.

References

1. M. Altenhofen, J. Dittrich, R. Hammerschmidt, T. Käppner, C. Kruschel, A. Kückes and T. Steinig: "The BERKOM Multimedia Collaboration Service", *Proc. 1st ACM Multimedia Conference*, Los Angeles, 1993, pp. 457-463.
2. A. J. Ballardie, P. F. Francis and J. Crowcroft: "Core Based Trees", *Proc. ACM SIGCOMM*, 1993.
3. D. Beaumont and J. Boucher: "ITU-T T12x Recommendations and Their Relevance to an API for ATM Networks", BT-Laboratories, 1994.
4. R. Bettati, D. Ferrari, A. Gupta, W. Heffner, W. Howe, M. Moran, Q. Nguyen and R. Yavatkar: "Connection Establishment for Multi-Party Real-Time Communication", *Proc. 5th International Workshop on Network and Operating System Support for Digital Audio and Video*, Durham, New Hampshire, 1995, pp. 255-265.
5. E. Biersack and J. Nonnenmacher: "WAVE: A New Multicast Routing Algorithm for Static and Dynamic Multicast Groups", *Proc. 5th International Workshop on Network and Operating System Support for Digital Audio and Video, NOSSDAV '95*, Durham, New Hampshire, 1995, pp. 243-265.
6. J. Boucher: "Draft Recommendation T.120: Transmission Protocols for Multimedia Data", Geneva, 1994.
7. R. Braudes and S. Zabele: "Requirements for Multicast Protocols", RfC: 1458, Internet Networking Group, TASC, 1993.
8. W. J. Clark and J. Boucher: "Multipoint Communications - the Key to Groupworking", *BT Technology Journal*, Vol. 12, No. 3, 1994, pp. 72-80.
9. D. E. Comer: "Internetworking with TCP/IP; Principles and Architecture." Vol I, Prentice Hall, Inc., Engelwood Cliffs, 1991.
10. A. Conta and S. Deering: "ICMP for the Internet Protocol Version 6 (IPv6)", Internet-Draft, draft-ietf-ipngwg-icmp-01.txt, Network Working Group, 1995.
11. S. Deering, D. Estrin, D. Farinacci, V. Jacobsen, C.-G. Liu and L. Wei: "An Architecture for Wide-Area Multicast Routing", *Proc. SIGCOMM '94*, London, ACM Press, 1994, pp. 126-135.
12. L. Delgrossi, R. G. Herrtwich, C. Vogt and L. Wolf: "Reservation Protocols for Internetworks: A Comparison of ST-II and RSVP", *Proc. 4'th International Workshop on Network and Operating Systems Support for Digital Audio and Video*, Lancaster, 1993, pp. 199-208.
13. H. Erikson: "MBONE: The Multicast Backbone", *Communication of the ACM*, Vol. 37, No. 8, 1994, pp. 54-60.
14. A. Gupta, W. Heffner, M. Moran and C. Szyperski: "Network Support for Real-Time Multiparty Applications", *Proc. 4th International Workshop on Network and Operating Systems Support for Digital Audio and Video*, Lancaster, England, 1993, pp. 37-40.

15. A. Gupta, W Howe, M. Moran and Q. Nguyen: "Resource Sharing for Multi-Party Real-Time Communication", *Proc. INFOCOM '95*, 1995.
16. A. Gupta and M. Moran: "Channel Groups: A Unifying Abstraction for Specifying Inter-stream Relationships", Technical Report, TR-93-015, Tenet Group, Computer Science Division, University of California at Berkeley & International Computer Science Institute, Berkeley, 1993.
17. M. J. Handley, P. T. Kirstein and M. A. Sasse: "Multimedia Integrated Conferencing for European Researchers (MICE): Piloting Activities and the Conference Management and Multiplex Centre", *Proc. JENC '93*, Rare, Netherlands, 1993, pp. 18-31.
18. G. J. Heijenk, X. Hou and I. G. Niemegeers: "Communication Systems Supporting Multimedia Multi-User Applications", IEEE-Network, Vol. 8, No. 1, 1994, pp. 34-44.
19. L. Henckel: "Multimedia Communication Platform: Specification of the Enhanced Broadband Transport Service (M26a)", GMD-Germany, 1993.
20. R. G. Herrtwich: "The HeiProjects: Support for Distributed Multimedia Applications", Technical Report, 43.9206, IBM European Networking Center, Heidelberg, 1992.
21. R. Hinden: "Internet Protocol, Version 6 (IPv6) Specification", Internet-Draft, draft-hinden-ipng-ipv6-spec-00.txt, Sun Microsystems, Inc., 1994.
22. R. Hinden: "IP Next Generation Addressing Architecture", Internet Draft,, draft-hinden-ipng-addr-00.txt, Sun Microsystems, Inc., 1994.
23. R. M. Hinden: "IP Next Generation Overview", Internet-Draft, draft-hinden-ipng-overview-00.txt, Sun Microsystems, Inc., 1994.
24. ISO/IEC_JCT1/SC6: "First Draft of Enhanced Communications Transport Service Definition", Beppu, 1995.
25. ISO/IEC_JTC1/SC6-N9161/IV: "Draft Text on the subject of "Multi-peer Taxonomy"", Beppu, 1995.
26. ISO/IEC_JTC1/SC6/WG4: "Meeting Report and Action Items, Interim Meeting (29 June - 4 July 1995)", Geneva, 1995.
27. ITU-T/SG7: "First Draft of ISO/IEC 7498 - 5 Multi Peer Communication Architecture", 1995.
28. ITU-T/SG7: "Liaison Statement from ITU-T SG 7 Regarding "Multicast Transport Service (X.tms)"", 1995.
29. ITU-T/SG7: "Liaison Statement from ITU-T SG 7 Regarding "Network Layer Multicast"", 1995.
30. ITU-T/SG7: "Report on Q23/7", Geneva, 1995.
31. P. T. Kirstein, M. J. Handley and M. A. Sasse: "Piloting of Multimedia Integrated Communication for European Researchers (MICE)", *Proc. INET '93*, 1993.
32. A. Kückes: "BERKOM Multimedia Collaboration Service, AVC Semantics", Project Report, Draft 0.1, The BERKOM Project, 1993.
33. L. Mathy, G. Leduc, O. Bonaventure and A. Danthine: "A Framework for Group Communications", *Proc. Third International Broadband Islands Conference*, Elsevier Science Publisher (North Holland), Hamburg, 1994, pp. 167 - 174.
34. A. Mauthe, D. Hutchison, G. Coulson and S. Namuye: "From Requirements to Services: Group Communication Support For Distributed Multimedia

Systems", *Proc. Second International Workshop, IWACA '94*, Springer-Verlag, Heidelberg, 1994, pp. 266-277.

35. D. J. Mitzel, D. Estrin, S. Shenker and L. Zhang: "An Architectural Comparison of ST-II and RSVP", Computer Science Department, University of Southern California, Los Angeles, 1994.
36. R. Ramanathan: "Multicast Support for Nimrod: Requirements and Solution Approaches", Internet-Draft, draft-ietf-nimrod-multicast-01.ps, BBN Systems and Technologies, 1995.
37. R. T. Sanders, B. J. Dempsey and A. C. Weaver: "XTP: The Xpress Transfer Protocol." Addison Wesly Publishing Company, Inc., Reading, 1992.
38. M. A. Sasse, U. Bilting, C.-D. Schulz and T. Turletti: "Remote Seminars through Multimedia Conferencing: Experiences from the MICE project", *Proc. INET '94*, 1994, pp. 251-259.
39. H. Schulzrinne: "Issues in Designing a Transport Protocol for Audio and Video Conferences and other Multiparticipant Real-Time Applications", Internet-Draft, Internet Engineering Task Force; Audio-Video Working Group, 1993.
40. H. Schulzrinne and S. Casner: "RTP: A Transport Protocol for Real-Time Applications", Internet-Draft, Internet Engineering Task Force; Audio-Video Transport WG, 1993.
41. R. Simon, T. Znati and R. Sclabassi: "Group Communication in Distributed Multimedia Systems", *Proc. IEEE-Distributed Multimedia Systems*, IEEE, 1994, pp. 294-301.
42. R. Steinmetz and K. Nahrstedt: "Multimedia: Computing, Communications and Applications" Prentice-Hall International, Inc, Engelwood Cliffs, 1995.
43. C. Szyperski and G. Ventre: "A Characterization of Multi-Party Interactive Multimedia Applications", Technical Report, TR-93-006, The Tenet Group, International Computer Science Institute, Berkeley, 1993.
44. C. Szyperski and G. Ventre: "Efficient Multicasting for Interactive Multimedia Applications", Technical Report, TR-93-017, The Tenet Group, International Computer Science Institute, Berkeley, 1993.
45. Talpade and Ammar: "SCE: an Architecture or Reliable Multicasting", Internet Engineering Task Force INTERNET-DRAFT, College of Computing, Georgia Institute of Technology, 1995.
46. C. Topolcic: "Experimental Internet Stream Protocol, Version 2 (St-II)", BBN, RFC 1190, 1990.
47. XTP: "XTP Protocol Definition", Santa Barbara, 1992.
48. XTP-Forum: "Xpress Transport Protocol Specification, XTP Revisions 4.0", 1394 Greenworth Place, Santa Barbara, USA, 1995.
49. N. Yeadon, F. Garcia, A. Campbell and D. Hutchison: "QoS Adaptation and Flow Filtering in ATM Networks", *Proc. Second IWACA '94*, Heidelberg, Springer-Verlag, Heidelberg, 1994, pp. 191-202.
50. L. Zhang, R. Braden, D. Estrin, S. Herzog and S. Jamin: "Resource ReSerVation Protocol (RSVP) - Version 1 Functional Specification", Internet Draft, 1994.

Object-Oriented Framework for a Scalable Multicast Call Modelling

M.I. Smirnov
GMD FOKUS
Hardenbergplatz 2, 10623 Berlin, Germany
smirnow@fokus.gmd.de

Abstract

The proposed approach for the multicast Call modelling framework considers logical entities participating in the Call process. A new mediator for the multicast Call establishment is introduced - the multicast agent (MCA), which is a software object moving across the internetwork and delivering to all members of the group all functionalities needed for reliable and flexible participation in a multicast session. The Call is totally distributed, and it is highly scalable due to the multicast tree decomposition into unicasts and fancasts. Performance aspects are discussed. MCA implementation sketch is proposed within CORBA 2.0

1 Introduction

Current progress in the multimedia multipoint-to-multipoint collaboration support from the internetwork could be seen both as the challenge for the application designers and the testbed for the design alternatives. New applications supporting collaboration are on their way to customers in various domains such as distributed production management, scientific collaboration, medicine, distributed computer aided design, and so forth. However, the complete understanding of the multicasting technology especially in the case of multimedia collaboration still is not achieved. Hopefully, the system model of that case can contribute to the desired understanding providing a sort of a framework for successful implementations. The framework model was designed to help new applications development and to provide for a broader spectrum of related research tasks. As in any object-oriented design one of the key success factors is object classes definition suitable for design and implementation within a particular domain and design targets. This paper introduces such classes for the multimedia multicasting with some reasoning behind the choice.

The paper is organized as follows. After observing a certain number of similar approaches (Section 2) and making a general conclusion about the need for the multicast specific framework for the Call modelling (Section 3), the paper defines some very general case of multicast collaboration (Section 4). Section 5 introduces a Multicast Agent - major mediator for the call establishment - from structural and functional viewpoints. The multicast tree development scenario and the definition of framework components could be found in section 6. Section 7 presents some performance aspects of the proposed approach. Section 8 deals with possible implementation of the Multicast Agent within the CORBA environment. The paper ends with a short conclusion.

2 Related Work

Traditional Call Establishment model emphasizes needed actions (functions); it is point-to-point oriented; it is dependent on existing network technology; it has data transfer orientation for the data transfer phase. On contrary to this, new smart connections require more complex behaviour, they need to be highly scalable to cope with the existing internetwork dimensions and to deal with objects rather than with pure data. The Call will be treated in this paper as a protocol which controls and allocates network resources required for a multipoint multimedia session; these requirements are to be confirmed by the service provider in a form of a contract. A snapshot of some recent works follows and gives an impression of development directions in the area.

The Multimedia Call Establishment [1] gives a *global view* of the Call Establishment process and deals with several objects, *medium* being among the object set, including also *relation* and *attachment*. The *relation* object is proposed to use for synchronization purposes between different information streams, while the *attachment* object is needed to relate a user to a certain medium. This approach could not be overestimated because of the potential it gives to designers. However after introducing these objects the authors deal only with service primitives.

From the signalling viewpoint, one of the trends in a call control area is separation of Call Control from a Connection Control; introducing some building blocks in the Call Control information structure and generic signalling flow mechanism based on atomic actions [2]. The authors of [2] argue also in favour of greater functional specialization among different sets of signalling entities. When defining mentioned above building blocks authors present suddenly a set of objects, such as *Calls, Parties, Services, Service Groups, Attachments, Access Channel Endpoints* and *Mapping Elements*. All these elementary call objects could be Confirmed, Local or Virtual. When this set of elementary objects is introduced they are to be "combined and organized into composite structures that logically represent a service"[2]. It should be mentioned that on contrary to the global view approach the authors represent objects only within the context of a *local view*.

The control domain for the Call Modelling in a broadband Intelligent Network [3] is partitioned into 3 domains - Service Control (modelled by the IN service logic); Connection Control (being essential for the B-ISDN) and Session Control domain, where the Call Model is considered to have 2 parts: Session Control State Model (SCSM) and the Connection Coordination Model (CCM). The SCSM provides control functions for one session (equal to one call) which actually can invoke more than one connection. The SCSM model consists of 11 Points In Call (PIC); where PIC is a distinguished state in the state transition diagram indicating the flow of operations. The CCM describes the connection topology and types of connections in some abstract way in order to be network independent. This description is based actually on 4 objects: Leg - unidirectional communication path with address, bandwidth, QoS, direction, etc. as attributes; Connection Point - interconnection in a switching matrix between two Legs; Virtual Subscriber - agent for all Call resources to initiate or terminate; Port - a source or delivery point address. The Call is modelled with the use of these objects and their attributes.

In [4], which is actually based on the same concepts and testbed as [2], the object oriented Call model is represented with the number of 12 objects grouped into 3 generic classes, namely Local Object (8 elementary objects), Shared Object (2 elementary objects) and Virtual Object (2 elementary objects). Notably that the number of messages is rather small (Request, Accept, Ack and Reject).

The most relevant work deals with the concrete multicast protocol description for the so called Protocol Independent Multicast - PIM [5]. One can also derive the object model out of this paper.

Another multicast implementation, Reliable Process Group Communication, deals with a group view maintained at each host (somewhat similar to a global view) and with a sort of multicast address hierarchy enabling group dynamics within a certain domain [6]. All group views are treated consistently due to the implementation of a "two-phase commit" scheme.

In [7] the author suggests the new metaphor of the network providing a sort of a universal service for the multimedia objects (Open Bearer Service); which services are also able to provide the multicast service.

In [8] authors of Conference Control Channel Protocol describe a scalable base for building conference control applications and list a number of urgently needed primitives for the multicast.

Sometimes, however, it is rather hard to define a correct set of call primitives used in different approaches; let us agree that Call Establishment capabilities will refer to some generalized primitives in each referenced model.

It is evident from this short survey that there is no approach available for object definition in general for this kind of model. Table 1 summarizes all observed approaches in purely quantitative manner. As a complexity measure we use a product of two numbers - the number of objects and the number of primitives. The average complexity of observed approaches is 34. Probably, any object-oriented approach for the Call establishment in order to be manageable will have its complexity near this value. However, reasonable complexity is not the only requirement. This complexity has to fit properly a certain framework enabling modelling and piloting.

Table 1:

Model	Objects	Primitives	Complexity
Multimedia Call Establishment [1]	3	8	24
Call Control Information Model [2]	8	9	72
Call in Broadband IN [10]	4	5	20
Object Oriented Call Model [3]	12	4	48
Open Bearer Service [7]	4	4	16
Protocol Independent Multicast [5]	4	6	24
Reliable Process Group Communication [6]	4	10	40
Conference Control Channel Protocol [8]	2	12	24

What the multicast designer would like to have is, according to the author's opinion, some framework which will provide the designer with typical objects (i.e. most general, suitable almost for any cases); their interrelations, some "method template" for every object, and implementation scenario. An attempt to provide such a framework for a multicast Call establishment modelling was undertaken in this paper based on a very few considerations, essential to a multicast support for collaborative work.

3 Design Approach

The standard approach for any network technology modelling (for the sake of its evaluation), according to one of the fundamental works in the area [9], is as follows. The model of the network technology, say, the protocol is to be represented with a set of its [distributed] functions and entities which are to be placed within the model of the network environment.

In this traditional model the communication environment usually is modelled by the limits of possible variations of major factors influencing the protocol. Normally the designer should be fairly happy if [s]he can prove via this model that the protocol (mechanism) under development operates satisfactorily under all given environment conditions.

Normally in a point-to-point case we have 2 communicating entities (objects) with one set of protocols in between facilitating any of required communication modes. This general model is also applicable for the Call Establishment. However in the [multi]point-to-multipoint case we have more complicated picture of communication. Introducing logical channels between the sender and each of the receivers of the multicast flow is probably a bad idea, especially for the Call model because it hides essential features of the multicasting. These features are: partial sharing of the multicast tree between receivers, negotiating custom quality of service (QoS) requirements for heterogeneous receiver applications, the receiver group dynamic properties. The latter means that we can't separate in time the Call establishment phase from the information transfer phase within the multicast. Multicast receivers are to be allowed to join and leave the session at any time. However, we can separate these two phases logically. Moreover, we can also take into consideration the possibility to advertise the multicast session for potential receivers during the Call Establishment phase.

The proposed approach for the Call modelling framework tries to separate further logical entities participating in the Call process as much as possible and specify their interfaces as well as object' methods behind them in such a way that real design could benefit from this separation in terms of model flexibility, completeness, feasibility and, mainly, manageability. The rationale behind this separation was perfectly introduced in [2]; it was stated there that such services as communication among heterogeneous terminal equipment; selectable quality of service; support of multiple connection calls; support of multiple party connections; individual control of mapping and presentation of information on each user's access, - "cannot be modified effectively by a monolithic call object or single state machine. It would be necessary to enumerate a very long list of service and capability variations and the resulting finite state machine would be so enormous that it would be unmanageable" [2, p.302]. Proposed logical separation of objects is essential for multicast. It distinguishes source - flow, network - flow, and receiver - flow interfaces.

4 The Multicast Case Description

The Internet philosophy it to provide the so called best-effort service. In order to prove the vitality of any networking approach it seems reasonable to show how this approach works within the Internet environment. If the approach shows all desired properties, such as robustness, scalability, reliability, etc. in this environment then one can expect even better behaviour in less hard conditions, which could be improved by additional management and control policies, resource allocation, QoS assurance scheme, etc.

In terms of communication mode parameters any multicast session could be [10]: planned or spontaneous, managed or unmanaged, with controlled or uncontrolled access; with different number of senders. In terms of network environment it could also utilize the same multicast tree for all possible senders or use different multicast trees, make use of the same applications and/or QoS requirements throughout the receiver group or different applications/QoS requirements, etc.

Let us describe some generic multicasting case, having in mind application collaboration demanding multicasting support rather than pure conference. The multicast session starts as a planned event according to some schedule with one sender and a certain number of receivers who agreed to participate in advance (*dedicated* receivers - DR). That means that this single sender (initial sender) will issue some request with the list of dedicated receivers in it. The request message has certain parameters requiring certain services from the network, i.e. from the network service provider (SP). The SP has to confirm/reject requirements via separate message to all/some of the multicast group members (e.g. to initial sender or/and to all dedicated receivers). Negotiation mechanism on service parameters with the SP is not considered here. When confirmed in a form of some *contract,* the multicast Call Establishment request has to follow the *multicast tree* (MCT) for the particular session, dedicated receivers being the leaves of this tree. We assume that MCT is a pre-calculated logical structure spanning all DR in most efficient way [13].

More formally a contract could be presented as follows:

```
multicast_session_contract(management_domain,
   initiator, QoS_requirements,
   [number_of_simultaneous_senders],
   [session_duration], MCT, [modality],
   [MCT_development_policy]).
```

The MCT could be seen as the main routing information for the Call establishment phase and as an association of network routers participating in the Call process at the moment. Each router, obtaining a contract, has an ability to send the *join request* message for more destinations than it was required in the dedicated receivers list or to more hosts adjacent to it, thus *advertising* the future session. Note, that the number and the set of advertised routers could be restricted by a *modality* parameter of a contract together with the *development policy* [15]. Normally, every dedicated receiver upon receiving the *request* issues its own *acknowledgement* message to the sender of its *join request* message. The initial sender can start sending multicast packets when the number of acknowledgements received from DR is sufficient. This decision making is out of the scope of the paper.

This paper proposes a highly scalable approach:

- only one member of the multicast group negotiates initially with the SP, namely the so called initial source;
- receivers send their *acknowledgements* only to the nearest *splitting point* of the multicast flow;

- all active senders share the same multicast tree (that's why the SP has to limit, probably, the maximal number of active sources);
- all QoS negotiations with the receivers could be provided at the splitting points.

The Call establishment thus continues during the information transfer phase of the session: acknowledgements are still coming to the initial sender and to splitting points; advertised receivers start joining the session; some of them can decide to leave the session; others can decide to request their own flow specifications different from the default flow specification. We also assume that any of the dedicated receivers can become a sender to the same session. This option could be enabled by a special *floor request* message sent to the group. The only difference of the floor request from the initial sender request is that it does not need to be confirmed by the service provider. We assume that for some period of time there could be several senders for the same multicast session sending in parallel.

For the sake of the design simplicity it is reasonable to decompose the whole Call process into three phases: 1. Session initialization; 2. Continuous Session; 3. Session End. The last phase could be introduced as follows. If the session is planned for a certain period of time, then it has obviously to have some agreement on session termination which is assumed to be fulfilled by all current senders stopping their transmission and informing the SP about the session end. Another possibility could arise when no one receiver is listening to the session. However, the latter case has to introduce the group control entity within the multicast.

5 Multicast Agents

5. 1 Structural Motivation

The most general idea of the approach comes out of the decomposition which in turn follows the general definition of the multicasting system [11] (MCS):

$$MCS = (MCAgents, MCConnectivity, MCProtocols, MCDescriptionTools).$$

The above mentioned decomposition affects two sides of the definition - the MC Connectivity, that is the Multicast Tree spanning all members of the multicast session; and the MC Agents which are defined in this paper as some objects capable to move across the network, clone, intercommunicate and perform a set of specified actions.

Topologically, any multicast tree could be seen as a composition of several serial interconnections (unicasts) and fanouts of nodes. Therefore, any communication supporting and providing multicast could be composed out of several unicasts and simple multicasts. The multicast could be called simple when the corresponding multicast tree is a fanout, for that reason a new term could be introduced - *a fancast,* i.e. inevitable multicast which could not be reduced. From these reasonings comes the idea of a scalable multicast. The multicast tree decomposition is presented in Figure 1, where simple composition components are put within smaller icons. This decomposition is complete, i.e. no more fanouts or unicasts could be found. However, this is not a single possible decomposition. Different structures which can arise during the call establishment will be addressed later. Out of this example it is evident that due to inevitable overlapping of unicasts and fanouts the functionalities needed to support unicasts and fancasts are to be placed within their intersection points, while these points are actually the *splitting points* of the multicast flow. A splitting point could be defined as a network router receiving the multicast flow, where packets of the flow have to be reproduced at least in 1 copy additionally. All over the paper we assume that receiving hosts connected to routers are not shown in the figures.

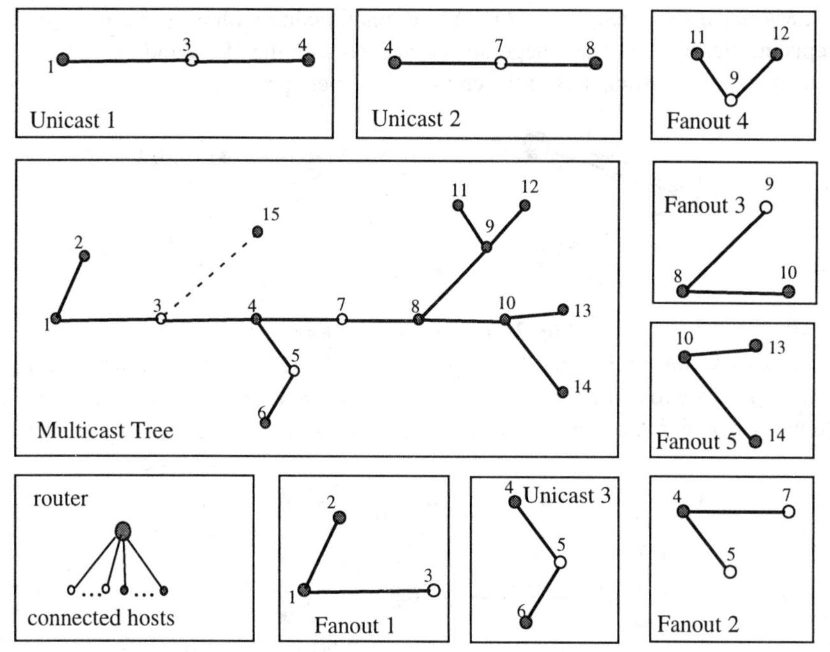

Fig. 1 A Sample Multicast Tree Decomposition

The desired scalability feature could be achieved via limited visibility out of each splitting point: we say that from each splitting point only a set of nearest reachable splitting points and receivers could be observed. This improves scalability and manageability of these partial multicasts. However, the need to provide the self-distribution of the multicast functionality over the multicast tree is the cost for these improvements.

5.2 Scalability

In this section the scalability feature for the multicast session will be introduced in some details. First we consider the mostly primitive case of the simplest fancast and a model for essential communication between involved entities; then we show how this scheme could be made scalable. In the following sections we'll try to show how more complicated interactions between components could be modelled.

The simple multicast case will consist of one sender and two receivers (Fig. 2). The only thing the sender has to do in this case is to copy packets to two destinations. The multicast flow direction is shown by solid arrows (Fig. 2(a)); the call establishment flows of messages and corresponding components are shown in dashed lines (Fig. 2(b)). We introduce here 3 generic functional components, namely:

1. *Src* - the source of the multicast flow (it could be an application or a composition of applications, each sending its own flow to a different media such as video, audio,

joint editing/drawing application space, etc.);

2. *Multicast Agent (MCA)* - a generic functionality within the multicast flow source network node which communicates with the service provider and actually sends multicast packets to the receivers;

3. *Multicast Receiving Agent (MRA)* - a functionality within the receiver router; it accepts multicast packets and negotiates if needed with the MCA and the receiver host applications; also participates in the call establishment process.

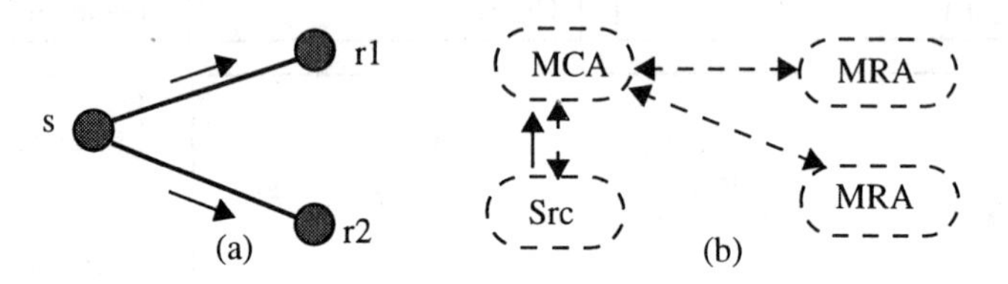

Fig. 2 The fancast development

Let us make this embryonic MCT grow a little bit (Fig. 3) and see what additional functionalities are to be introduced for the new case. When the intermediate node (*n*) is introduced between the sender and receivers we see that in relation to receivers this

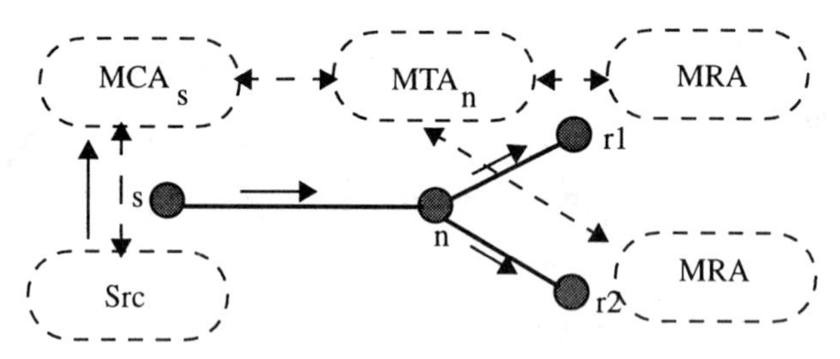

Fig. 3 Grown up multicast case

node plays exactly the same role as the sender *s* did in Fig. 2 except the acceptance of the multicast flow from Src. It looks like one MCA has been split into two agents - the initial MCA and the network agent (let us call this new object MTA - the *Multicast Transmission Agent*). It is useful to note that the transition from Fig. 2 to Fig. 3 could be made with the insertion of an additional node to the left of the initial source and with artificial movement of the multicast flow source to this new node. The functionality of MCA in Fig. 2 has been split into two parts with ordinary point-to-point connection (unicast) between them.

From this object presentation it could be seen that there are two different interfaces for the initial and the intermediate senders of the multicast flow; optionally the initial MCA can have also the MRA interface as well. If every receiver will exchange messages (for patching of losses of packets, for QoS negotiation, for receiver group dynamics control, for floor and application control, etc.) with the initial source throughout the whole MCT then this scheme will not scale well. Evidently, sooner or later the source node will become a bottleneck. In the proposed approach an exchange of messages between agents exists only during the call setup/disconnect phases. However, this exchange between receiver applications and MRA can exist at any time.

In this paper we are especially aware of the case of the so called sparse group multicast [12], when the number of receivers is very large comparatively to the number of possible senders; receivers could also be distributed across several domains. Current multicast systems are composed with applications and network support, sender and receiver applications communicating via the network. The network support consist of special address scheme, host extensions and routing support. Many authors argue in favour of more and more support for future multicast multimedia applications form the network. However, networks become more and more high speed, i.e. tend to provide less functionality during the data transfer. Hence, one of the requirements for a successful multicast implementation - to keep it simple for the network, e.g. following the Internet best-effort philosophy. The second reasoning comes from the fact that one can not change the placement of a multicast flow (source) for obvious reasons (human participants, cameras, data bases, etc.), however the *starting point* for the multicast tree could be placed *anywhere* within the network, provided that this anywhere place has a good connection to the source. Out of these two considerations comes the proposed scenario for a scalable multicast Call establishment, which was named the *multicast tree development.*

6 The Multicast Tree Development

6.1 Scenario

To obtain an object model we'll use the term *agent* for some instance inside the sender host having all interfaces described above. The multicast session starts with the agreement between the network service provider and the multicast agent, the list of dedicated participants being one of the parameters of this agreement. We assume that somehow is calculated the MCT [13] which spans the initial source and all the dedicated receivers. The initial source (i.e. MCA) sends its *request* and after *confirmation* from the SP starts the process of the MCT development. Shortly, the process could be seen as follows: the MCA clones itself and places its embryonic objects in every node of the MCT in a step by step mode. Each embryonic multicast agent initially has only one (unicast) interface to communicate with the parent MCA (i.e with that MCA instance which had created it). During the MCT development each copy agent activates needed number of copies of needed interfaces. From the receiver viewpoint there is no difference to communicate with the remote MCA (placed near the multicast flow source) or with the nearest MCA. We assume that from the functional viewpoint each copy of the MCA provides the same service for the receiver[s].

This scenario will have the following script for the MCT shown in Fig. 1:

- Initial source node (MCA entity in node 1 or MCA(1) for short) establishes the contract with the SP; obtains confirmed contract;
- MCA(1) clones itself and sends copies to nodes 2 and 3;
- MCA(2) recognizes from the contract that it has only receivers attached to its current node (2) and invokes needed number of MRA interfaces;
- MCA(3) recognizes that it has no receivers attached, optionally invokes the *adv* interface and sends advertising messages to its hosts, but, first retransmits its copy to node 4;
- MCA(4) recognizes that it has different types of attached nodes and clones itself in two copies, then sends them to nodes 5 and 7; and so forth.

Note that the MCA copy could be placed in nodes 3, 5, 7 with the only *unicast* interface invoked; however we suppose that this ordinary unicast could be supported by normal network service. In some future, when a new receiver 15 (say, attached to node 3) will try to join the multicast session it will need to join with the nearest shaded node. If the MCA functionality would be provided at all MCT nodes, including those which had no initially attached dedicated receivers, then the MCA(2) obtaining the *join* request could simply invoke needed MRA interface. Note that within the proposed scenario only the source node makes the contract negotiation with the service provider.

What we have as the result of the MCT development process is a tree spanning all confirmed participants of the group; each being provided with the needed multicast functionality. In terms of a multicast flow distribution this tree could be seen as a hierarchy; however from the multicast call establishment viewpoint this tree is a distributed association of peer agents. These agents share the multicast flow being sent to a group by the source.
The embryonic MCA is shown in Fig.4. Within the Object-oriented programming ter-

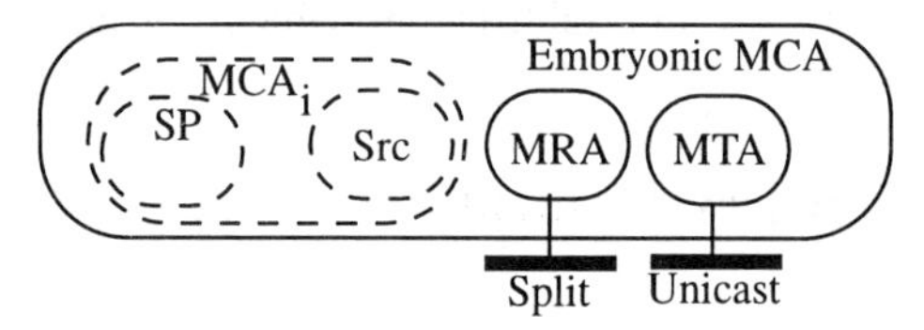

Fig.4 Embryonic MCA

minology [14] the Embryonic MCA could be seen as a complex object with SP, MTA. MRA, Src being its methods, however we'll call them also objects with their own methods (only *split* and *unicast* methods are shown here). The methods are invoked by messages according to the multicast call establishment scenario.

6.2 The Multicast Call Establishment Components

The Multicast Call establishment scenario development is not the purpose of this paper, however we show here what components are essential with the use of the MCT Development concept and the Multicast Agents object presentation. Actually we consider the MCT Development scenario as a framework for the Call Establishment Model design. Following the above mentioned MCTD scenario and [1, 5, 6, 8] let's start with the source node.

The *Src* method of the complex MCA object is an object with methods:

- attach/add/detach/remove media;
- define medium-to-medium relations (for multi-media multicasting);
- include/exclude applications;
- floor control;
- give current multicast state (optional).

The *SP* method of the complex MCA object is an object with methods (parameters):

- enable/disable multicast session (contract);
- quality of service parameters (DR list; Media list).

The *MRA* method of the complex MCA object is an object with methods:

- filtering (Media list; QoS requirements list);
- include/exclude applications;
- attach/add/detach/remove media;
- patching (Medium, badly received packets);
- advertising (optional);
- floor control;
- participation control;
- give current multicast state (optional).

The *MTA* method of the complex MCA object is an object with methods:

- clone (MCA);
- send copy (MCA);
- invoke (MCA_method);
- prune (MCA_instance);
- split, i.e send copy (multicast packet).

We assume that when the SP method is activated it starts negotiation with the service provider for a multicast session with required dedicated receivers set and required QoS parameters for a certain media involved. This contract is assumed to be assured by the service provider all over the MCT. If different receiver hosts will require changes to the contract defaults then they have to negotiate them with their MRA.

Actually the MTA entity realizes a protocol for distributed multicast management; this particular method has to realize essential multicast functionality to achieve scalability of the sparse multicast groups.

The complete set of objects introduced to support the object-oriented framework for a scalable multicast is presented in Figure 5. This set is called a framework because it is not elaborated yet in the sense of object messages, also some methods are optional and others could arise during the multicast implementation. These building blocks can support the whole lifecycle of the multicast session, including Call Establishment and Call Modification.

The Call Establishment phase needs to invoke the Src and SP objects to establish primary contract with the service provider and MTA object to clone, transmit and self-install the MCA complex object within the nodes of the MCT. Note that we suppose to clone not an abstract MCA complex object but its instance customized according to a pre-negotiated set of session parameters (dedicated receivers list; media list; default QoS parameters; maximum number of active senders, etc.). The Call Establishment phase continues with invoking of MRA objects and their local negotiations with local receivers, session advertising, etc.

The Call Modification phase can start even during the Call Establishment. For example some of the dedicated receivers could become unreachable; crossing the domain boundary may cause changes in the default session parameters, one of possible active senders may ask for a floor even before the start of the initial flow, etc. We assume that this could not cause any problem because any MCA instance after the self-installation is ready to accept any of the messages which could take place within the entire multicast session. The shortcoming of this approach is that we have no opportunity to change all global parameters of the multicast session (DR list, Media List and default QoS for the reserved Media). The only possibility to change these parameters is to renegotiate with the service provider via the SP object. This object could be invoked only by any of the dedicated receivers (each of which have a potential to become a sender). The only possibility to change the DR list is to apply to the initial source.

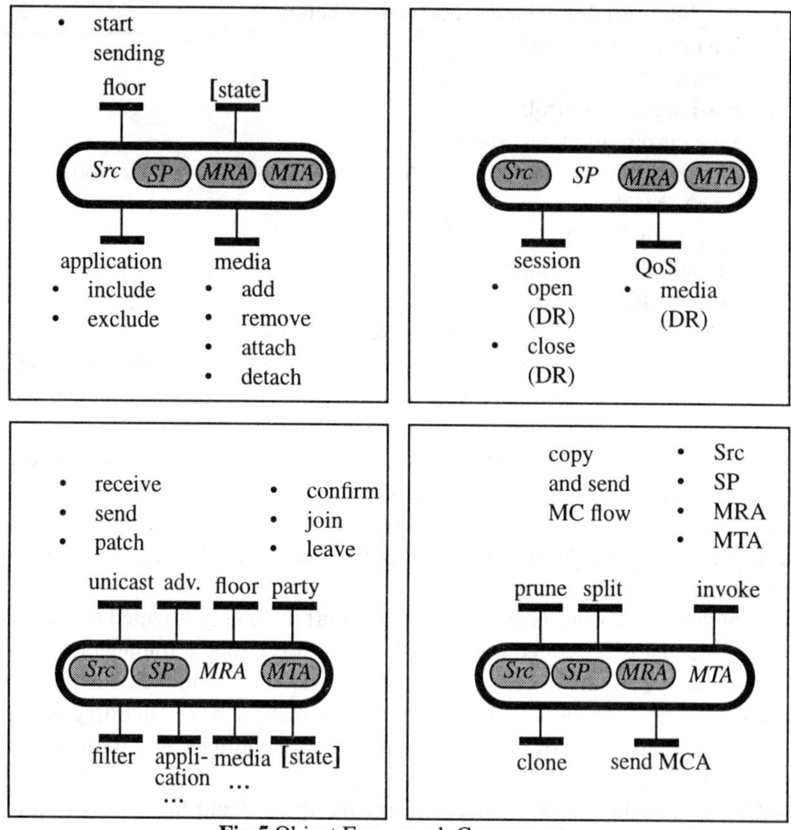

Fig.5 Object Framework Components

Note, that if the number of dedicated receivers is too large for a service provider to permit them to send in parallel, then the number of active sources is to be introduced. However this number is to be somewhat less than the actual possible number of concurrent sources because we assume no information exchange between the instances of the MCA. Consequently the only way for them to know the number of actual active sources is to listen to the multicast. However, due to a delay introduced by the network several sources could start sending simultaneously based on their local permission. To prevent this situation we assume simply that the MCA instance nearest to the last source has to stop sending when recognizes that actual number of active sources is greater than default maximum.

7 Performance aspects of the MCT Development

As long as the proposed agent technology for the MCT development is totally distributed we have to pay special attention to some potentially dangerous events during the development. Main features which will be addressed here are: scalability; manageability and global state. The global state problem exists because no one MCA has a complete knowledge of the global state, however MCA at all splitting points notify all receivers joined within their fanout in a bottom-up manner. Finally, the decision making on total MCT development quality is made by initial MCA. All reasonings will be made regardless the network structure; we require for the multicast agents only the knowledge of initial MCT. First, we assume that exchange of information between MCA entities exists only within each fanout.

Let us start with the formal description of the MCT under development which will facilitate some taxonomy of possible changes during the call. The multicast tree structure could be presented as

$$MCT = \langle L, \{S_l\}, \{R_s\} \rangle,$$

where L is a number of levels in the MCT, i.e. the largest receiver-hop distance from the initiator to the tree leaf; $\{S_l\}$ is a set of sets of splitting points at each level $l, l \in \overline{1, L}$; $\{R_s\}$ is a set of sets of receivers which are to be connected to particular splitting point s, where $s \in S_l, l \in \overline{1, L}$.

That is we can define two mapping functions $f(S_l, R_S) \to N$, $\varphi(S_l, S_{l+1}) \to N$, where N is a set of integers. Assuming that MCT is a tree structure we can use simply $f(S_l)$ and $\varphi(S_l)$ to denote the number of receivers and the number of splitting points to be attached to the given splitting point. Where it will be possible we omit the second index for a splitting point, however a complete depiction should be $S_{l,t}$, where $l \in \overline{1, L}$, $t \in \overline{1, \varphi(S_{l-1,i})}$, i is a parent splitting point index.

The MCT under development is assumed to be a pre-calculated Steiner tree structure which may have or have not other network nodes (s-nodes) in it except dedicated routers [13]. During the development of this MCT by multicast agents a number of events may happen which will change pre-calculated structure, effecting different components of the MCT. Table 2 gives a summary of possible events during the development and shows how these events influence the notification process. Most of events could be caused by current network situation and the fact that splitting point routers are able to advertise future session for new receivers, which actually were not considered when the MCT was calculated.[i]

Splitting of a splitting point is considered as any of two possible cases: add new splitting point and remove [partly] receivers from the old one; or add two new splitting points and delete an old splitting point. When s-nodes are considered within the MCT (due to obvious reasons for sparse multicast groups), the operations of add/delete s-node are to be considered at the internetwork unicast level. Other cases (when s-node becomes a receiver or even a new splitting point) could be found in the Table 2. Note that only one event (Delete receiver) could be notified only at the very top of the MCT

---

i. However, according to the heuristic proposed in [13], the MCT development algorithm may try to make effective use of those sub-optimal MCT branches which were not pruned.

(when all of the nearest fanout acknowledgements will arrive). All the rest of these structural changes for initial MCT could be successfully resolved in a distributed manner. Note also that some cases may change the number of MCT levels (Degrade/Delete splitting point, and Delete receiver). And, finally, it should be mentioned that notification points presented in Table 2 will be valid if all migrations will take place at the

Table 2:

Changes to	Case	Definition	Notified by
R	Migrate receiver within one level	$S_{l,i} \rightarrow S_{l,j}$	S_{l-1}
	Migrate receiver between levels	$S_{l,i} \rightarrow S_{h,j}$	S_l if $l < h$
			S_h if $l > h$
	Add new receiver	$f(S_l) = f(S_l) + 1$	S_l
	Delete receiver(s)	$f(S_l) = f(S_l) - 1$	S_0
S	Migrate splitting point	$\{S_l\} \rightarrow \{S_h\}$	S_l if $l < h$
			S_h if $l > h$
	Add new splitting point	$\varphi(S_l) = \varphi(S_l) + 1$	S_{l-1}
	Delete splitting point	$\varphi(S_l) = \varphi(S_l) - 1$	S_{l-1}
	Degrade splitting point	$\varphi(S_l) = 1$	S_{l-1}

same branch of the MCT. In opposite case almost all of the events could be resolved only at the very root of the tree. However, we assume that the MCT (and its unpruned branches being the most probable candidates for the advertising) covers the network topology in a way somewhat similar to the existing multicast connectivity in the Internet [16]. That is why we assume that probability of moving any receiver from one branch to another branch of the MCT is negligibly small.

For the sake of the MCTD manageability it is reasonable to set some timer within the initial MCA, estimating the waiting time for the whole MCT development. The same type of timer should be set in every copy MCA, placed within splitting points to estimate the development time of their fanouts. Let us denote as t_d - the average one hop delay for the MCA messages; t_i - the maximal MCA installation time; τ - the timeout to be set at the splitting point as estimated time for developing its own fanout only, and $\tau = 2 \cdot t_d + t_i$. Let also *dN(h)* denote the number of positive acknowledgements (*acks*) from the fanout under development according to the MCT. However, at some level *h* of the MCT we may achieve the following number of positive *acks*:

$$dN(h) = \sum_{i \in \{S_{h-2}\}} [f(S_{h-1,i}) + \varphi(S_{h-1,i})] + dN(h+1) + aN(h) ,$$

where *aN(h)* stands for unknown number of *acks* from routers being advertised to join the session, and *dN(h+1)* - for expected *acks* from the rest of the MCT branch. The addenda of the above sum should refer to three *different rates* of the MCT development as viewed from the current MCA. The first addendum refers to the greatest rate, while the second - to the slowest development. The rate of *acks* for advertised routers is unknown and depends on their number, structure and contract parameters. Even when the multicast session is planned beforehand there is no hope that all routers will have exact knowledge of the MCT being part of the contract. That is why among these new receivers could be those dedicated to join the session via another splitting point.

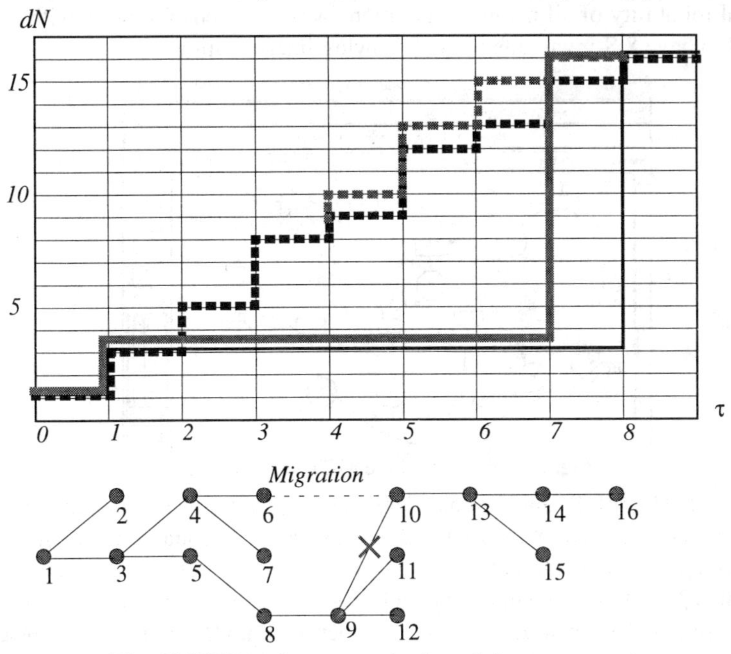

Fig. 6 MCT Development and acknowledgement process

Figure 6 shows a typical picture of the MCT development; dashed line depicts a distributed development process which takes place in reality, however the knowledge about the progress of development is shared among unknown number of MCA; bold line depicts *acks* receiving process for the initial MCA if it goes as planned. Thick grey line shows another possible arrival process, when splitting point migrates from receiver 9 to receiver 6.

Let us now examine how this performance could be improved with additional messaging to the initial MCA. According to Table 2 the most critical message to be forwarded backwards to the initial MCA is the negative acknowledgement (*nack*) from a dedicated receiver; this nack could be generated after expiry of the timeout by appropriate MCA in the nearest splitting point. However, in order to assure that this receiver is unreachable, the MCA has to wait for the whole hierarchy of timeout periods spanning its branch. This long delay prevents from early detection of failed receiver within the totally distributed scheme of the MCT development. From the other hand, if the *nack* will be sent after a single timeout it may confuse the initial MCA in case when this failed receiver has joined the MCT via another MCA (receiver migration). Also if *all acks* and *nacks* will be sent to the initial MCA this will cause a high risk of MCA flooding by these messages and, again, may cause conflicts in detection of the MCT state. Here we propose another solution: every MCA reports *acks* on new receivers to initial MCA only in case if this receiver is dedicated one but joining the MCT not in planned manner.

To make clear major behavioural aspects of the MCT development with the MCA, a simulation study of this protocol was undertaken with the use of XPetri [22] (Fig. 7) and PTOLEMY [23] tools. The Timed Petri net (TPN) model [24] of the simplified protocol (total reliability of all network operations was assumed) for MCT with 3 levels was found to have 8 S-invariants of the following interpretation:

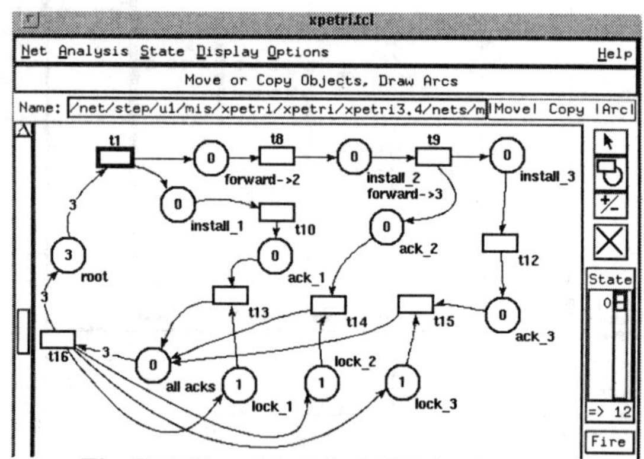

Fig. 7 TPN model of the MCT development

Invariant 1: The MCT could be considered developed if only the total number of acknowledgements received by the root node for this MCT is equal to the number of receivers mentioned in the contract;

Invariants 2 - 8: If the total number of acknowledgements received by the root at any instance of time is less then the contract' number, then nodes (immediately reachable from the root and those reachable via several hops) with pending acknowledgements could be in one of the following states only: waiting for MCA; installing MCA; sending acknowledgement. Note, that numbering of invariants follows the numbering of MCT levels being involved.

Though, the number of invariants (N_{inv}) as a function of the number of MCT levels (L) is growing according to $N_{inv}(L) = 1 + \sum_{i=1}^{i=L} \frac{L!}{i! \cdot (L-i)!}$, the model with $L=3$ provides for some typical protocol behaviour. Additionally, the model could be applied in a hierarchical way, assuming each branch of the MCT as a subtree.

Major timing parameters of the protocol are:

1. T - MCA copies distribution for a current fanout;

2. τ - MCA installation at splitting points and receivers (including customization at root);

3. t - acknowledgement to the sender of actual MCA copy;

4. Δt - acknowledgement processing at the MCA sender (including root).

According to a TPN analysis methodology suggested by J. Sifakis [25] each TPN's invariant (I_i, ($i = 1, \ldots, \rho$), ($\rho = 8$)) is responsible for a single inequality of the form $\sum_{j, k \in I_i} z_k \cdot f_j \le p$, where k is an integer defined by initial marking (here, for $L=3$ for all invariants, $p=3$), f_j is unknown frequency of firing of the *j-th* transition, and z_k is one of the timing parameters associated with the *k-th* position of the TPN.

This set of inequalities can give a decision for unknown firing *frequencies* assuming that all parameters are constants. However, all real life timing parameters have some essential fluctuations. That is why some randomization of timing parameters was used when solving the set of inequalities. The model output was averaged then. To simplify the routine matrix operations we had used the matrix tool within the synchronous data flow domain of the PTOLEMY [23]. Modelling results are presented in Figures 8 and 9 for the initial distribution of the MCA and for pre-installed MCA(i.e. for the case of $T/\tau = 1$) respectively. All timing parameters for Fig.9 were assumed equal and are shown at the horizontal axis. For the case of MCA distribution (Fig.8) the horizontal axis shows numbers of data sets in accordance with Table 3. The vertical axis (labelled *freq*) shows for each invariant (marked 1 - 8 in both diagrams) some performance measure [25] combined out of protocol parameters and involved firing frequencies (see left part of the inequality above). The best performance could be achieved when for the given protocol parameters *all* inequalities shown above hold with a minimum difference from equality (i.e. $p=3$ marked with '*' in the diagrams). Not surprisingly, that the results show greater dependence on the timing parameters of that protocol invariants which have larger numbers (i.e. relying on receivers which are far away from the root). On contrary, invariant 1 is almost parameter - insensitive. These simulation study shows clearly, how the combination of protocol timing parameters could be optimized for a particular network environment.

Other performance issues include the following.

Support for the Multidomain Networks. When the multicast Call traverses several domains (where domain could be represented by a single service provider and/or network management) then at the border node of a new domain the MCA has to invoke it's SP method to renegotiate all needed parameters and services (media, etc.). At the same time the whole new domain is visible to the source domain as a single receiver, that's why the border node within the sender domain has to invoke the MRA method and to communicate with the Src method of the MCA copy inside a new domain.

Sender Substitution. One of the main advantages of the proposed scheme is that any network node which belongs to the MCT possesses the embryonic MCA, that is why any node can invoke needed interface, thus enabling a new active sender or joining the current MCT by a new receiver, etc. We assume that when the SP method starts negotiation with the service provider it sends a multicast session request message with the list of dedicated receivers as it's parameter; that means for the service provider that any member of the d-nodes list can become a sender. As an option we can also use another parameter, say, maximal number of simultaneously active senders.

Quality of Service. The default quality of service with regard to each media mentioned in the contract with the service provider is to be guaranteed by the service providers in each domain. However, due to different hardware equipment, applications

Table 3:

Data set No	T [s]	τ [s]	t [s]	Δt [s]
1	10.0	2.0	0.2	2.0
2	10.0	2.0	0.2	0.1
3	5.0	1.0	0.1	1.0
4	5.0	1.0	0.1	0.05
5	2.5	0.5	0.05	0.5
6	2.5	0.5	0.5	0.025

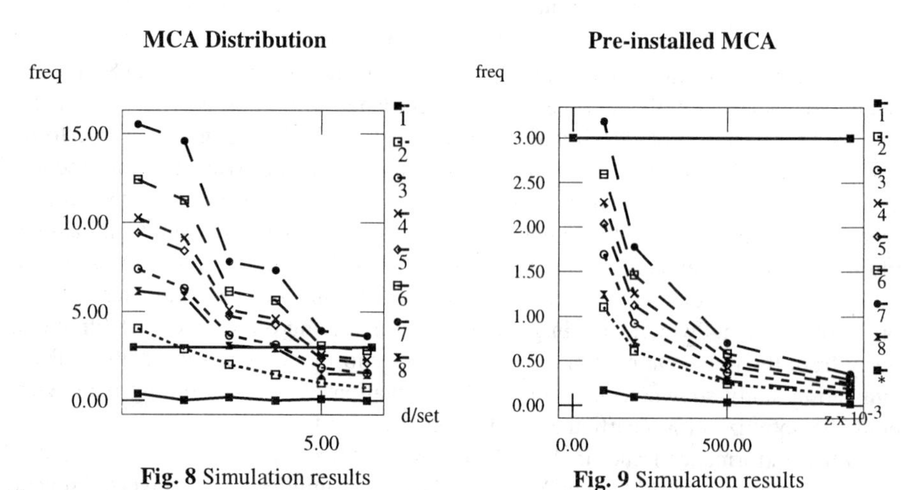

Fig. 8 Simulation results

Fig. 9 Simulation results

and types of host connection to the internetwork actual receivers may have a willingness to change the default QoS parameters. This can be done by the filter interface of the MRA object for every pair (MRA, Receiving Host).

Reliability. This feature could be achieved at the network level with the use of the same technology as proposed in [21] for a single application. However, reliable integration of applications could be achieved only with application independent reliability, i.e. within the underlying transport service including multicast.

8 Implementation Remarks

The technical availability of the migrating multicast agent could be explained within the Common Object Request Broker Architecture (CORBA) specification as shown in Figure 7. Two disjoint Object Request Brokers (ORB) can exchange any objects via the internetwork with the use of Replication and Externalization/Internalization Services of CORBA 2.0. The Externalization Service assumes that the Stream Service object holds externalized objects as files. To solve the problem of unique object references the CORBA suggests [17] that the Interoperable Object Reference (IOR) should be externalized together with the externalization of the copy object. Consequently, two proxies are to be installed within the naming services of two collaborating ORBs. This paper suggests that a special Contract Service should be considered within CORBA specification set, and, additionally, that no reference should be created for a copy MCA, except the IOR. Moreover, locking of the destination suggested in [15] actually locks the copy MCA as well, preventing from confusion with multiple objects within single ORB.

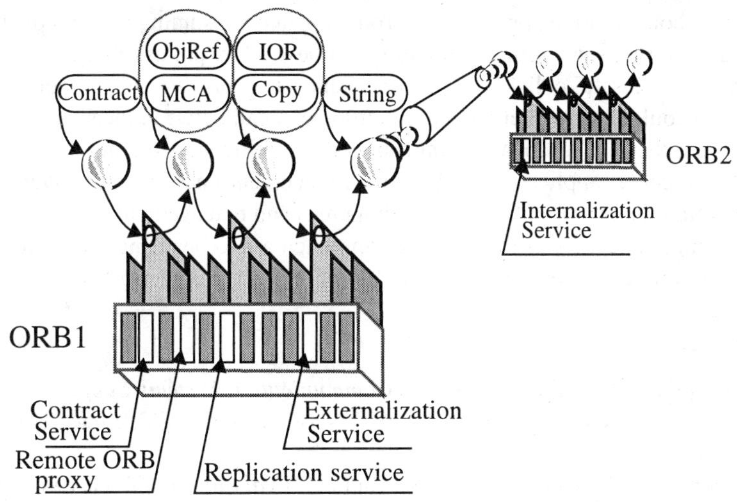

Fig. 10 MCA migration in CORBA environment

The CORBA specification was chosen for the following reasons:

- CORBA implementations provide efficient interoperability across heterogeneous software platforms;
- being a software development platform CORBA has an application viewpoint;
- CORBA deployment is rather fast in the direction for a cross-compatibility with ODP (TINA) [18]; OpenDoc, OLE [19] and CAD's Portable Common Tool Environment [20];
- adopted UNO approach [17] enables flexible distributed systems architecture.

The following implementation approaches seem to be possible:

- all nodes of the internetwork could be equipped in advance with customizable copies of the MCA ('pre-installed MCA' case) to save network resources needed to retransmit and self extract the MCA;
- MCA distribution at the Call Establishment phase could be made as proposed without any further exchange of information between the entities of MCA;
- providing for MCA distribution with some information exchange between these peer entities.

Future research in this area will consider quantitative scalability evaluation via different modelling approaches, as well as functional extensions of the presented framework.

9 Conclusion

This paper aimed to demonstrate that the *agent paradigm* could be extremely successful not only in the area of Artificial Intelligence and personal communications but for a network technology as well, e.g. for collaborative multicast call establishment.

The *multicast connectivity on demand* is a big challenge for telecommunication service providers. However, existing approaches for the call establishment don't fit special requirements of the multicast session establishment, or, when modelled, appear to be enormously complicated.

Presented approach tries to benefit from the logical separation of three major players in the area: collaborative applications, producing and consuming multimedia multicast flows; the internetwork with its structure and best effort philosophy, and, finally, the mediator of successful collaboration - the multicast agents dynamic association.

If there could be any intelligence in setting up a call between the sender and the set of receivers, then the multicast agent brings this intelligence to a number of positions nearest to receiver applications. This MCA placement enables to receivers effective QoS negotiation, error recovery, group dynamics and management.

Performance aspects of the proposed approach are to be refined in more details as well as the multicast agent implementability within existing platforms.

References

1. Heijenk X.H., Niemegeers I.G. *Communication Systems Supporting Multimedia Multi-user Applications.* - IEEE Network, January/February 1994, pp.34 - 44
2. Minzer S., Bussey H., Porter R., Ratta G. *Evolutionary Trends in Call Control.* - Proc. XV International Switching Symposium, ISS'95. Vol.2, pp. 300-304
3. Maastricht van C., Schalk E. *Call modelling in a Broadband IN architecture.* - In. Proc. XV International Switching Symposium, ISS'95. Vol.2, pp. 340 - 344
4. Grebenö J., Hanberger N., Ohlman B. *Experiences from implementing an Object Oriented Call Model.* - Proc. XV International Switching Symposium, ISS'95. Vol.2, pp. 77 - 81
5. Deering S., Estrin D., Farinacci D., Jacobson V., Liu Ch.-G., Wei L. *An Architecture for Wide-Area Multicast Routing,* In: Proc. SIGCOMM 94 -8/94, pp.126 - 135
6. Murata S., Shionozaki A., Tokoro M. *A Network Architecture for Reliable Process Group Communication.* - In Proc. 14th Int. Conf. on Distributed Computing Systems, June, 1994, Poznan, PL, ICDCS'94 (Piscataway, N.J.: IEEE), pp. 66 - 73
7. Covaci S. *Towards the Global Provision and Management of a New Open Bearer Service*, In: Proceedings of the First International Workshop on High Speed networks and Open Distributed Platforms (HSN&ODPl), June, 1995, St.Petersburg, Russia, 16 p.
8. Handley M., Wakeman I. *CCCP: Conference Control Channel Protocol. A scalable base for building conference control applications*. v.1.4. An html link was provided at http://www.research.att.com/cgi-bin/bibsearch_html?CCCP

9. Reiser M. *Performance Evaluation of Data Communication Systems.* - In Proc. IEEE v.70, No 2, February, 1982, pp.171-196
10. Szyperski C., Ventre G. *Efficient Multicasting for Interactive Multimedia Applications*, TR-93-017, International Computer Science Institute, Berkeley.
11. Smirnov M.I. *Multicast Routing: State of the Art and Research Ideas*. Presentation at GMD Fokus, STEP group, March, 1995; URL ftp://ftp.fokus.gmd.de/pub/step/papers/Smir9503:MCSofA.ps.gz
12. Deering S., Estrin D., Farinacci D., Jakobsen V. *IGMP Router Extensions for Routing to Sparse Multicast-Groups*, Internet Draft, available e.g. at http://karin.elch.lu.se/pub/stan... ft-ietf-idmr-igmp-sparce-00.txt
13. Smirnov M.I. *Efficient Multicast Routing in High Speed Networks.* - In: Proceedings of the First International Workshop on High Speed networks and Open Distributed Platforms (HSN&ODPl), June, 1995, St.Petersburg, Russia, 17 p.
14. Prabhat A. K., Gretzinger M.R. *Distributed Object-Oriented Data-Systems Design*, Prentice Hall.
15. Smirnov M.I. *Object-Oriented Framework for the Multicast Applications Integration.* - In: Proceedings of the DOOC'95 Workshop, Object World Conference, Frankfurt/Main, 10-11, October, 1995, 12 p.
16. Casner S. *The MBONE Frequently Asked Questions,* at: http://www.research.att.com/mbone_faq.html
17. ORB 2.0 RFP Submission *Universal Networked Objects*, Sept.,1994, OMG TC Document 94.9.32
18. *Comparison of the OMG and ISO/CCITT Object Models. The Report of the Joint NM Forum/OMG Task force on Object Modelling*, OMG Document 94-12-30, 27 p.
19. Adler R.M., *Emerging Standards for Component Software.* - In Computer, March, 1995, vol.28, No. 3, p. 68-77
20. Start K, Patel A. *The Distribution Management of Service Software.* - Computer Standards and Interfaces, 17(1995):291-301
21. Floyd S., Jacobson V., McCanne S., Liu Ch.-G., Zhang L. *A Reliable Multicast Framework for Light-weight Sessions and Application Level Framing.* - to appear in Proc. SIGCOMM, Cambridge, Mass., Sept.1995
22. Kahn, B., Noel, R., O'Keefe, S. *XPetri - Graphical Petri net simulator for XWindows*, Boston University, May, 1993.
23. Buck, J., Goei, E., Huang, W.-J., Kamas, A., Lee, E. *PTOLEMY: Users Manual*, U.C.Berkeley, Department of EECS, Apr., 1995.
24. Smirnov, M.I., Iljin, V.P. *Timed Petri-nets with inhibitor arches for system modelling*. - In: Electronic Modelling, 1990, vol. 12, No 2 (in Russian).
25. Sifakis, J. *Use of Petri Nets for Performance Evaluation.* - In: Acta Cybernetica, Vol. 4, No 2, pp. 185 - 202, 1978.

M-Connection Service: A Multicast Service for Distributed Multimedia Applications

José F. de Rezende[1], Andreas Mauthe[2], David Hutchison[2] and Serge Fdida[1]
{rezende,fdida}@masi.bp.fr, {andreas,dh}@comp.lancs.ac.uk

[1] Laboratoire MASI - CNRS
Université Pierre et Marie Curie
4, Place Jussieu - 75252 Paris Cedex 05
France
[2] Computing Department
Lancaster University
Lancaster LA1 4YR
UK

Abstract. New high speed networks and the requirements of users for the support of interpersonal communication in a computerised heterogeneous environment demand for new communication services. These services have to provide support for multimedia group communication. In this paper we introduce a multicast transport service for multimedia group applications called M-Connection Service. It is a protocol independent service which provides an interface to different transport protocols. Apart from data transfer this service offers functions to establish, join, leave, change properties of and terminate multicast connections. Multicast connections are characterised by QoS and integrity conditions. We propose distinct QoS negotiation and group management schemes suitable for multicast multimedia communication. We describe the service elements of the M-Connection Service in detail. To demonstrate the feasibility of the service its service elements are mapped onto XTP 4.0 protocol mechanisms.

1 Introduction

Emerging high-speed networks provide enough bandwidth to support the exchange of both discrete and continuous media. This is met on the user side with an ever-increasing demand for interpersonal communication in a computerised heterogeneous environment. In order to support this communication, new services have to be provided. Only a few protocols and services exist which utilise the features of high-speed networks optimally *and* provide the required service to the user. In particular there is a lack of services supporting multimedia *group* communication.

Multimedia group applications place a multitude of requirements on such services [1, 2]. The services offered have to be flexible, efficient and dynamic, i.e. it must be possible to change characteristics, such as topology and QoS. These stringent requirements make the design of supporting services complex and difficult.

Some proposals to support group communication at different levels of the system architecture have been introduced in [3, 4, 5, 6]. However, these proposals only address particular aspects or problems of group communication. Many suffer from a high degree of complexity, are too general in some issues, whereas in others they are very specific.

To deal with the high demands and complexity of group communication a *comprehensive* approach towards the problem is necessary. To provide efficient and flexible services we propose a group communication service architecture built from independent service modules. In this architecture all tasks are clearly defined and performed once by just one service module where this can be done most efficiently. Thus, redundant service definitions are avoided. Policies and mechanisms are separated, and any user dependent policy decisions are taken close to the user whenever possible.

In this paper we introduce the M-Connection Service which is a fundamental module of this architecture. It provides an end-to-end communication support between one sender and multiple receivers. It is a transport service that interfaces multiple transport protocols which support multicast (1:N) and a certain degree of QoS. In the overall group communication architecture it offers the *communications* support needed by multimedia group applications. The M-Connection Service is an independent module that can be either used by other service modules or directly by the application.

The service offered by the M-Connection Service is independent of any specific protocol. However, two protocols have so far been considered as potential candidates for our service, viz. XTP 4.0 and a proprietary multimedia transport protocol developed at Lancaster University [7]. To show the feasibility of the M-Connection Service and how it can be provided we map it onto XTP calls and mechanisms. XTP was chosen because it provides the required features and an implementation is available [8].

This paper is organised in five sections. An overview on XTP 4.0 is given in section two. In section three we introduce the M-Connection Service. Section four describes the service elements of our service in detail and maps them onto XTP mechanisms. Finally we discuss open issues and summarise the paper in section five.

2 XTP 4.0 "Xpress Transport Protocol": Overview

XTP 4.0 is a transport protocol designed to be used by all kinds of applications on top of emerging high speed networks [11]. It allows different modes of communication, viz. connection-less, transaction and connection-oriented. The fundamental design principle of XTP is to provide mechanisms rather than to implement any particular policy. In the view of the protocol designers only the application has enough information to optimally configure the control procedures for the data exchange, therefore XTP is a set of mechanisms whose functionalities are mutually orthogonal. The control procedures can be turned on or off to tailor the protocol according to specific requirements. Bitflags in the XTP header are used for this purpose. There is no service interface specified for XTP. Thus, it is up to the implementor of the protocol to define his own service interface adapted to the specific needs of the user.

Certain features of XTP, for instance the provision of multicast and QoS support, make it suitable for use in multimedia group communication. We decided to exploit these features and to use XTP as basic transport protocol for the M-Connection Service. A detailed discussion of XTP is beyond the scope of this paper. In the following we discuss those aspects which are relevant to the M-Connection Service.

2.1 XTP Multicast

XTP multicast is an integral part of the protocol and not an addendum. It provides a simplex data flow from one transmitter to an arbitrary number of receivers. Before a multicast association can be established, a set of users who wish to communicate must request that their contexts are placed into the *listening* state. A multicast association is established when one or more listening multicast contexts receive and accept a FIRST packet from the multicast transmitter. A FIRST packet is accepted if address, traffic specification and options match the ones specified during the listen. In the options, the kind of control procedures employed are specified. The MULTI bit in the option field indicates that this association is a multicast association. After accepting a FIRST packet the context moves into the *active* state.

The receivers reply with a TCNTL packet to the FIRST packet if the SREQ bit was set. The TCNTL packet contains all necessary information about the receiver including its traffic specification. Only then does the multicast transmitter know the number and identity of receivers in the receiver group. Different user reliability semantics are proposed in the specification. These include k-reliability, a specific sub-set of members or a hybrid reliability semantic.

The traffic specification, i.e. QoS values, are negotiated during the establishment phase (FIRST packet) or during the course of an on-going association (TCNTL packet). This negotiation may be initiated by the user and can involve all users, XTP contexts and the network provider. Also, the implementor can construct XTP in such a way that the contexts can renegotiate without user intervention. This is for instance useful when conditions arise where the traffic specification is no longer satisfied.

Receivers can join an on-going association by sending a JOIN packet to the multicast group. If the join is accepted the multicast transmitter answers with a JOIN packet, else a DIAG packet is sent to the joining context. To leave an association the receiver sends a control packet with the END bit set to the multicast transmitter. The multicast transmitter can force a receiver to leave by sending it a CNTL packet with the END bit set.

Once all data from the multicast transmitter has been sent, the multicast data stream can be closed. A multicast association closure may be achieved with two degrees of gracefulness, viz. graceful close and abortive close. In a graceful close, the transmitter initiates the termination by setting the WCLOSE bit in an outgoing packet. The transmitter then collects responses from all active multicast receivers carrying the RCLOSE bit. After that it sends a packet with all three bits WCLOSE, RCLOSE, and END set to end the association. In an abortive close, the multicast transmitter sends the END bit at any time which closes the association immediately, regardless of the state of its multicast data stream.

XTP multicast is a collection of mechanisms that support group communication; it does not impose how these mechanisms should be used. In general, it is up to the user to determine how multicast mechanisms are used and how changes in the membership are accomplished. Therefore, the XTP service interface must provide its users with the ability to determine the policies to be used in a multicast association.

2.2 Group Management

Group management in XTP 4.0 is list-based. The receiver group list is created from the replies of the receivers. The multicast transmitter gathers control information during the communication from the receiver group to up-date this list. At regular intervals, a multicast transmitter must send DATA or CNTL packets with the SREQ or DREQ bit set. These intervals can be determined by various, user specified, policies. The transmitter resolves the information in its receiver group list to determine the association's current state according to the replies of the receivers. This is done after a packet with the SREQ or DREQ bit set is sent. If for some reason a CNTL response from any of the receivers does not arrive for some time, the association enters into a synchronising handshake during which the transmission of data is stopped.

The active group is made up of active receivers which are receivers whose control information is used by the multicast transmitter when it runs its control algorithms. List-based multicast offers reliable multicast by tracking the explicit membership of the receiver set designed for reliable delivery.

2.3 Error, Flow and Rate Control in XTP Multicast

The multicast transmitter sends all data packet to the multicast group address. In case of an error in the transmission the lost or corrupted data might be retransmitted to the multicast group. Either positive or negative acknowledgment is used to indicate errors. Retransmission may be either go-back-N or selective. The NOCHECK bitflag indicates if the checksum is calculated over the header fields only or over the whole packet. NOERR is a mode bit that enables or disables error control. When set, it informs the receiver that error correction processing shall be disabled because the sender will not retransmit data. In the multicast case, the multicast transmitter retransmits all data reported lost or corrupted in response to a SREQ bit.

To control data transmission XTP offers two independent mechanisms: flow and rate control. For flow control an end-to-end windowing flow control mechanism is employed. The NOFLOW bitflag indicates to the receiver that the transmitter will not be constrained by flow control. For multicast transmission flow control is done according to the slowest receiver in the active group. Rate control values are determined during the traffic specification negotiation.

3 M-Connection Service

The M-Connection Service is a basic module which supports multimedia multicast communication at the transport level. It offers a protocol independent service interface to transport protocols that provide multicast (1:N) connections and support QoS. This service is provided by a restricted set of service elements which support all functionality needed for communication between one sender and multiple receivers. The service offered by the M-Connection Service allows service users to establish, join, leave, change properties of and terminate *m-connections*. M-connections are multicast transport connections established for the purpose of transmitting multimedia data. Unicast connections are considered as a special case of multicast connections. M-connections are characterised by certain properties, such as QoS and policies to determine the success of the communication, which can be dynamically changed by the service user. All service primitives that comprise the M-Connection Service are shown in table 1.

service element	*primitives*
M-Connection set-up	listen_mconn, open_mconn.req
	open_mconn.ind, open_mconn.rsp
	open_mconn.cnf, open_mconn.ack
Leave M-Connection	leave_mconn.req, leave_mconn.ind
	leave_mconn.cnf
Join M-Connection	join_mconn.req, join_mconn.ind
	join_mconn.rsp, join_mconn.cnf
Change M-Connection Properties	chg_mconn_qos.req, chg_mconn_qos.ind
	chg_mconn_qos.rsp, chg_mconn_qos.cnf
	chg_mconn_qos.ack
	chg_mconn_policy.req, chg_mconn_policy.cnf
Terminate M-Connection	close_mconn.req, close_mconn.ind

Table 1. M-Connection Service primitives.

The aim of the M-Connection Service is twofold, i) to provide the user with a set of clearly specified service elements for multicast communication and ii) to hide the details and specific characteristics of different transport protocols.

3.1 Service Classes and QoS Schemes

The M-Connection Service is designed to support continuous as well as discrete media data transfer. Three distinct service classes to accommodate the requirements of different kinds of traffic are provided:

A - best-effort
B - group integrity and/or QoS
C - group integrity and data transfer reliability

Class A provides a best-effort service without QoS negotiation. However, the sender can specify the QoS of the data stream only to inform the receivers about the traffic type and its QoS. This service class will be used by dissemination services like radio or television broadcast.

The second class, Class B, offers group integrity and the possibility to negotiate QoS. Group integrity in this context refers to the possibility to specify integrity conditions for the m-connection. QoS can be negotiated among all participants; three different negotiation schemes are considered:

- receiver-selected: The sender specifies the maximum QoS it can provide. Each receiver selects the most appropriate QoS which can be supported through mechanisms like filtering [9].
- imposed-QoS: The sender specifies a distinctive QoS which can not be negotiated. Individual receivers can only accept or reject it.
- common-denominator: This has to be agreed on a common QoS among all participants.

The common-denominator case is distinguished from the other cases in two ways. First, the sender might give a range of QoS rather than a clearly specified value. Receivers can propose their QoS within this range. Second, a third message is required to inform the receivers of the common QoS. The QoS parameters we consider are throughput, end-to-end delay, delay jitter, and error rate. A flow specification like that proposed in [10] can be used to fully describe the QoS requirements in such a case.

If QoS negotiation without group integrity is required the parameters determining group integrity are set to nil. If only group integrity is required, no QoS will be specified. Service class B is intended to be used for continuous data transfer and for discrete data transfer where no data transfer reliability is required.

Class C provides group integrity as well as data transfer reliability. Data transfer reliability is provided through retransmission of any lost or corrupted data. The specified QoS in this class gives only an indication of bandwidth and timing requirements for the m-connection. Any reservation of network or end-system resources are done according to this QoS specification[3]. No guarantees

[3] Note that error rate in this class will always be zero since the data transfer is reliable.

for the timely transmission of correct data units are given. Data reliability in this class has always priority over timing constraints.

3.2 Group Communication Policies

The M-Connection Service is characterised by a minimal set of policies which are related to the way a multicast communication is considered successful. Compared to the unicast case it is much more difficult in multicast communication to determine the success or failure of establishment and data transfer. Certain conditions have to be defined which specify the number and/or identity of participants necessary to consider the establishment and/or data transmission successful. We call these conditions establishment and communication integrity conditions (M-AGI, Active Group Integrity) respectively. These conditions might be different for each m-connection.

We distinguish between establishment and communication integrity conditions for two reasons. First, establishment conditions are specified by the initiator of a group communication and will never be negotiated. Communication can only commence if these conditions have been fulfilled. In contrast, communication integrity conditions can be changed and it is left to the service user when and under what conditions this is done. Second, different conditions might be specified for the establishment and the data transfer. A connection set-up, for example, is usually quite costly and therefore it might only be worth proceeding if a certain number of users agree to participate. During the communication, on the other hand, it might be sufficient if only one participant is left who bears the cost of transmission.

The following conditions are considered:

- *key-member* (identity) and/or
- *k* (number)
- *all*
- *quorum*[4].

We are currently considering an upper-bound on the maximum number of receivers permitted. This might be necessary for several reasons; for instance the available resources that can be allocated to a specific m-connection might be restricted.

4 M-Connection Service Elements

The M-Connection Service provides five service elements to set-up, manage and terminate a multicast connection. For each service element the primitives and their parameters are defined in this section . Their function is explained and they are mapped onto corresponding XTP mechanisms whenever appropriate.

[4] For establishment condition *quorum* and *all* refers to a known group, i.e. the number and/or identity of group members is known by the service user or the M-Connection Service. For the communication integrity conditions *quorum* and *all* refer to the number of positive replies during connection establishment.

4.1 M-Connection Establishment

M-Connections are established between a service user who acts as sender and one or more receivers. The initiating user is called *master*.

The service primitives and parameters used during the establishment phase are shown in table 2 and 3, respectively.

primitives	*parameters*
listen_mconn	(@mgroup, service)
open_mconn.req	(@master, @mgroup, QoS, mconn_policy);
open_mconn.ind	(@master, @mgroup, QoS);
open_mconn.rsp	(@user, @master, mconn_id(, QoS));
open_mconn.cnf	(mconn_id, QoS, status);
open_mconn.ack	(mconn_id, QoS);

Table 2. M-Connection Establishment primitives.

parameters	*meaning*
@master	address of the master
@mgroup	address of the receiver group
@user	address of a receiver
QoS	quality of service specification
mconn_policy	policies applied to the m-connection
mconn_id	identifier of the m-connection
status	returns success or failure and a list of the receiver group
service	service class (A, B or C)

Table 3. M-Connection Establishment parameters.

Before an m-connection can be established, all users who wish to participate must issue a *listen_mconn* primitive to be placed in listening state. The parameters carried by this primitive inform the M-Connection Service to which address the user wishes to listen and the type of service expected. This primitive is directly mapped onto an XTP listen. A XTP listening context is established where the address and service value are specified. The XTP traffic specification is left empty.

To establish an m-connection a potential sender issues an *open_mconn.req* primitive. This is mapped onto a FIRST packet which is sent to the multicast group address. The FIRST packet is encoded according to the parameters specified in the *open_mconn.req* primitive. The option bits in the FIRST packet are set according to the profiles shown in the table 4. For each service class a specific profile exists. In order to gather the information from the receiver the SREQ bit must be set. In the classes B and C the SREQ bit is set when integrity conditions and/or QoS have been specified.

Class	*NOCHECK*	*NOERR*	*MULTI*	*NOFLOW*	*SREQ*	*RCLOSE*
A	ON	ON	ON	ON	OFF	ON
B	(i)	ON	ON	(ii)	ON	ON
C	OFF	OFF	ON	OFF	ON	ON

(i) - according to the QoS parameters.
(ii) - according to the kind of traffic.

Table 4. FIRST packet bitflags.

When receiving a FIRST packet an *open_mconn.ind* is issued to each user listening on the specified *mgroup* address. This primitive conveys the QoS specified by the master and the policy to be used for QoS negotiation. Receivers respond either positively or negatively by issuing an *open_mconn.rsp* primitive. In case the SREQ bit has been set in the FIRST packet, a positive response provokes the emission of a TCNTL packet. This is encoded according to the parameters specified in the *open_mconn.rsp* primitive. In case of a negative response, a CNTL packet with the END bit set is issued. If no SREQ bit was set no packet is sent back to the master.

At the master side all replies are gathered and a data structure is built that identifies the receivers and captures their current state. Establishment conditions are validated using these replies. If the establishment is considered successful an *open_mconn.cnf* is issued to the master including the list of the receiver group. If for QoS negotiation the common-denominator scheme is used a TCNTL packet specifying the final QoS is sent to the receivers. The reception of this packet causes the emission of a positive or negative *open_mconn.ack* primitive at the receiver side. If the agreed QoS is equal or "weaker" than the QoS selected by the user a positive *open_mconn.ack* primitive is issued. Otherwise, a negative *open_mconn.ack* primitive is used. In this case, the user will not take part on the m-connection.

A *close_mconn.ind* primitive is issued to the master user if the establishment conditions are not met. Further, a CNTL packet with the END bit set is sent to the receiver group. At the receiver side this packet generates a *close_mconn.ind* primitive.

The complete time sequence diagram for a successful m-connection establishment with the corresponding mapping onto the XTP protocol is shown in the figure 1. Figure 2 depicts a non-successful m-connection establishment time sequence diagram.

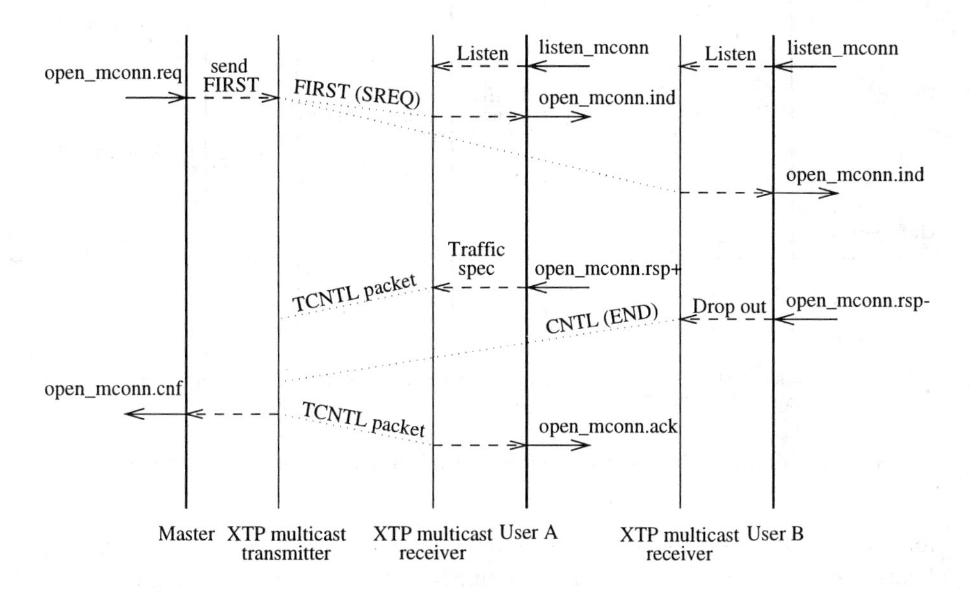

Fig. 1. Successful M-Connection Establishment time sequence diagram.

Group Management. During the establishment phase integrity conditions are specified according to which the establishment and data transfer are deemed successful. The M-Connection Service validates these conditions. In XTP 4.0 the already discussed group management scheme is proposed to accomplish this task. Semantics for what is called *group reliability* are also proposed, but we found the way these semantics are supported are too restrictive and not suitable for our purpose. In particular the proposed synchronising handshake after one or more receivers fail to respond to an SREQ bit does not suit our requirements. During a synchronising handshake all data transmission is stopped. In such a case there will be unacceptable gaps in the media play-out. Lost or corrupted continuous-media data can generally not be retransmitted in time. Therefore, data reliability is not required. Only group reliability or integrity conditions are validated *a posteriori*. The following group management scheme accommodates our requirements better.

Our proposed group management scheme is also list-based. A list of receivers is constructed from the replies to the FIRST packet. This list contains both static and dynamic receiver information. The former contains receiver address

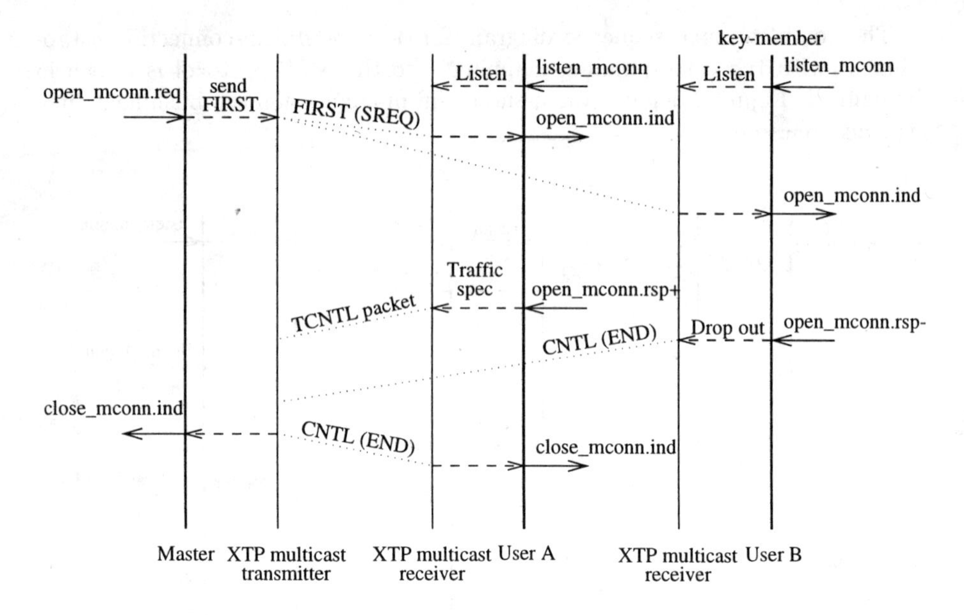

Fig. 2. Non-Successful M-Connection Establishment time sequence diagram.

and return key. The latter, i.e. dynamic part of the list contains the following fields: the latest *echo* value (return to an SREQ), *rseq* (sequence number of the next in-sequence byte expected) and *alloc* (credit left at the receiver).

Packets (CNTL or DATA packets) with the SREQ bit set are sent periodically (SREQ interval). Every reply from a receiver up-dates the dynamic receiver information. After a clearly specified number of SREQ intervals the list is checked and the integrity conditions are evaluated. If a key member has not replied for some time and/or the necessary number of receivers are not present, the m-connection will be terminated.

Group management policies determine how multicast mechanisms are used and how changes in the membership are accomplished. The M-Connection Service provides its users with the ability to specify these policies. For each class different group management policies are used. The following policy areas specified in the mconn_policy parameter are relevant to our service:

- establishment condition - all, k, quorum or/and key-member.
- communication integrity (M-AGI) - all, k, quorum or/and key-member.
- ejection[5] - gives a policy according to which a receiver is forced to leave the m-connection because it breaches certain conditions (e.g. QoS).

4.2 Leave M-Connection

The leave operation is used to drop a corresponding user from an m-connection.

[5] This case might not apply to continuous media where receivers who can not keep up with the pace of the communication do not disturb other receivers.

The service primitives and parameters used in this operation are shown in table 5 and 6, respectively.

primitives	*parameters*
leave_mconn.req	(@user, mconn_id, reason);
leave_mconn.ind	(@user, mconn_id, reason);
leave_mconn.cnf	(@user, mconn_id, status, reason);

Table 5. Leave M-Connection primitives.

parameters	*meaning*
@user	address of the receiver
mconn_id	identifier of the m-connection
reason	reason to leave
status	returns success or failure

Table 6. Leave M-Connection parameters.

This operation may be initiated by three different entities:

1. master user
2. receiver user
3. master provider

In the first case the master user initiates a leave operation by issuing a *leave_mconn.req* primitive in order to drop a specific receiver user from a particular m-connection. The service provider checks the integrity conditions before the operation is executed. It refuses to execute the leave if this would breach the M-AGI. Then, a negative confirmation, *leave_mconn.cnf*, is sent to the master user. This policy was adopted to prevent a master user from releasing an m-connection accidently. In highly dynamic groups this can happen when the master user is not aware of the current state of the receiver group. It should be ultimately a user decision if an m-connection is released or if instead integrity conditions are changed.

If the requested leave does not breach the integrity conditions, this primitive provokes the emission of a CNTL packet with the END bit set to the unicast address of the respective receiver. At the receiver side this packet causes a *leave_mconn.ind* to be issued to the receiver user. The master user receives a positive *leave_mconn.cnf* primitive indicating that the user has been dropped.

Figure 3 shows the complete time sequence diagram for the leave initiated by the master user. The corresponding mapping onto the XTP mechanisms is also depicted in this figure.

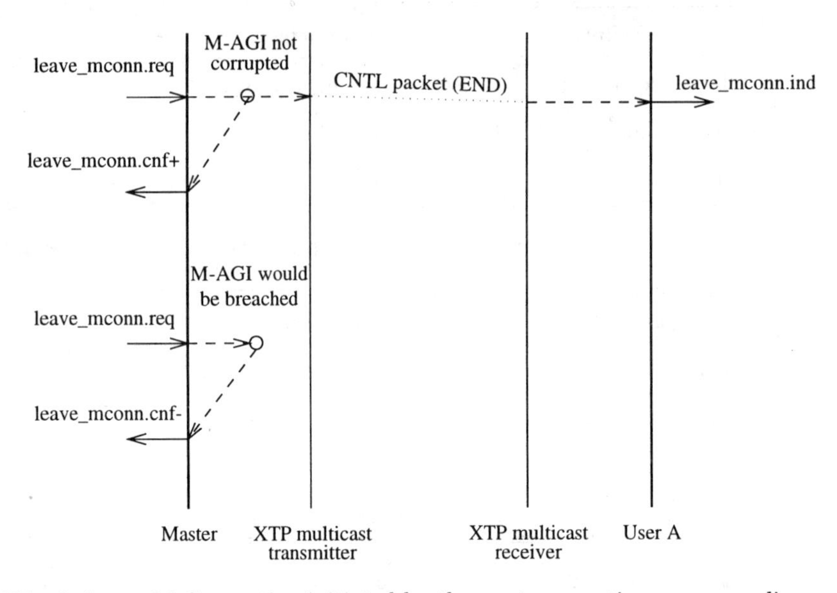

Fig. 3. Leave M-Connection initiated by the master user time sequence diagram.

In the receiver initiated case the receiver user issues a *leave_mconn.req* to quit the m-connection. This primitive is mapped onto a CNTL packet with the END bit set which is sent to the unicast address of the multicast transmitter. After receiving this packet the master verifies the M-AGI conditions. When the M-AGI is not corrupted the master user receives a *leave_mconn.ind* primitive that indicates that the corresponding receiver has left the m-connection. Otherwise the m-connection must be terminated. Figure 4 shows the time sequence diagram for a leave operation initiated by the receiver user.

During the establishment phase, the master user has to specify under which conditions a receiver must be dropped from the m-connection. For example, if the receiver falls too far behind in a reliable multicast data transfer this might jeopardise the success of the data transmission for other receivers. In this case this user should be ejected from the m-connection. Hence, the third case of the leave operation is related to the group management policy. The ejection policy must determine if the service provider has the right to drop certain endpoints even if the M-AGI is breached. This would result in the termination of the m-connection. As above, two cases must be considered. In case the M-AGI is not corrupted, the master provider sends a *leave_mconn.ind* primitive to the master user. This primitive informs the master user that a receiver has been removed and gives the identity of this receiver including the reason. In parallel, it issues a CNTL packet with the END bit set to the receiver that has to be dropped. At

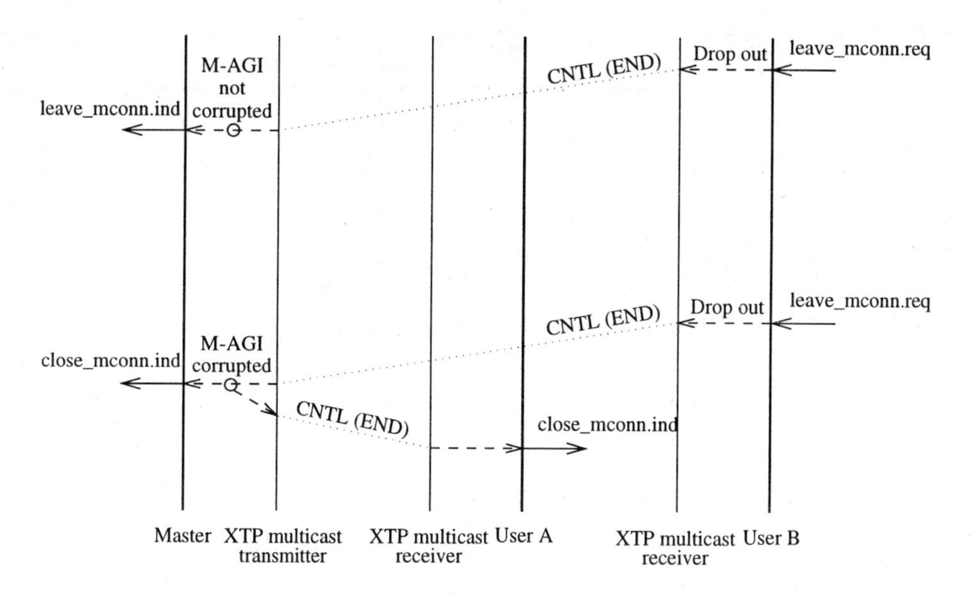

Fig. 4. Leave M-Connection initiated by the corresponding user time sequence diagram.

the receiver side this packet causes the emission of a *leave_mconn.ind* primitive. If the M-AGI is corrupted the master provider terminates the m-connection. The first case is depicted in figure 5.

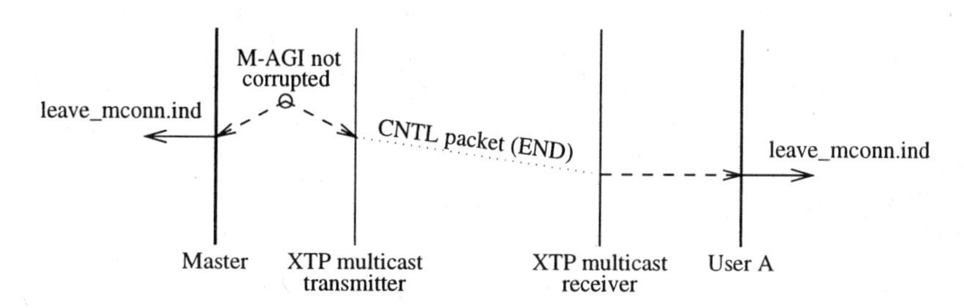

Fig. 5. Leave M-Connection initiated by the master provider time sequence diagram.

4.3 Join M-Connection

The join operation is used to add a new receiver to an m-connection. It is an unconditional service element of type confirmed. The primitives and parameters used to join an m-connection are described in table 7 and 8.

The join is performed in a two-way handshake. The joining user issues a *join_mconn.req* in which it gives its address, the id of the m-connection it wants to join and the required QoS. The *join_mconn.req* is mapped onto an XTP JOIN

primitives	*parameters*
join_mconn.req	(@user, mconn_id, QoS);
join_mconn.ind	(@user, mconn_id, QoS);
join_mconn.rsp	(@master, @user, mconn_id, QoS, status);
join_mconn.cnf	(@master, @user, mconn_id, QoS, status);

Table 7. Join M-Connection primitives.

parameters	*meaning*
@user	address of the joining user
@master	address of the master
mconn_id	identifier of the m-connection
QoS	QoS required by the joining user
status	returns success or failure

Table 8. Join M-Connection parameters.

packet. An incoming JOIN packet is passed as a *join_mconn.ind* to the master user. In the *join_mconn.rsp* the master gives his address, the QoS (in case the policy requires a common QoS for all) and the status, i.e. if the join was successful or not. The response is mapped onto a JOIN packet in XTP. If the join was unsuccessful a DIAG packet is sent. This might be for instance the case when the receiver and/or network can not handle the required QoS. At the receiver side the JOIN packet is indicated as a positive *join_mconn.cnf.* A DIAG packet will be passed as a negative *join_mconn.cnf* to the service user. The time sequence diagram for the join operation is shown in figure 6.

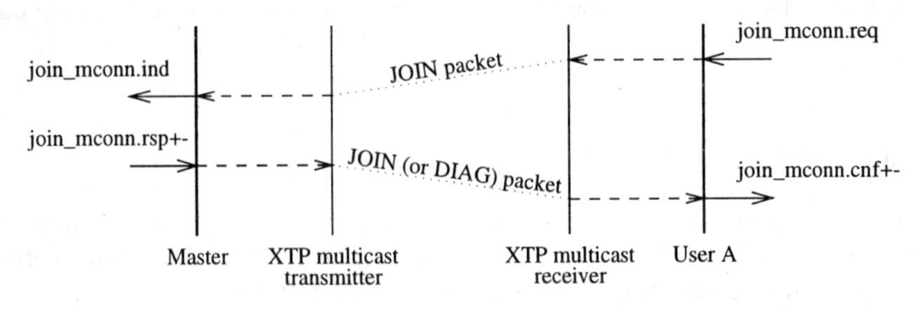

Fig. 6. Join M-Connection time sequence diagram.

4.4 Changing M-Connection Properties

Policies (i.e. integrity and ejection conditions) and QoS of an m-connection can be dynamically changed. The M-Connection Service Interface offers two generic sets of service primitives to change QoS and policies. Integrity conditions can only be changed by the master. The same is true for QoS which is valid for the whole m-connection. A receiver may request to change its QoS in the receiver-selected case. Further, the service provider, i.e. the network might indicate a change in QoS, which would result in an indication to the m-connection service user.

Table 9 shows the primitives provided to change m-connection properties and table 10 describes the parameters of this operation.

primitives	*parameters*
chg_mconn_qos.req	(@user, mconn_id, QoS);
chg_mconn_qos.ind	(@user, mconn_id, QoS);
chg_mconn_qos.rsp	(@user, mconn_id, QoS);
chg_mconn_qos.cnf	(@user, mconn_id, QoS);
chg_mconn_qos.ack	(mconn_id, QoS);
chg_mconn_policy.req	(mconn_id, mconn_policy);
chg_mconn_policy.cnf	(mconn_id, status);

Table 9. Changing M-Connection Properties primitives.

parameters	*meaning*
@user	requesting user address
mconn_id	identifier of the m-connection
QoS	new QoS
mconn_policy	new policy
status	returns success or failure

Table 10. Changing M-Connection Properties parameters.

In the case of receiver-selected QoS scheme a receiver can change its QoS by issuing a *chg_mconn_qos.req*. This is mapped onto an XTP TCNTL packet with the SREQ bit set which is addressed to the multicast transmitter. New resource reservations will be made according to the QoS specification and filters will be re-instantiated if necessary. The multicast transmitter will answer this TCNTL packet with a TCNTL packet addressed to the requesting user stating the agreed QoS. At the receiver this TCNTL packet is indicated as a *chg_mconn_qos.cnf*. The change is only successful if the specified QoS matches the one stated in the *chg_mconn_qos.req*.

In the two other cases (i.e. common-denominator and imposed-QoS) only the master can issue a *chg_mconn_qos.req* primitive. This will be mapped onto a TCNTL packet with the SREQ bit set sent to the receiver group. At the receiver side a *chg_mconn_qos.ind* is passed to the service user containing the QoS that can be provided by the network. The user responds with a *chg_mconn_qos.rsp*. Again, this is mapped onto a TCNTL packet sent to the multicast transmitter. A *chg_mconn_qos.cnf* is issued to the master user. In the common denominator case a last TCNTL packet is sent to the the multicast group stating the new common value. This is passed as a *chg_mconn_qos.ack* to the users.

The time sequence diagrams for changing the QoS initiated by the master and by the user are shown in the figures 7 and 8, respectively.

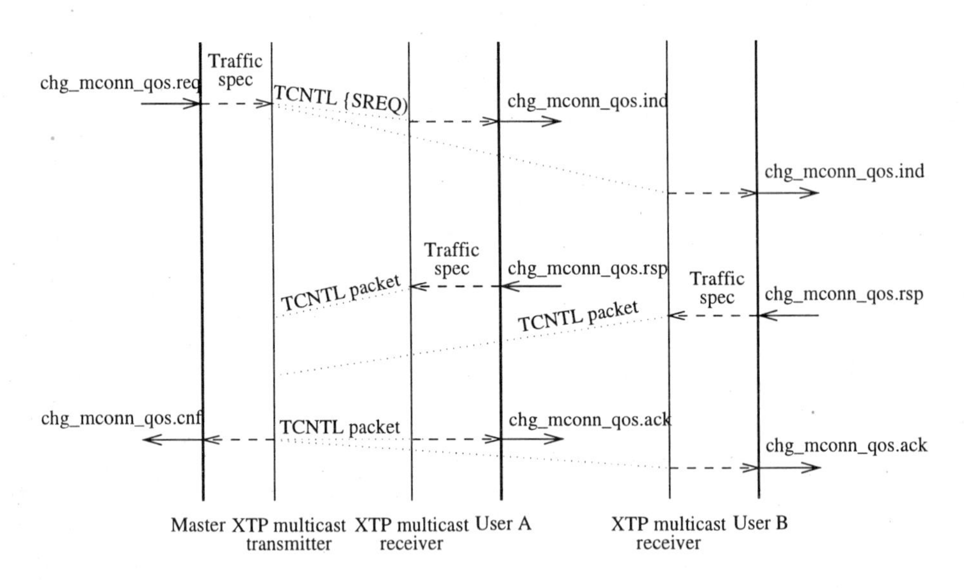

Fig. 7. Master initiated change of M-Connection QoS.

To change the policy associated with an m-connection, a *chg_mconn_policy.req* primitive is issued by the master user including the new policy to be applied. Following this, the master user receives a confirm from the service provider that indicates whether the operation was successful or not. This does not involve any interaction with the receivers and/or the XTP protocol. Therefore nothing has to be sent over XTP.

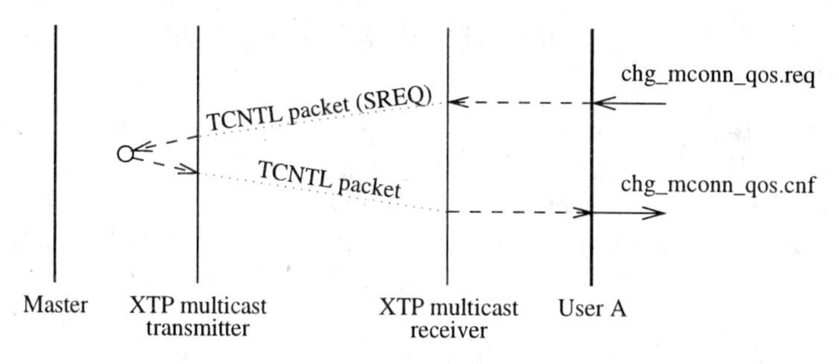

Fig. 8. Receiver initiated change of M-Connection QoS.

4.5 M-Connection Termination

Using this operation both the master user or master provider may terminate an m-connection. It is an unconfirmed service element. The service primitives and their parameters used during the termination are shown in table 11 and 12.

primitives	*parameters*
close_mconn.req	(mconn_id, mode, reason);
close_mconn.ind	(mconn_id, mode, reason);

Table 11. M-Connection Termination Primitives.

parameters	*meaning*
mconn_id	identifier of the m-connection
mode	mode of the closure (graceful or abrupt)
reason	reason to terminate the m-connection

Table 12. M-Connection Termination Parameters.

The M-Connection Service provides two ways to terminate an m-connection, *m-release* and *m-abort*. With the former it is ensured that the m-connection is only terminated when all data was correctly received according to the specified integrity conditions. With the latter an m-connection is aborted immediately.

This can be easily mapped onto XTP abbreviated graceful close and abortive close respectively.

To terminate an m-connection an *close_mconn.req* primitive is issued by the master user indicating whether it is an m-release or an m-abort operation and the identifier of the m-connection. In case of an m-release, this primitive causes the XTP protocol to enter into a standard graceful close procedure for multicast as described in section 2. In the case of an m-abort a packet with the END bit set is sent to the multicast group, the multicast association is terminated immediately. After receiving a packet with the END bit set a *close_mconn.ind* is issued, the m-connection does not exist any longer. Figure 9 depicts the time sequence diagram for the m-release or m-abort operation.

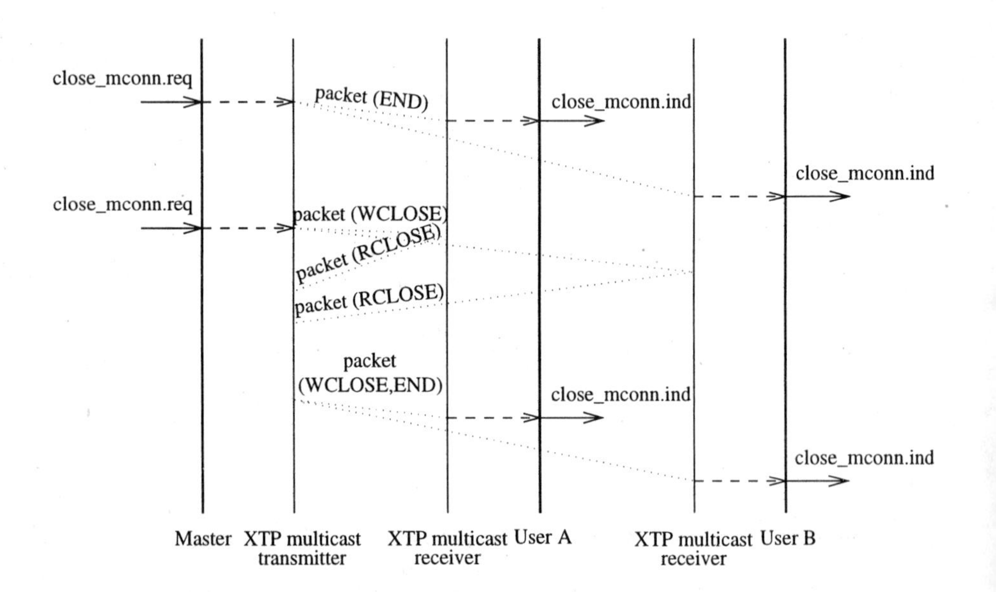

Fig. 9. M-Connection Termination time sequence diagram.

An m-connection can also be terminated through an m-abort initiated by the service provider. This is for instance the case when M-AGI conditions are breached. The master service provider issues a *close_mconn.ind* to the master user and sends a packet with the END bit set to the receiver group. At the receiver side, a *close_mconn.ind* primitive is sent indicating the reason why the m-connection has been aborted.

The m-release will be only used with service class C where data transfer reliability is provided. In any other case an m-connection will be terminated using m-abort.

5 Conclusion

The M-Connection Service is part of a service architecture to support multimedia group communication. It is a protocol independent service that provides an interface to different transport protocols. A set of service elements to support multicast communication is defined. Using this service interface, a user can establish, join, leave, change properties of and terminate multicast connections. These connections are characterised by QoS and integrity conditions. The service specification includes the definition of QoS negotiation and group management schemes. Also, different group management policies are defined which can be specified by the user through the service interface.

The aim of this service is twofold, to provide the user with a set of clearly specified service elements for multicast communication and to hide the details and specific characteristics of different transport protocols.

To show how the M-Connection Service can be provided with an existing transport protocol each service element is mapped onto XTP mechanisms. The corresponding XTP functions are discussed in detail for each service element.

XTP does not dictate in every detail how to use the offered protocol mechanisms. On the contrary, every mechanism and function can be adapted to specific user needs. It gives a high degree of freedom how to use its mechanisms. A drawback of this approach is that descriptions of mechanisms are very often unclear and ambiguous. For instance the way a response to a SREQ bit is generated is considered as an implementation aspect. Sometimes user intervention is required, for example during the establishment of an association. The protocol does not clearly specify if the reply to a SREQ bit is generated by the user or the receiver context. However, the text in the protocol specification indicates that the response is generated by the receiver context without user intervention.

The group management scheme proposed by XTP was apparently designed for reliable multicast. This can be seen in the way a failure to respond to an SREQ bit is treated. In this case the multicast transmitter initiates a synchronising handshake during which the entire data transmission is stopped. This is not suitable for multimedia applications where continuous, uninterrupted data streams are required. Moreover, this actually breaches the design philosophy of XTP not to impose any policy. To overcome this problem we specify our own group management scheme using similar mechanisms.

In the M-Connection Service specification an m-connection is always terminated when the integrity conditions are breached (hard AGI). It is still an open issue if it should be possible to suspend an m-connection in some cases (soft AGI). This is for instance proposed in [12]. However, in the case of soft AGI a problem arises with XTP since it does not provide a way to suspend a multicast association and resume it later. It is left for further research if the support of soft AGI conditions is required and how this could be accommodated by the lower layers.

Another open issue is if join invitation is needed at this level. It is not needed if the availability of the potential user and his willingness to participate is checked before the actual join. On other hand, it is required if a user is called to participate in a telephone like model. In this case, it has to be clear if the service provider has sufficient resources before calling the user. This problem will be also addressed in our future research.

References

1. A. Mauthe, D. Hutchison, G. Coulson and S. Namuye: "From Requirements to Services: Group Communication Support for Distributed Multimedia Systems", in *Proc. 2nd International Workshop on Advanced Teleservices and High-Speed Communication Architectures (IWACA94)*, Heidelberg, Germany, Sept. 1994.
2. C. Szyperski and G. Ventre: "Efficient Support for Multiparty Communication", in *Proc. of Multimedia Transport and Teleservices, International COST237 Workshop*, Vienna, Austria, Nov. 1994.
3. L. Henckel: "Multipeer Transport Services for Multimedia applications", in *Proc. of HPN'94, 5th IFIP Conference on High Performance Networking*, Grenoble, France, pp. 165–183, June 1994.
4. W. J. Clark and J. Boucher: "Multipoint Communications - The Key to Groupworking", *BT Technology Journal*, vol. 12, no. 3, pp. 72–80, July 1994.
5. G. J. Heijenk, X. Hou and I. Niemegeers: "Communication Systems Supporting Multimedia Multi-user Applications", *IEEE Network Magazine*, vol. 8, no. 1, pp. 34–44, Jan. 1994.
6. M. Altenhofen, et.al: "The Berkom Multimedia Collaboration Service", in *Proc. First ACM International Conference on Multimedia*, Anaheim, CA, May 1993.
7. F. Garcia: "Continuous Media Transport and Orchestration Services", PhD Thesis, Lancaster University - Computing Department - Lancaster, UK, May 1993.
8. W. T. Strayer, S. Grayer and J. Raymond E. Cline: "An Object-Oriented Implementation of the Xpress Transfer Protocol", in *Proc. 2nd International Workshop on Advanced Teleservices and High-Speed Communication Architectures (IWACA94)*, Heidelberg, Germany, Sept. 1994.
9. N. Yeadon, F. Garcia, A. Campbell and D. Hutchison: "QoS Adaptation and Flow Filtering in ATM Networks", in *Proc. 2nd International Workshop on Advanced Teleservices and High-Speed Communication Architectures (IWACA94)*, Heidelberg, Germany, Sept. 1994.
10. F. Garcia, A. Mauthe, N. Yeadon and D. Hutchison: "QoS Support for Video and Audio Multipeer Communications", in *Expert Contribution to SC6 WG4 and SC21 - ISO-Meeting*, Beppu, Japan, 1995.
11. XTP Revision 4.0: "Xpress Transport Protocol Specification", XTP Forum, Santa Barbara, CA, Mar. 1995.
12. P. Cocquet: "4th Draft on Multipeer Taxonomy", ISO/IEC-JTC1/SC6/N, Mar. 1995.

GLASS:
A Distributed MHEG-Based Multimedia System

H.Cossmann, C.Griwodz, G.Grassel, M.Pühlhöfer,
M. Schreiber, R. Steinmetz, H.Wittig, L.Wolf[1]

IBM European Networking Center, Vangerowstraße 18, D-69115 Heidelberg

Abstract: This paper is about GLASS[2], a distributed multimedia system that is currently under development. The multimedia department of the IBM European Networking Center in Heidelberg, Germany, participates in the project along with other industrial and university partners. GLASS is an acronym for GLobally Accessible ServiceS, expressing the system's wide range of possible applications. Interactive TV scenarios can be presented as well as multimedia applications for interactive learning and games for entertainment. Clients for standard telecommunications services like FAX and E-mail can be included as well as presentation components providing access to hypertext-oriented internet services like WWW. The system's components can be highly distributed using networks with TCP/IP. An application is driven by MHEG-encoded presentations that allow for the definition of sophisticated presentations. The system comprises multiple server and client components. All components can be operated on a mix of different platforms. Although running on different operating systems, all clients have the same look and feel which is defined completely by the MHEG presentation.

1 Introduction

The GLASS project started in January 1994 and will be finished in December 1995. The goal of the GLASS project is the development of a distributed multimedia system based on the ISO/IEC MHEG standard. GLASS comprises the following components of an interactive distributed multimedia system: client systems, application server, video server and management systems. Gateways to existing services (e.g. TV, radio, World-Wide-Web, electronic mail, FAX, BTX) are currently under development.

The MHEG standardization activity is motivated by the following consideration: Standardization only at the level of monomedia information (e.g. bitmaps, text, video and audio) is not sufficient to guarantee application portability. Monomedia standardization does not address the interchange of multimedia and hypermedia information. By using the MHEG standard, the same presentation can run on different platforms providing the end user with an identical look and feel. In the GLASS project user end-systems are running on DEC Alpha, Intel, Motorola 680x0, Power and Sparc processors under the AIX, Linux, MacOS, DOS-Windows, OS/2, OSF/1 and Solaris operating systems.

---

1. E-mail: {hc,griff,grassel}@heidelbg.ibm.com, mpuehl@vnet.ibm.com, mschreib@heidelbg.ibm.com, steinmet@vnet.ibm.com, {wittig,lars}@heidelbg.ibm.com

2. This project is supported by DeTeBerkom.

This paper intends to give a brief overview of the GLASS project. It presents the major components of the system that were available at the time when the article was written. Their interaction is demonstrated by an exemplary presentation. The components allow the presentation of basic scenarios that make use of some discrete media types as well as continuous ones. User interaction is already supported but there is the possibility to add many more features to the current system. A short outlook to future extensions is given at the end of this paper.

The paper is structured in the following way: Chapter 2 introduces the MHEG standard and the MHEG Object classes. Chapter 3 describes the client and server components which comprise the MHEG run-time environment. Chapter 4 clarifies the interaction of the components by presenting a walk through the system. Chapter 5 gives a conclusion of the work presented in this paper.

2 Overview of MHEG

Primarily, MHEG is the name of the "Multimedia and Hypermedia information coding Expert Group". This group is organized as working group 12 of the ISO/IEC Joint Technical Committee 1/ Sub-Committee 29 [5].

MHEG is also the common title of the document 13552, which comprises the work of this group. The standard supports presentation, representation and manipulation modelling of multimedia and hypermedia applications. By placement with the information interchange standardization efforts, MHEG is related to the Joint Photographic Expert Group (JPEG) [2], the Moving Pictures Expert Group (MPEG) [3] and the Digital Storage Media (DSM) Group [6], which are in the same sub-committee of ISO/IEC JTC1.

This section briefly introduces the role and overall concepts of the MHEG standard.

2.1 General Concept

The MHEG standard provides an interchange format for multimedia and hypermedia information and specifies its machine-independent encoding [8]. It provides generic multimedia information structures which are suited for real-time multimedia applications, synchronization and real-time interchange of applications. Temporal and spacial relationship of monomedia objects can be expressed, timer mechanisms can be applied. User interaction with the presentation objects can be specified in detail, as well as reactions to these interactions. MHEG is also a container and description format for various monomedia formats.

To clarify the semantic information contained in a presentation coded in MHEG, a comparison to monomedia content encoding can be applied. Figure 1 depicts an analogy between the multitude of conventional text processing systems and multimedia editors. Distribution of text documents in electronic form has at the moment settled for PostScript, while in the multimedia and hypermedia domain, a multitude of players are required, one for each presentation coding. Similar to PostScript as the standard page description language, MHEG represents the common coding for multimedia and hypermedia applications. Since an MHEG presentation is not human-readable, it is also called "final form".

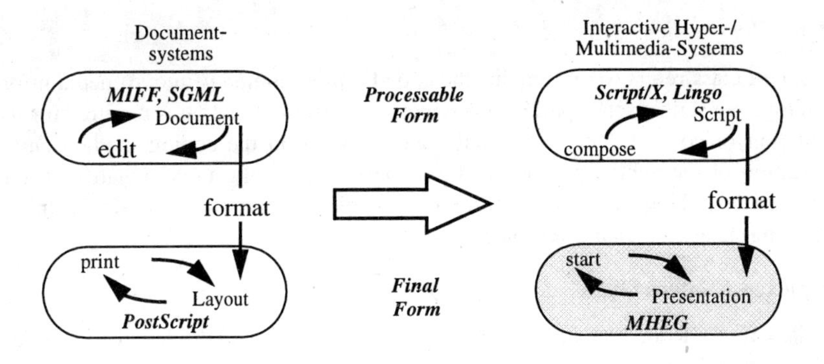

Fig. 1. MHEG as Multimedia/Hypermedia Presentation Format

The MHEG framework consists of five parts. The focus of the GLASS project is on MHEG-1 which defines the "Coded Representation of Multimedia and Hypermedia Objects (ASN.1)", and MHEG-5 which is the MHEG-1 sub-profile for interactive and digital television. The experiences of the MHEG system design and implementation within the GLASS project has influenced the ISO/IEC MHEG-1 and MHEG-5 standardization activities.

2.2 MHEG Classes

MHEG makes an object oriented approach towards presentation interchange. It describes a presentation by means of class descriptions. From these descriptions, classes may be set up in the MHEG interpretation service to form so-called interchanged objects, which can in turn be instantiated to run-time objects. Figure 1 shows

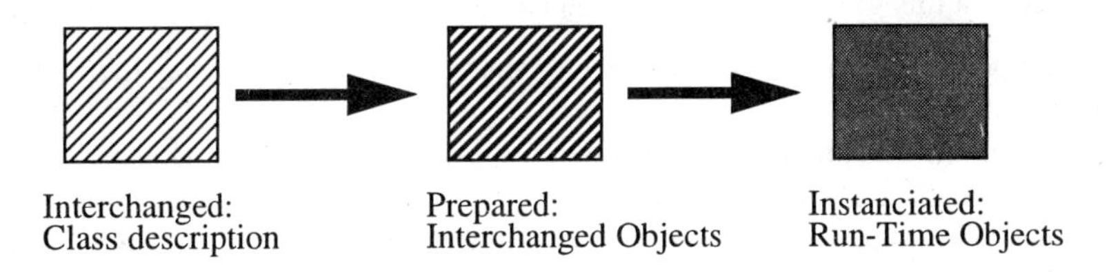

Fig. 2. Object-Oriented Modelling

the three steps that are required by the MHEG interpretation service to create a run-time object. An effect of this modelling principle is that presentations can not be started by the presentation provider, but that the interpretation service must always take the initiative and retrieve at least a single initial object to start a presentation. This start-up mechanism is not described in the MHEG documents and is private to each MHEG presentation system.

Content Class

The content class refers to or contains the coded representation of mono-media information together with a set of parameters containing information that is required for the presentation of content. This set contains information on the coding method and a specification of the application-oriented parameters (e.g. color table, Quality-of-Service parameter). They form the description of object classes that are set up in the MHEG Engine as Interchanged Objects.

Multiplexed Content Class

This class refers to or contains the coded representation of multiplexed media data together with a description of each multiplexed stream.

Container Class

The container class provides a container for grouping multimedia and hypermedia data in order to interchange them as a whole set.

Composite Class

The composite class provides support for multimedia and hypermedia objects that are processed as a single entity. This mechanism provides a consistent approach to the synchronization in time and space. It also provides short-cuts by allowing action objects to be applied to groups of objects.

Action Class

The MHEG standard defines an initial behavior for each MHEG Object. It also describes how to modify the initial behavior of each object by defining a list of elementary actions to be applied to the objects. The actions are used within a link object to describe a link effect. The MHEG standard defines the following types of behavior:

- Preparation: loading of objects into the system and removing them (Prepare, Destroy)
- Creation: instantiation and deletion of run-time objects (New, Delete)
- Presentation: control of the progress of the run-time objects (Run, Stop)
- Rendition: control of the projection of the run-time objects (e.g., Set Speed, Set Size)
- Interaction: control of the result of an interaction (e.g., Set Selectable, Set Modifiable)

Actions are also provided to retrieve a current value of an object's attribute for further processing in Links or Actions.

Link Class

The Link Class defines a structure which specifies a set of relationships. Each relationship is defined between one or more sources and one or more targets. The relationship is composed of conditions associated with the sources (link condition) and the actions to be applied to the targets (link effect). The actions are to be applied to the targets when the conditions become true. An exemplary link object is shown in Figure 3.

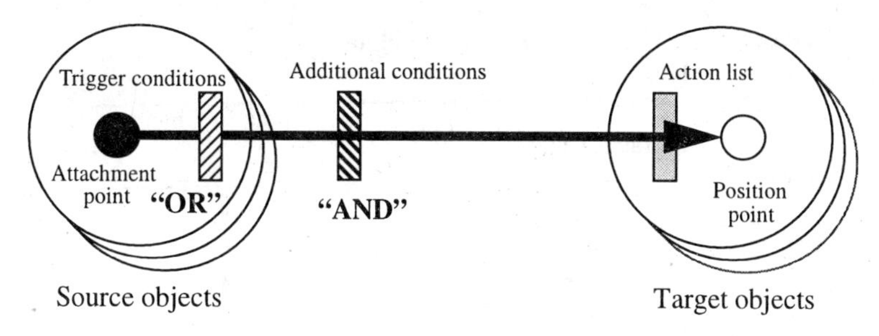

Fig. 3. Link Class

Script Class

The script class defines a container for complex relationships between MHEG Objects, defined by a language not defined within the first part of the MHEG standard. Within the standardization body, the MHEG-3 effort is undertaken in order to provide an exchange format for the scripting language.

Descriptor Class

The descriptor defines a structure for the interchange of resource information about a single object or a set of multiplexed objects. For example, this information is used to facilitate a correspondence between the resources required to present the objects and the resources available to the system.

3 Architecture of the GLASS system

The GLASS system consists of various components that work together to make an MHEG presentation run and to provide the user with visual and audible output as well as with devices for interaction. The components can be distributed over different machines on interconnected networks and they are able to exist in a heterogeneous environment. This is possible because they use project specific application protocols that are based on a standard internet protocol suite that includes TCP/IP. A couple of the protocols will be described along with the components that communicates with each other.

The components can be subdivided into three categories:

- Client Components, consisting of the User Interface Agent (UIA), Presentation Objects (POs), the MHEG Engine (Engine) and the Control Agent (CA).
- Management Components, consisting of the Session Management Agent (SMA) which, in turn, includes the Authoring Agent, the Security Agent, the Locator Agent, the Directory Agent, and the Data Distribution Agent.
- Stores consisting of MHEG Object Stores and Content Data Stores.

Figure 4 gives an overview of the system´s architecture.

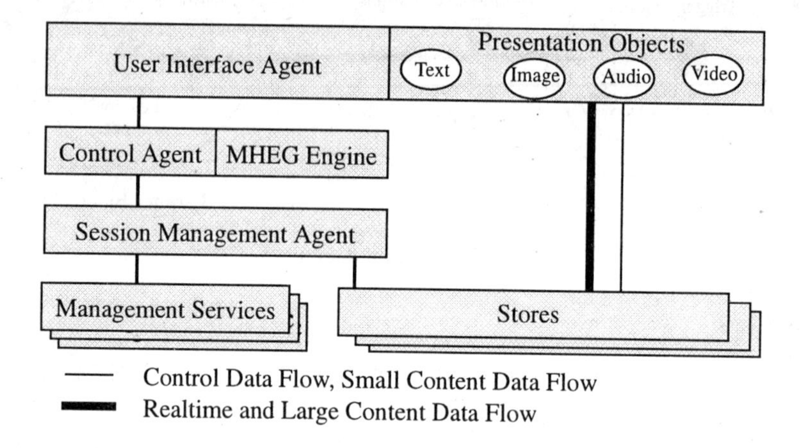

Fig. 4. Architecture of the System

The User Interface Agent is responsible for the management of Presentation Objects that autonomously present the audible, visible and interactive parts of a presentation. The MHEG engine is the heart of the system. It interprets MHEG presentation objects and drives the UIA and its POs. It's communication front-end is the Control Agent. It is connected to the Session Management Agent that manages subcomponents for data distribution, security and accounting purposes. Stores are components that are responsible for content delivery of either discrete or continuous media. The components will be described in more detail in the following paragraphs.

3.1 Client Components

User Interface Agent

The User Interface Agent (UIA) is a client component that is responsible for the creation, maintenance and destruction of Presentation Objects. It is completely driven by the MHEG engine with which the UIA communicates through an intermediate component - the Control Agent. Communication with the Control Agent is done using the User Interface Control Protocol (UICP). This protocol has primitives for session establishment, control and accounting, as well as primitives for Presentation Object control. A typical transaction consists of a Request that is asynchronously answered with a Response. Most requests originate from the MHEG engine that wants the UIA to perform another presentation task. Most of the UICP PDUs (Protocol Data Units) represent atomic MHEG actions that are combined to perform the presentation. Nevertheless there are PDUs from the UIA that either request an SMA service or signal an event to the engine. The Control agent routes these accordingly.

The UIA design is thread-based to allow for concurrent Presentation Objects that will act independently from each other without blocking the UIA from its management tasks. Figure 5 shows the UIA's architecture.

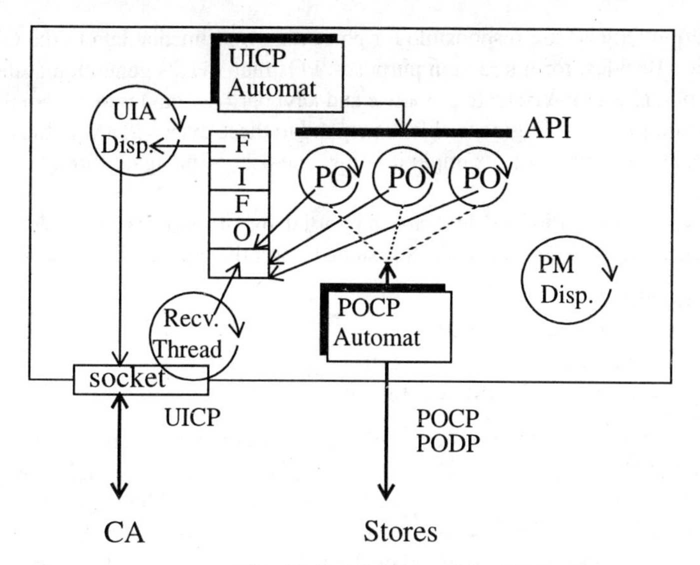

Fig. 5. Architecture of the UIA

All incoming and outgoing PDUs are collected into a queue that guarantees the correct execution sequence. A protocol automaton is used to build an internal object oriented representation of the MHEG presentation´s elements that are currently in use. A set of Presentation Interchange Objects (PIOs) for atomic presentation elements like JPEG Images, MPEG Video, MPEG Audio and Text is constructed. PIOs allow for efficient resource management to prevent the POs from multiple buffering of content data. They also reflect MHEG´s concept of presenting contents in two steps. The first step is used to prepare POs for running, e.g allocating resources, retrieving data and initializing devices. It is completely transparent to the user and ensures that the prepared PO is ready to run. The second step starts the presentation of a PO that then becomes visible (or audible) to the user. The UIA only tells the appropriate PO to start its presentation, the presentation process itself is out of the UIA's scope and up to the PO. When the MHEG Engine tells the UIA to do so, a PO or a PIO with all POs belonging to it will be deleted from memory.

Presentation Objects

Presentation Objects (POs) handle the presentation of content data objects as well as the user interaction with these objects. POs are created, modified and destroyed by the UIA on demand of the MHEG Engine (Engine). Content data objects, managed by Presentation Objects, are either extracted from an MHEG Object, if included, or retrieved from the Content Store, if referenced in the MHEG Object. The Presentation Objects handle the necessary communication in the latter case. They communicate with the store using the Presentation Object Control Protocol (POCP) and the Presentation Object Data Protocol (PODP). The POCP establishes and controls the PODP which performs the actual data transmission. While the POCP provides primitives such as “open data connection”, “start/stop data streaming”, “set stream speed” etc., the PODP transmits raw data of different media types.

Presentation Objects are responsible for presenting multimedia data to the GLASS system's user. Besides, for interaction purposes, POs map events generated at the proprietary graphical user interface (e.g. mouse and keyboard actions) into a form that the Engine can interpret. The Engine is able to respond to these events in a way the presentation author has intended, for example it allows the user to navigate through the presentation.

Media types like audio and text are supported by different types of POs. This allows for easy extension of the GLASS system. Currently supported media types are

- Video (MPEG I),
- Audio (MPEG-Audio and WAVE),
- Images (JPEG) and
- Text (plain text and a GLASS text format)

which are associated with the respective POs. This set of POs can be separated into two subsets of POs according to the contents' media types: discrete and continuous. Video and audio are examples for continuous media types. They may neither be cached nor instantiated more than once at a time from an interchanged object. They require real time flow-control and are played out directly on the local system. Except for this, they work like discrete media. Examples of those discrete media that can be preloaded when the PO is instantiated are pictures and texts. The GLASS architecture for example allows the ImagePO to receive an entire JPEG image, to decode it and to store the decoded image for other ImagePO instances' usage.

MHEG Engine and Control Agent

The MHEG Engine (Engine) is the driving and controlling force of MHEG presentations. Its task which is central to the system is to control the presentation. To fulfil this task, the MHEG Engine interprets the interchanged MHEG Objects. As a result, it issues requests to the presentation system.The interpretation process might also result in requests for retrieval of further objects.

The Engine requests MHEG Objects asynchronously from the Control Agent (CA) which in turn makes use of the MHEG Stores through the Session Management Agent to retrieve these objects. The result of this retrieval process is an asynchronous response as an input to the Engine. Another source of input for the Engine are events. These events originate in the presentation system and are passed to the Engine through the User Interface Agent and the Control Agent. MHEG Objects and events are processed by the Engine. This may result in a request to the presentation systems or trigger the retrieval of further MHEG Objects. Internally, the Engine manages a single event queue which serializes the events received from the presentation system, the notifications of MHEG Objects provided by the CA, and actions that are triggered by internal state changes which are the result of processing such incoming information. Actions will in turn result in the retrieval of new objects, or sending of requests to the presentation system.

The Control Agent (CA) is a service instance which handles all communication with remote or local system components for the MHEG Engine. During a session setup phase, which is always initiated from the client side, the Control Agent establishes a connection to the Session Management Agent and initializes the MHEG Engine.

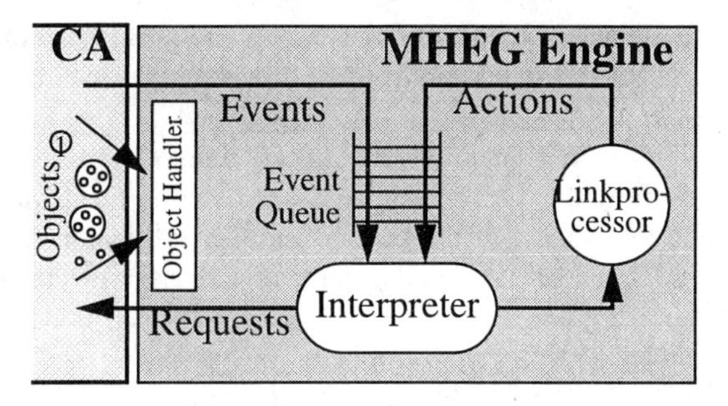

Fig. 6. Architecture of the MHEG Engine

Afterwards, it processes PDUs from the three components according to the automaton described in [1]. Packets can be either translated to another protocol and forwarded or they can be dropped.

As a switching component, the CA is connected to three other components, and it requires up to three different communication mechanisms. While the connection to the SMA is always networked and uses TCP sockets, and the communication with the MHEG engine is always local and uses message queues, the connection between the CA and the UIA uses either of them depending on the way the UIA connects to the CA.

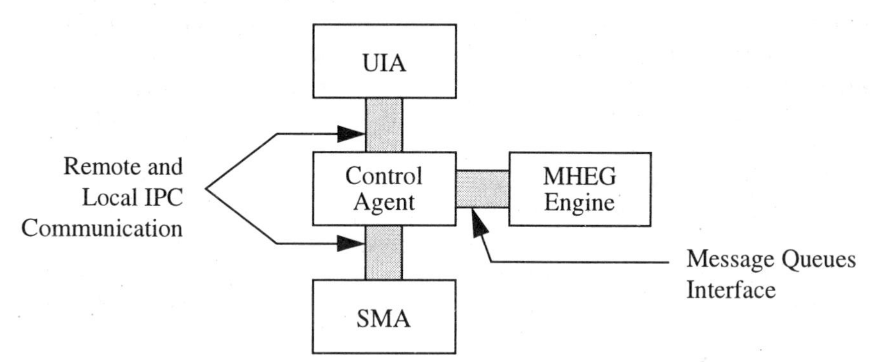

Fig. 7. Control Agent's Interfaces

Besides handling the communication for the Engine, the Control Agent services the timer interrupts required for the timer management of the Engine. For a distributed environment, the Control Agent is implemented to work on both client and server sides of the system. Basically, putting the CA and Engine on the client achieves better performance, while putting them on the server reduces resource requirements on the client system and achieves greater flexibility. This flexibility has major effects on the design of the component, in that it must allow an efficient implementation on the clients' operating systems that can also be applied to a server operating system without major changes.

The state information held in the CA does not contain information about the activity state of a particular presentation or presented object. Rather than that, the state information is only concerned with the connection itself. Because of this, messages can arrive during the run of the session from all three neighboring entities in no particular order as far as the knowledge of the CA is concerned.

In order to service all different kinds of connection, the CA consists of two threads, one of which exists initially. These two processes have no need for communication except for common knowledge of the state of the CA automaton. Thus, this state is kept in shared memory and accesses to the state variable are protected by semaphores.

At start-up time, only one thread, the CA-Server-Thread, is active. The MHEG Engine is uninitialized. When a UIA connects to this server and sends a Connect Request message, the CA-Server-Thread establishes a connection to a Session Management Agent (SMA), initializes the MHEG Engine and splits into two threads, namely the CA-Sender-Thread and the CA-Receiver-Thread. It is advisable that the MHEG Engine runs as another thread in the same process as the CA, allowing for message passing in shared memory.

3.2 Server Components

Session Management Agent

The Session Management Agent (SMA) organizes the retrieval of MHEG Objects and content data objects using the Locator Agent, Accounting Agent, and Security Agent. The Control Agent requests MHEG Objects from the Session Management Agent that retrieves these objects from the MHEG stores and passes them back to the Control Agent. This mechanism is not used for the transfer of referenced content data objects but only for MHEG Objects and control information. The referenced objects are transferred between the Content Data Stores and the Presentation Objects by individual transport mechanisms. However, the transport is initialized by the Session Management Agent.

The communication between SMA and Stores is performed using the Store Control Protocol (SCP). This protocol contains primitives for the initiation of data transfers between Presentation Objects and Stores/Gateways, for the retrieval of MHEG Objects from MHEG Stores, and for accounting purposes.

There are several subsystems and sub-agents in the SMA that will not be discussed in this document.

Content Store

The Store is responsible for the storage of content data and the transmission of this on request towards its clients. As such, it provides:

- mechanisms to control the exchange of content data with Presentation Objects,
- transport system mechanisms for transfer of continuous-media data.

In later stages of the implementation the Store will offer additional transport system mechanisms for transfer of continuous-media data with varying transmission characteristics depending on the Quality of Service (QoS) requirements of clients, functionality to pass accounting information back to the Session Management Agent and protocols for the communication with Locator Agent and Data Distribution Agent.

The Stores communicate via exchange of PDUs with Presentation Objects (POs) and the Session Management Agent (SMA).

The Store consists of two components (Figure 4):

- a Video on Demand (VoD) Server including appropriate stream handlers (SH) to perform the handling of multimedia data, i.e., the storage and transmission of continuous-media data, and
- a Control Process (*ctrlproc*) which communicates via GLASS protocols with other GLASS system components (i.e., POs and SMA) and instructs the *VoD server* accordingly to deliver the required multimedia data stream.

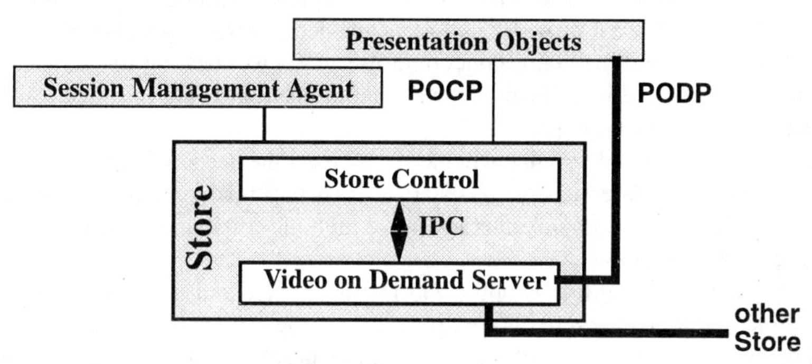

Fig. 8. Store Interfaces

Control Process

The *ctrlproc* consists of and uses various subcomponents:

- Communication Interface to exchange information with other GLASS components
- Handler modules (pocphdlr, scphdlr, credhdlr) for the control protocols POCP and SCP,
- Peer, the internal representation of a peer, i.e. the receiving Presentation Object on the receiving side,
- Key, the internal representation of an SMA key,
- cm_stream, the internal representation of a content data stream transferred to a Presentation Object
- server_conn, the module which controls the communication with the *VoD server*

To set up a stream, two prerequisites are required. Firstly, the SMA must provide a key to the control process, secondly, the peer must authenticate itself with this key for content retrieval. If this authentication is successful, the stream is opened. Further requests from the peer are required to start streaming of data.

Processing within the *ctrlproc* is done within one thread. Since it is not responsible for the transmission of large content data such as continuous-media data but performs control operations only, this incurs no performance drawback and simplifies implementation and testing.

Video on Demand Server

The Video on Demand server used within GLASS is a server designed for IBM RISC System/6000 workstations running under IBMs AIX Version 3 operating system. To provide flexibility, the server allows dynamic addition of hardware- and protocol-specific stream handlers.

The server is designed to provide high quality, guaranteed, on-time delivery of continuous-media data such as video, audio, or animation from a server to desktop computers located around an existing network. It allows delivery of *streams* of multimedia data (e.g. video) to desktop computers for applications wishing to "play" multimedia data. At the same time it provides non-real-time access through existing network filesystem interfaces and maintains concurrent use of the workstation for standard applications. This is possible by guaranteeing smooth play-out using resource reservation mechanism as described in [9].

The filesystem can determine whether it can meet the needed quality of service for a specific stream before the stream is started. It ensures that it can meet these requirements by tracking the complex interactions of multiple simultaneous streams with different qualities of service.

It provides a format that is optimized for the storage and retrieval of multimedia data while traditional filesystems are optimized for the storage of small data files.

This is enhanced by a set of generic multimedia services and resource reservation for UNIX systems. It utilizes the facilities provided by the filesystem to provide an integrated continuous-media server and to provide generic resource control for the various subsystems. They also provide interfaces to easily load new modules to support different types of networks for customers with special requirements.

4 Interaction of the components

The last chapter introduced the components of a GLASS system in a very abstract manner. This one wants to clarify their functionality and the communication among them from a more practical perspective. A very simple presentation will be used for a walk through the system.

The elements of a simple presentation are shown in Figure 9. The first page consists of a background image and two small images labelled 'Video' and 'Next'. Both labelled images are defined to be selectable, which makes them buttons. In the center of the background image is a frame that is used to overlay a video clip. If the user presses 'Video' the video clip will be displayed in the overlay area. Clicking the 'Next' button will lead to page 2.

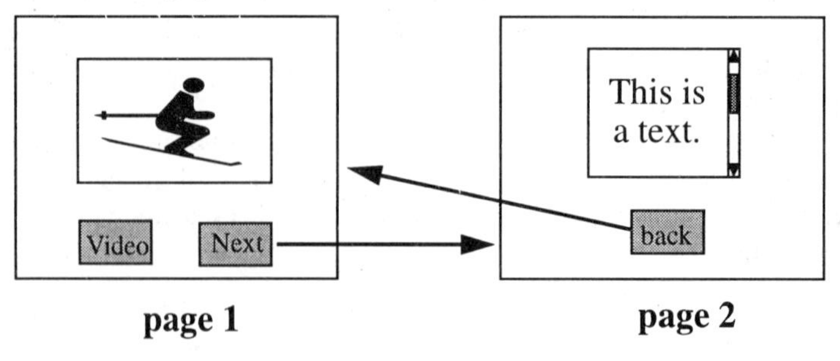

Fig. 9. Simple Presentation Example

Page 2 consists of a background image that is overlaid with a smaller text window. The user can scroll through the text using a scroll bar and up and down buttons. A small image labelled 'back' is positioned below the text window. Button functionality is assigned to it. If it is clicked it leads back to page 1.

Now let us start a walk through the components. All MHEG Objects that belong to the presentation are stored in an Object Store. All contents, the discrete as well as the continuous, are stored in a Content Store. We assume that the SMA is set up to receive requests (e.g. running or automatically starting by the inetd) and that the MHEG Engine is waiting for our presentation's initial object. Figure 10 shows the connection setup as it is requested by the UIA. When the user starts the UIA, e.g. by clicking on its icon on the client machine, the presentation begins. A UICP Connect Request PDU is sent from the UIA to the CA. The CA forwards the request to the SMA and responds to the UIA after the connection between SMA and CA is set up. Then, the CA retrieves the initial object and feeds it into the Engine which starts interpreting it.

We assume that in the described case, the initial composite object contains all MHEG Objects required to present the first page. Any further object requests from the engine would be satisfied by the CA that communicates with the SMA using the SMCP protocol. With each object request, the SMA checks from which store the MHEG Objects are available and retrieves them. The retrieval mechanism and the communication between the SMA and the media stores are out of the scope of this document.

While interpreting the initial composite, the MHEG engine is usually required to prepare monomedia contents that are referenced in the presentation. A presentation without monomedia content does nothing user-detectable, and a presentation consisting purely of included content is not feasible for a retrieval system like the GLASS

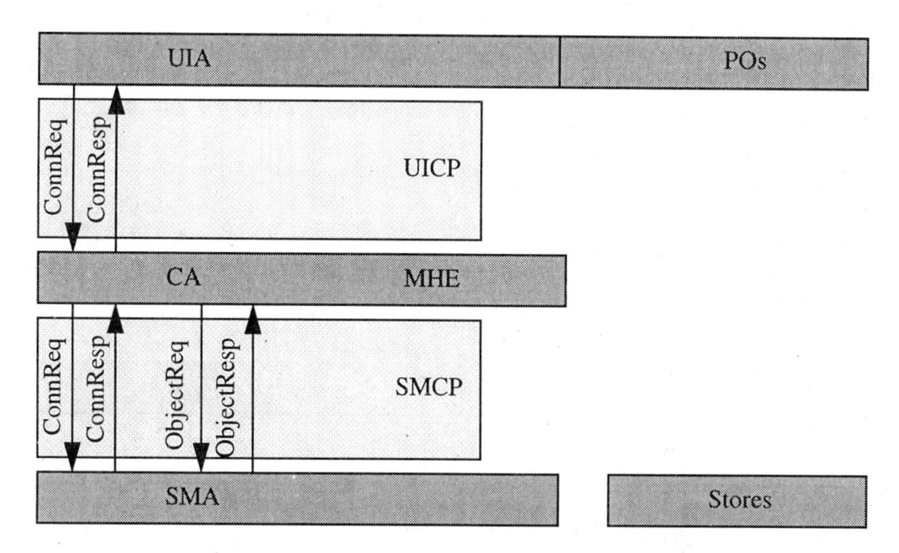

Fig. 10. Communication among the components: startup

system. It is advisable to transport small data portions as included data because of the transportation overhead, whereas big data quantities should be transmitted as referenced data because all included content data is copied into the MHEG engine once.

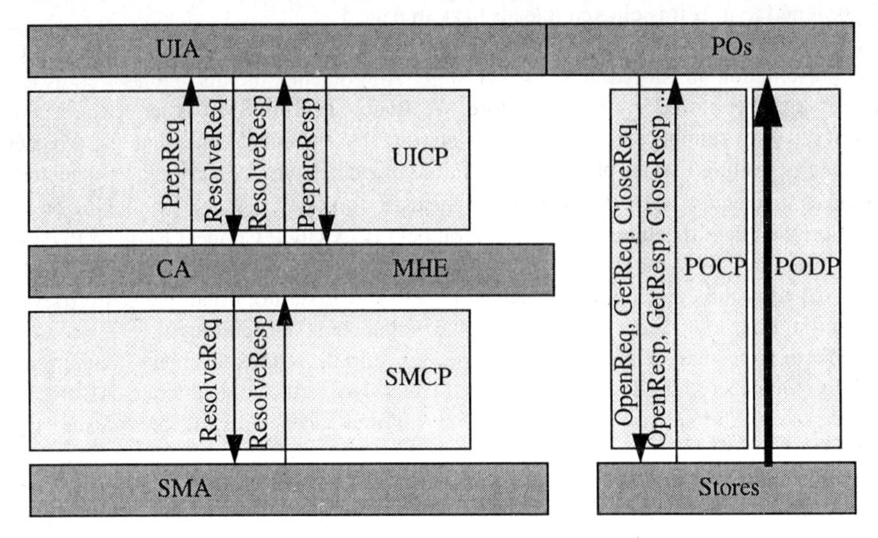

Fig. 11. Communication among the components: content preparation

Figure 11 shows how the MHEG engine uses the UICP primitive PrepareRequest to tell the UIA which Presentation Object is to be prepared for execution. In our example a PrepareRequest PDU for each of the three images on the first page is sent to the UIA. The UIA builds internal representations for these objects that are called Presentation Interchange Objects (PIO). Such a PIO has to communicate with the SMA for resolution of the references to monomedia content (ResolveReq and ResolveResp primitives). The resolution is provided as a uniform resource locator (URL) indicating the protocol required (POCP), machine name, optional port number, and an additional string resembling a Unix-like path to identify the content.

The PIO then uses the POCP protocol to build a connection to the store that is pointed to by the URL's machine name. When a connection has been established the PODP protocol is used to retrieve the data. As content is stored and transferred in a standard format, e.g. JPEG for images, it is necessary to convert it to the client machine's local content type. After this is done, the UIA answers the successful preparation of the objects with a PrepareResponse.

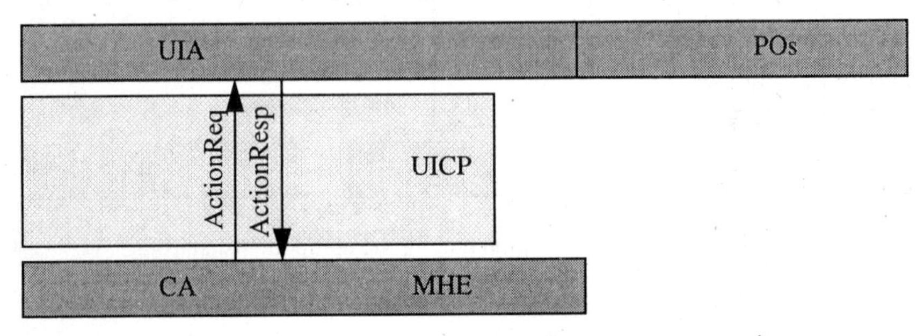

Fig. 12. Communication among the components: actions

After the prepare phase has finished, the engine can execute actions that are instances of the Behavior Class on the prepared objects. An action is considered finished after it has sent an ActionReq PDU to the UIA, been synchronously executed there, and an ActionResp PDU has been sent back to the engine as shown in Figure 12. The first action sent to a Presentation Interchange Object (PIO) has to be the New action. For each prepared image, one New action is sent from the engine to the UIA. The UIA creates for each New action a Presentation Object (PO). Each PO derives from a PIO whose content it inherits and shares with optional other instances of the same PIO. Thus the images have to be stored in the client's address space only once. After instantiation, a PO is ready for presentation to the user.

The MHEG engine now instructs the UIA to perform a Run action on each Presentation Object to show it on the screen. In parallel the engine interprets some Behavior objects that enable the images labelled 'Video' and 'Next' to be selectable and installs some Link objects that start video playback and creation of the second page, respectively, when one of the (now selectable) images is clicked. These are actions propagated to the UIA using the UICP protocol as well. The UIA itself tells the two POs to be clickable and to send back a message when being clicked.

When the user clicks the 'Video' button, the UIA informs the engine of the event. The engine knows that the button click triggers the video playback and sends a Run action on the video Presentation Object to the UIA. A video clip is a special type of referenced data object, a continuous one. It can not be downloaded in a burst like an image. The control (POCP) and data (PODP) connections between the store and the PO as presented in Figure 13 have to remain open during the existence of the video PO. Data is transmitted continuously as a real-time data stream until the end of the clip is reached or another event in the presentation results in preliminary destruction of the video PO. Then the connection has to be closed. In our simple example the user has to

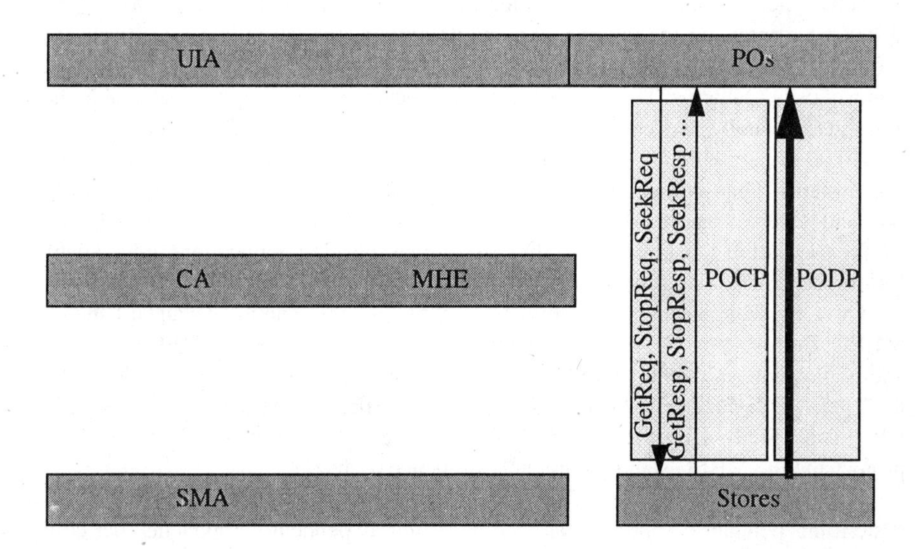

Fig. 13. Communication among the components: video playback

wait for the end of the video clip until he is able to interact with the presentation again. Nevertheless, it is possible to express a VCR like functionality in an MHEG presentation using the SetSpeed and SetPosition actions.

Clicking the 'Next' button will lead to the second page. The MHEG engine will (in our example) clean up all resources used by the first page and will instruct the UIA to do so as well. The Stop action is used to hide a Presentation Object from the screen. The Delete action is the counterpart to the New action and will delete a Presentation Object. The Destroy action cleans up all resources that were allocated with the Prepare action.

The second page is created in the same way as was explained for the first page. Clicking the 'Back' button will lead back to the first page that has to be created again as described above. This procedure can be sped up considerably by not destroying downloaded contents in a page transition that are expected to be reused. As we have seen our simple presentation example is a kind of endless loop. It can only be ended by stopping the UIA. In that case, a DisconnectReq PDU is sent to the CA that answers with a DisconnectResp PDU. After that the MHEG engine is re-initialized and waits for a new connection from the same or another UIA to start the presentation again.

5 Conclusion

The following chapter summarizes the results of the paper and gives an outlook on future extensions to the system.

5.1 Summary

Before coming to a conclusion about the results of the GLASS project as presented in this article, it may be helpful to think about the situation of today's multimedia applications.

Today there are only few distributed multimedia systems available. Most of the systems are intended for running on a single system, though multiple platforms are supported. These are kiosk systems and presentation systems for Point-of-Information/ Point-of-Sale scenarios. There are a lot of multimedia authoring systems available that allow user friendly and easy construction of those applications. However, there is a lack of interoperability and interchangeability of the presentations among the presentation systems. Thus, content providers have to decide which systems they want to support and which to ignore. This decision will be risky for them as well as unsatisfactory for the user in the consumer market because only a subset of applications are available for him depending on his decision for a system.

Those limitations can be overcome by sticking to an international standard such as MHEG. In this sense GLASS is a pioneer project. It makes use of MHEG's concepts and implements an engine that is capable of interpreting MHEG Objects as defined in the MHEG DIS document. On the other hand the MHEG standardization process from DIS to the final IS profits by the GLASS designers' experiences. So the GLASS project and the MHEG standard mutually influence each other.

The GLASS architecture defines a distributed system that allows coexistence of system independent components as well as system dependent components that communicate using protocols based on TCP/IP. System independent components can be easily ported to different platforms and system dependent components are shielded to allow a heterogeneous system. Since Presentation Objects make use of standard media types like JPEG images and present them without applying features that are specific to a graphical user interface, the same look and feel is maintained across all platforms.

Coming to a conclusion one can say that the GLASS system is demonstrating the successful use of a standard multimedia and hypermedia exchange format like MHEG to guarantee both the content provider's and the end user's investments without losing flexibility.

5.2 Future Outlook

The GLASS system as described in this article provides all the basic means for a flexible and extensible multimedia system. These means can be used in many ways to support the user with global services that really demonstrate the power of the architecture.

Currently under development are gateways that build interfaces to standard services like TV and radio broadcast, BTX, World Wide Web (WWW) and Electronic Mail. Some gateways make use of facilities that allow real-time compression of continuous media using standards like MPEG. Others convert hypertext information and semantic actions to MHEG Objects that are used to build MHEG presentation parts dynamically.

Another important task will be to develop tools that allow designers to quickly and easily produce MHEG-based multimedia presentations. Those tools would help to increase the number of available presentations rapidly. Until then presentation designers could be provided with converters from other multimedia authoring tools to MHEG.

Since the MHEG standardization is still in process, it will be necessary to modify the engine when the official MHEG International Standard document is passed.

As outlined, there is still some work to do but the existing GLASS system provides a good starting platform for further research activity.

Acknowledgements

The authors thank all partners of the GLASS consortium, namely DEC CEC, DeTe-Berkom, GMD Focus, Grundig Multimedia Solutions, TU Berlin PRZ for the co-operative work. Many thanks to DeTeBerkom for sponsoring and supporting the project. The authors express their gratitude to Hans Werner Bitzer, Thorsten Illies, Stefan Koenig, Thomas Meyer-Boudnik and Olaf Rehders for their discussions and contributions.

References

1. DEC, GMD Fokus, Grundig Multimedia Solutions, IBM ENC, Technical University of Berlin PRZ: *BERKOM Globally Accessible Services: System Specification 1.0.* BERKOM, Berlin, May 1994.

2. ISO/IEC IS 10918:1992: *Information Technology – Digital Compression and Coding of Continuous-Tone Still Images* (*JPEG*). 1992.

3. ISO/IEC IS 11172:1992: *Information Technology – Coding of Moving Pictures and Associated Audio for Digital Storage Media up to about 1.5 Mbit/s (MPEG).* 1992.

4. ISO/IEC IS 11544:1992: *Information Technology – Digital Compression and Coding of Bi-level Images (JBIG)*. 1992.

5. ISO/IEC WD 13818-6:1994: *Information Technology – Coded Representation of Multimedia and Hypermedia Information Objects (MHEG) – Part 1: Base Notation (ASN.1)*. June 1993.

6. ISO/IEC CD 13552-1:1993: *Information Technology – MPEG-2 Digital Storage Media Command and Control Extension (DSM-CC)*. November 1994.

7. ITU-T Draft Recommendation T.170: *Audiovisual Interactive (AVI) Systems – General Introduction, Principles, Concepts and Models.* Second Revision, Geneva, CH, 16-25 November 1993.

8. T. Meyer-Boudnik, W. Effelsberg: *MHEG - An Interchange Format for Interactive Multimedia Presentations.* Accepted for IEEE Multimedia Magazine, 1995.

9. Carsten Vogt, Ralf Guido Herrtwich, Ramesh Nagarajan: HeiRAT: The Heidelberg Resource Administration Technique - Design Philosophy and Goals. Tagungsband Kommunikation in Verteilten Systemen, Munich, Germany, March 3-5, 1993

Performance Evaluation of the CSCW Application JVTOS

Espen Klovning[†], *Olivier Bonaventure*[‡]
[†]*Telenor Research and Development, P.O. Box 83, N-2007 Kjeller, Norway*
[‡]*Institut d'Electricité Montefiore, B-28, Université de Liège, B-4000 Liège, Belgium*
Email: Espen.Klovning@fou.telenor.no, bonavent@montefiore.ulg.ac.be

Abstract

The primary contribution of this paper is an experimental evaluation of the architectural design of an application for computer supported cooperative work (CSCW) as well as an investigation of the traffic pattern generated by the application. The evaluation is done in a local area ATM network based on instrumentation of the main data path combined with resource monitoring of the end-system. The evaluation shows that the current architecture of the studied CSCW application JVTOS under X11R5/SunOS will not scale to a larger number of users. To summarize, JVTOS works nicely as either a PicturePhone or for cooperative work. Combining them will in some cases be a problem since the performance of the application sharing will suffer due to the CPU intensive PicturePhone. Our views of problem areas including transport layer protocol issues are underlined and potential solutions are discussed.

1 Introduction

The expected large scale deployment of high speed metropolitan and wide area ATM networks will provide an excellent opportunity for bandwidth demanding multimedia teleservices. One type of teleservice we think has potential especially among commercial users is computer supported cooperative work[1]. With ATM networks it will be possible for users located several thousand kilometers apart to communicate almost regardless of the distance between them. Thus, we are confident that cooperative work will bring computer supported video conferencing one step further. While audiovisual communication can not be replaced, computer supported cooperative work can make daily work easier, provided the performance of the system is good enough. It is obvious that user acceptance is very unlikely, unless the performance is acceptable. The performance of a CSCW system depends on a number of different factors including the chosen system architecture and of course the actual implementation. What is really important for CSCW applications is that the performance perceived by collaborating users for shared applications is comparable to the performance they are used to locally. In addition, the CSCW application should allow at least 3-4 collaborating partners with reasonably good performance. Another important issue is the traffic generated by the CSCW application.

In future ATM networks possibly spanning several different administrative domains, compliance with the requested traffic contract will be essential. The policing functions in the networks are used to protect the network from misbehaving users. Violations of the ATM level traffic contracts will normally lower the performance of the end-to-end transport layer protocol significantly. Thus, detailed knowledge about the basic characteristics of the generated traffic will be necessary for any CSCW application.

Our paper is organized as follows. Section 2 gives a short introduction to the different components of the CSCW application JVTOS, and an overview of the overall architecture. Section 3 includes an overview of the experimental environment and our evaluation methods. Sections 4 and 5 include the results from our experimental evaluation. The end-system utilization is presented in Section 4 while Section 5 presents an overview of the traffic pattern generated by JVTOS. The paper closes with some concluding remarks in Section 6.

2 Joint Viewing and Tele-Operation Service

Joint Viewing and Tele-Operation Service (JVTOS)[1] is a CSCW application designed for an environment with heterogeneous computers and networks sharing standard single user applications (e.g. spreadsheets, desktop publishing, multimedia applications) which are usually used locally with homogenous systems (hardware and operating system) running cooperation-aware applications. With JVTOS, it is possible to share a standard single-user application between heterogeneous systems over long distances[2]. Typical scenarios for JVTOS are heterogeneous desktop conferencing, distance learning, cooperative engineering and telecommuting.

Applications supporting CSCW are sophisticated. An important part of these applications is control and management of all collaborating participants. In JVTOS, the control and management of the collaboration are organized in sessions which are controlled and managed by a session manager process. Each session is started and managed by one of the participants hereafter denoted chairman. The chairman decides the admission policies, and has power to invite and exclude session partners.

JVTOS is built as a collection of nearly independent processes which interact to a limited extent. The three main components are; 1) PicturePhone, 2) Application sharing and 3) Telemarkers. These independent components are managed and controlled by a common session manager which is the core of the JVTOS application. JVTOS uses UDP (User Datagram Protocol) and TCP (Transmission Control Protocol) for communcation of data and control information between remote machines. The application has been implemented for Macs and PCs as well but in this paper we focus exclusively on the X11R5 implementation for SunOS.

2.1 Picture Phone

The Picture phone (PP) provides audiovisual communication between the participants in a session. All participants can export their own audio (A) and video (V) stream and import the incoming AV streams from other participants at will.

The audio part uses the available audio device in the workstation and offers two audio qualities based on 16 bit samples. The sampling frequencies in our Sun Sparc 10 stations are 22khz for the high quality or 11 khz for the low quality. Thus, the bit rate of the audio stream is either 352 kbps or 176 kbps respectively. Compared to the bit rate of other audio conferencing tools [3] which use standard audio compressing algorithms, JVTOS demands a high bandwidth for the audio channel.

The video stream requires a high-end video board, Parallax Power Video, to provide the requested video quality. This video board offers JPEG (Joint Photographic Expert Group) compression on an embedded C-Cube microprocessor on the video board. The JPEG compression done by the C-Cube processor is controlled by a software configurable quantization factor which decides the compression quality and thereby the image size. The video quality available to JVTOS users is split in three different cathegories, high, medium and low. The frame rates of these qualitites are user configurable but the default rates which we have used in this evaluation are 20 frame/s, 7 frames/s and 1 frame/s respectively. The Parallax video board generates interrupts upon digitalization of a frame, and after compression or decompression is done. Both the audio and the video quality can be changed at any time during a JVTOS session. Hereafter we use the acronyms HQA (High Quality Audio), LQA, HQV(High Quality Video), MQV and LQV for the different audio and video qualities respectively.

2.2 Application Sharing

Most application sharing solutions under X-windows use an X-multiplexor between the X-client application and the X server. The multiplexor intercepts the X traffic to the client application in order to distribute it to the other sites. In JVTOS, the implementation is based on a distributed approach where the multiplexor is divided in a pseudo server and pseudo client modules. These two modules communicate using an extension of the X protocol denoted the X' protocol which supports multicasting of X requests. The application sharing component in JVTOS uses a centralized architecture with single-instance application execution. The shared window system is distributed which means that sharing functionality is distributed among all participating nodes for reliability and scalability. X11R5 events are collected and distributed from each node. Figure 1 illustrates how the different modules communicate when sharing an X client application. The pseudo server on the local machine running the application distributes the X requests and the resource information to the local and remote pseudo clients. These clients interact with their corresponding X server which, in this case offers a video extension suitable for the Parallax Video Board. The remote pseudo server is idle until a shared application is started on that machine. All the participants in a session can start applications for sharing. JVTOS offers several different floor control policies (i.e. the policy controlling user input to shared applications) including implicit mode, chaired mode, baton mode, FCFS mode. The participant who started an application controls the floor control policy of that particular application.

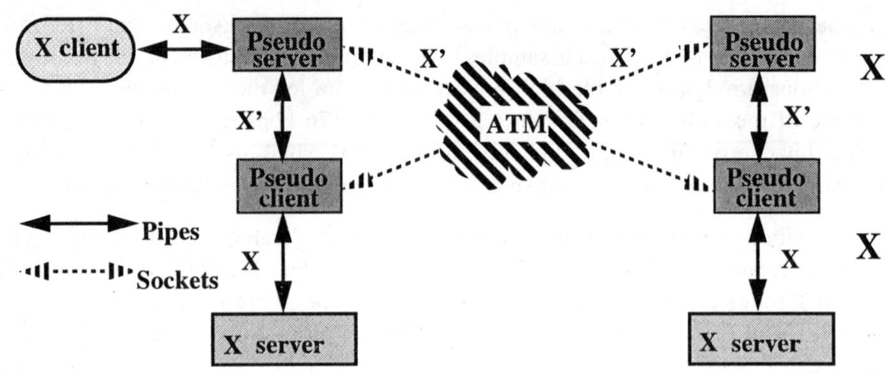

Figure 1: *Application sharing components*

The pseudo server keeps track of applications and their use of X resources which makes it capable of answering some requests without consulting the X server. The pseudo clients map the information to local X server characteristics. This resource monitoring makes it possible to allow dynamic user participation. If the sharing modules are active, applications will connect to the X server via the pseudo server. Unless the application is shared, the pseudo server will only forward the information to the X server. In heterogeneous environments with different window servers, the pseudo server or the pseudo client will translate the graphical information to a format understandable by the local window server.

Limitations under X11R5

One problem with JVTOS under X11R5 is the use of pseudo color maps which are only capable of supporting 256 colors at a time. Each X11R5 application is assigned a range of this color map which can be used with the corresponding X server. The problem is that each X server uses its own color mapping which makes it necessary to map every pixel of bitmapped data from a shared application to the correct color maps on the remote X server. Thus, sharing bitmapped applications can be slow. According to the designers of the application sharing module, there is no easy solution for this problem. The performance penalty will disappear when new display hardware which allows 24 bit true coloring is used, and the X server is configured to use it. Another problem the application sharing must solve is that different X11R5 applications use different event handling mechanisms. Some applications poll for event information while others rely on asynchronous event indications based on requested event types. This makes it more difficult for an application sharing module to be completely independent of the application. Other limitations for sharing applications under X11R5 are discussed in [2]. This reference presents an evaluation of the application sharing under X11R5 for SunOS 4.1.3, and the conclusion is that it provides acceptable performance. It includes a thorough description of the application sharing component of JVTOS and an extensive reference list of previous work done concerning application sharing.

2.3 Telemarkers

An additional feature available in JVTOS is the telemarkers which can be used by the different collaborating partners to point to objects in shared windows which is useful while doing cooperative work. The telemarker facility in JVTOS is flexible and makes it possible for each participant to use/control several telemarkers. The location of the telemarkers are updated with a regular 300 ms time interval. Telemarkers do not influence the end-system utilization significantly nor generate a lot of traffic. Therefore, telemarkers will not be discussed later in this paper.

3 Experimental Environment and Methods

In this experimental performance evaluation we have mainly been interested in two aspects of JVTOS. The first aspect we have studied is the end-system utilization and resource usage while running JVTOS. As with anything else, the performance of the JVTOS application is essential and will eventually decide potential acceptance among users. The other aspect we have focused on is the traffic generated by JVTOS during a session. Traffic characterization is interesting from a wide area ATM network perspective where resource reservation, traffic contracts and traffic policing [4] are used. An overview of the traffic generated by JVTOS with focus on the ATM level has been presented in [5]. In that paper, a composite source model[6] with exponential state sojourn times is used to model the ATM level traffic generated by JVTOS. Our contribution focuses on the end-system architecture and implementation and the effect it has on end-system utilization and traffic pattern.

3.1 Experimental Environment

The performance evaluation described in this paper was done with JVTOS version 3.1 for X11R5 under SunOS 4.1.3 in a local area ATM network based on an ASX-200 [7] ATM switch from Fore Systems Inc. All JVTOS sessions used in this evaluation were done with only two participants using Sparc 10 workstation clones (Axil 311/51/SunOS 4.1.3) equipped with both an SBA-200 [8] adapter with a 140 Mbps TAXI physical interface and a Parallax Power Video board for SBus. The SBA-200 adapter is an SBus master device using DMA in both the send and receive path. It has an embedded Intel i960 processor which controls the data transfers including the AAL5 CRC. The adapter is configured to provide an AAL5 end-of-frame interrupt on receive. The TCP implementation in SunOS 4.1.3 is similar to most BSD implementations [9] [10]. The maximum TCP window size which we used in our evaluation is 52428 bytes. The most interesting features of this TCP implementation and similar implementations are discussed in detail in [10], [11], [12], [13].

3.2 Experimental Methods

Evaluating the performance of a CSCW application is not a simple task. The necessary user interactions in a real JVTOS session are not repetetive which makes the measurements very difficult. The traffic pattern from the application will depend not only on the application which is actually used, but also heavily on the user. Even though the X11R5 events could be traced while running an application, the implementation complexity and the limited timer granularity of SunOS prevent an exact replication anyway. Thus, we have instead focused on examples of usage patterns and in some cases stress-testing of a limited set of applications.

Test Applications

In this evaluation we have used three different test applications. The first application, is the SunOS command */usr/ucb/head* which was used to print the first 400 lines of an 800 line long file containing */usr/ucb/vmstat* output. The second application is the MPEG viewer *mpeg_play* which is available via anonymous ftp from internet file servers. This application was used to play the MPEG file *skydiving.mpg*. The third and last application we tried was *Framemaker* version 3.0 which is a desktop publishing program. This application was started and a 10 page long *Framemaker* document was opened. The document contains a slide presentation of JVTOS with a lot of graphics. The operations we did on the document were simple. We scrolled to the end of the document and back again. Afterwards we zoomed the first document page 7 levels and then back again. Both these operations were repeated twice. Thus, the stress testing included 8 operations.

Operating System Monitoring

The two performance aspects we are interested in require different measurement techniques and methods. The end-system performance can only be found by interacting with the monitoring mechanisms of the SunOS 4.1.3 operating system. Thus, the resource utilization during the JVTOS sessions was captured with a slightly modified version of */usr/ucb/vmstat*. This application uses the kvm functions in SunOS to get the necessary information from the real-time resource monitoring in SunOS. The accuracy of the SunOS resource monitoring is not 100%, at least not for the CPU utilization due to the 10 ms sampling interval, but is adequate for our purpose. The record interval was set to 1 second to avoid influencing the results too much. The percentage of the CPU time which is used to process kernel code, i.e. system calls, interrupts, protocols, etc, is hereafter denoted *system time*. Similarly, we denote the percentage of the CPU time used to process application code in user space for *user time*. When the CPU is idle, we denote it *idle time*. Our modified *vmstat* captures, in addition to the usual statistics, more detailed information about autovectored interrupts.

Communication System Instrumentation

For Ethernet, FDDI and other similar technologies, powerful montoring stations operating in network promiscous mode can capture the traffic transmitted on the network. In an ATM network this is not possible unless the communicating workstations are connected through an external protocol analyzer. Unfortunately, we did not have any suitable protocol analyzer to monitor the ATM and IP level traffic, thus we instrumented the communication system (figure 2) with small optimized

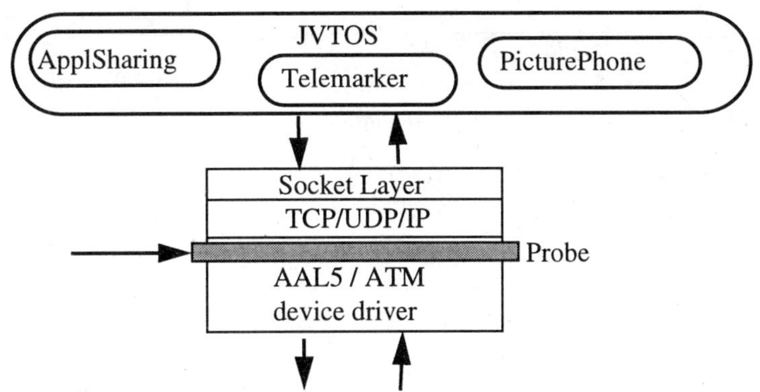

Figure 2: Communication system instrumentation

probes in the main send and receive data path. The main advantage of this approach is that the traffic generated from any CSCW application can be monitored without modifying the application source code.

The probes in the ATM device driver parse the header of all Internet Protocol (IP) packets in both directions to get the requested information, which then is stored in a large table in non-pageable kernel memory by calling a logging function. By instrumenting the communication system in both end-systems, we can monitor all the traffic generated by the end-systems. The UDP logging function records three different parameters; a timestamp, the destination or the source portnumber (configurable) and the length of the IP packet. The TCP logging function records in addition the announced window size and the acknowledgement or sequence number. The timestamp is recorded via a kernel function (*uniqtime()*) which accesses the microsecond hardware clock. The measured overhead of the logging mechanism is less than 5 μs which gives sufficient accuracy.

4 End-system Utilization

In this section we present measurements of the end-system utilization. Although the utilization was monitored in both workstations, we have only included the results from the workstation used by the session chairman. The results for the other workstation are not presented due to the fact that the results are for all practical purposes identical to the results presented here.

4.1 Reference Measurements

In a single user SunOS 4.1.3 environment, the end-system utilization depends on the applications and demons running on the particular machine. Thus, the user(s) can to some extent control the CPU utilization themselves. When a user joins a JVTOS session, several JVTOS processes which consume resources and CPU time are started.

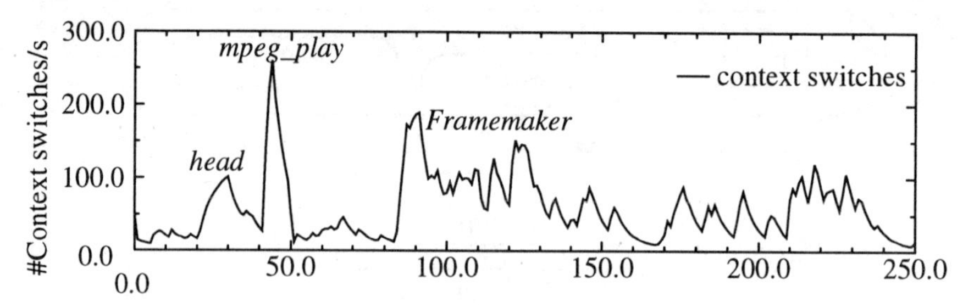

Figure 3: Context switch rate of single user environment

As a reference, we executed the three test applications in a single user environment under X11R5 as well. These reference tests showed that the CPU is fully utilized while the SunOS applications *head* and *mpeg_play* were executed and when major operations were done while editing the reference document in *Framemaker.* Otherwise, the CPU was idle. In the evaluation we have also been interested in other key parameters, e.g. context switch rate per second (CSR), interrupt rate per second (IR), etc,. Figure 3 shows the CSR while running the applications in the reference environment. The average CSR during the entire 250 second long test period is 60. The peaks occur when the applications are run or when major operations are done in *Framemaker.* The system time while the applications were running never exceeded 30%.

4.2 Influence of the PicturePhone

To evaluate the impact of the PicturePhone on the end-system utilization, we ran JVTOS with only the PicturePhone component enabled. The results of these measurements are presented in Table 1. This table presents average values of some of the most important factors influencing the end-system utilization and thereby the user perceived performance for different settings of the PicturePhone. We have only presented the average numbers due to the limited deviation during these measurements.

As these measurements illustrate, running JVTOS even with all components disabled requires a significant amount of resources and CPU time compared with the reference environment. Although there are hardly any interrupts when both the PicturePhone and the application sharing are disabled, the CSR is still over 220 and the system time is 75%. Thus, the background JVTOS processing is demanding compared to the reference measurements. However, with the PicturePhone enabled, the utilization gets even higher. As expected, the highest utilization of the end-system occurs with the HQV/HQA. In that case, the average CSR is 733 and the IR is 284. When only the HQA is enabled, the relatively low CSR and IR levels confirms our belief that the video part of the PicturePhone is responsible for the high resource usage and end-system utilization. The reason for the context switch level is the design of the JVTOS application which uses a set of processes. Some of the basic operating system mechanisms including context switching (CS) do not scale with an increase in processor speed[14]. Thus, using too many processes which increases context switching might be a problem even after a hardware upgrade. Combined with the need for additional pipe-based IPCs and data movements, the performance of the CSCW application degrades.

PP setting	CSR	IR (non-clock)	IR SBA-200	IR PVB	System time
-/-	227	11	2	0	75
HQV/HQA	733	284	33	80	59
MQV/HQA	527	139	20	28	68
LQV/HQA	415	69	13	4	72
HQV/-	607	253	22	80	65
MQV/-	361	105	9	28	72
LQV/-	246	35	3	4	75
-/HQA	232	30	13	0	76

Table 1: System parameters

Nearly half of the interrupts are caused by either the SBA-200 adapter or the PowerVideo board (PVB). The amount of interrupts from the SBA-200 corresponds to the expected interrupt rate for all the different PicturePhone settings. In addition to the interrupts from the audio or video traffic, the simple propriatary signalling protocol SPANS[1] [15] used by the ATM switch transmits two "Are-you-alive?" signalling messages to the connected workstations every second contributing to two additional interrupts. The PVB interrupts the end-system when video images are digitized and on (de)compression. The interrupt rate is, as shown, four times the requested frame rate. The remaining interrupts are caused by the other devices interrupting the end-system, e.g. audio device, mouse, keyboard, etc,. Later in this section we will illustrate that the IR level influences the CSR level significantly. This is expected since interrupts are asynchronous events which initiate context switches.

The results in Table 1 also illustrate that the system time is quite high when the PicturePhone is enabled. However, the significant amount of user application code which is necessary to process HQV lowers the system time by approximately 20 %. The system time varies with the video quality of the PicturePhone. The higher the video quality is, the more user application code processing is required which introduces a corresponding system time drop. The effect that execution of local applications has on the end-system utilization during a JVTOS session is illustrated in figure 4. It shows the CSR and the total IR level as well as the IR levels of the main interrupt sources, i.e. the SBA-200 and the PVB. The figure shows clearly that there is correlation between the IR and CSR levels. In this figure we have indicated with greyscale coded lines in the upper part of the figure when the test applications were run. The length of these lines illustrate the length of the execution

1. The SPANS protocol was specified by FORE as a temporary solution awaiting the UNI standards from ATM Forum. The protocol can establish switched VCs, but without solid QoS guarantees.

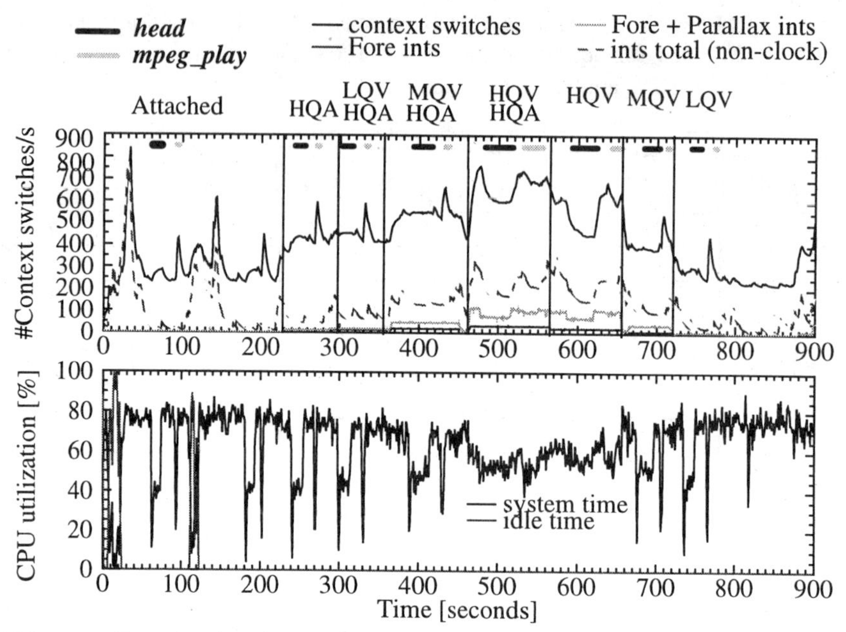

Figure 4: *Running head and mpeg_play locally with PicturePhone enabled*

time. Above the figure itself we have indicated the configuration of the PicturePhone. The figure illustrates that the execution time increases significantly if the HQV is enabled. This increase in execution time is also significant when the MQV is enabled. However, the CSR and IR levels are extremely high compared to the reference environment. The significant number of context switches is a problem for JVTOS.

There are some additional effects to notice. As long as the HQV is not enabled, running the *mpeg_play* application increases the CSR by roughly 175 CS. The increase can be explained by the additional context swiches caused by this application. However when the HQV is enabled, the CSR and the IR drop. The latter effect can be seen for the other application as well. For other settings, the CSR does not increase significantly. The drop in the IR and the CSR can be explained by a lowered video frame rate. The interrupts from the Parallax Video board drop in the same periods, which indicates that the video module is not capable of sustaining the HQV frame rate. The IR from the SBA-200 board does not drop since this interrupt rate is controlled only by the peer workstation and its utilization. Thus, the drop in Parallax video board interrupts are the (de)compress interrupts initiated by the workstation when handling the video stream. Another effect is that the user time increases when the applications are running. This corresponds to the system space processing drop depicted in figure 4.

Scenario	*head* (s)	*mpeg* (s)
X-windows, local	8.7 [1.00]	2.6 [1.00]
JVTOS user attached, local	11.0 [1.26]	2.8 [1.08]
Session, HQA, local	13.8 [1.59]	3.1 [1.19]
Session, LQV, HQA, local	14.7 [1.69]	3.2 [1.23]
Session, MQV, HQA, local	23.2 [2.67]	4.1 [1.58]
Session, HQV, HQA, local	35.3 [4.06]	23.0 [8.85]
Session, HQV, local	31.0 [3.56]	12.7 [4.88]
Session, MQV, local	20.1 [2.31]	3.5 [1.35]
Session, LQV, local	12.7 [1.46]	2.9 [1.12]
Session, LQV, HQA, shared	15.2 [1.75]	7.5 [2.88]
Session, MQV, HQA, shared	25.1 [2.89]	6.8 [2.62]
Session, HQV, HQA, shared	35.5 [4.08]	11.8 [4.54]
Session, HQA, shared	15.1 [1.74]	6.4 [2.46]
Session, shared	11.8 [1.36]	5.6 [2.15]

Table 2: Application execution times

The execution times of the *head* and *mpeg_play* applications for different settings of the different components of the JVTOS application, including the reference environment, are shown in table 2. Numbers presented in brackets are the relative execution time compared to the reference measurements. The results presented in table 2 illustrate that whenever the PicturePhone is used with medium or high quality, the performance of running additional applications will degrade quickly. This is of course expected, but a relative execution time more than four times higher than the execution time in the reference environment is too much. It is also obvious that the PicturePhone is the dominant bottleneck in the CSCW application JVTOS. Although the PicturePhone works nicely for pure video conferencing, using it while running applications locally will lower the performance of the application significantly.

4.3 Influence of Application Sharing

Included in table 2 are also measurements when the test applications were shared across the ATM network. When the PicturePhone is disabled, the application sharing provides good performance in line with the evaluation presented in [2]. However, the relative performance is better for the *head* application than for the *mpeg_play* application. Naturally, the relative performance of these applications is not comparable at all, but it illustrates that the performance of the application sharing depends highly on the application itself. In this case, the *mpeg_play* application must allocate a new window and do other time consuming operations under X11R5. We still think that the user perceived performance is good enough. Sharing the same application with the PicturePhone enabled, regardless of its configuration, will reduce the performance of the shared application. However, the conclusions in the last subsection still hold. The performance degradation for the MQV and HQV options is much higher than for the LQV option or when only the HQA option is enabled. Figure 5 shows the CSR and IR levels when the test applications are shared between the JVTOS partners. The PicturePhone options used while doing these tests are indicated above the figure. The CSR and IR levels are nearly the same as when the applications were only run locally. The application execution times are still much longer when the PicturePhone is enabled which is also illustrated by the results presented in table 2. The interrupt levels are different though, due to the application sharing traffic. The number of SBA-200 interrupts (i.e. due to the network traffic generated by the application) increases only slightly when the *head* application is run but increases a lot when the *mpeg_play* application is run. Since this workstation is distributing application sharing information between its pseudo server and the remote pseudo client, the increase in the incoming SBA-200 interrupt rate indicates that the X' protocol works on a request-response basis

Figure 6 shows the end-system utilization while running the *Framemaker* application. We have indicated when the test operations were performed with small circles in the upper part of figure 6 and the PicturePhone settings above the figure. The time between these operations, i.e. the circles, indicates the completion time and therefore the performance of the application sharing. The test operations we did can be divided in 4 clusters. In the first cluster, i.e. group of 8 operations (circles), the Picturephone was run concurrently with the LQV/HQA options enabled. The second and third cluster were done with the MQV/HQA and HQV/HQA options respectively. The last cluster was done with only the HQA enabled and no video component. This figure confirms that the CSR and IR levels are high whenever the MQV or HQV option is used. At the same time, the system time drops due to the increased processing requirements in user space by the PicturePhone mod-

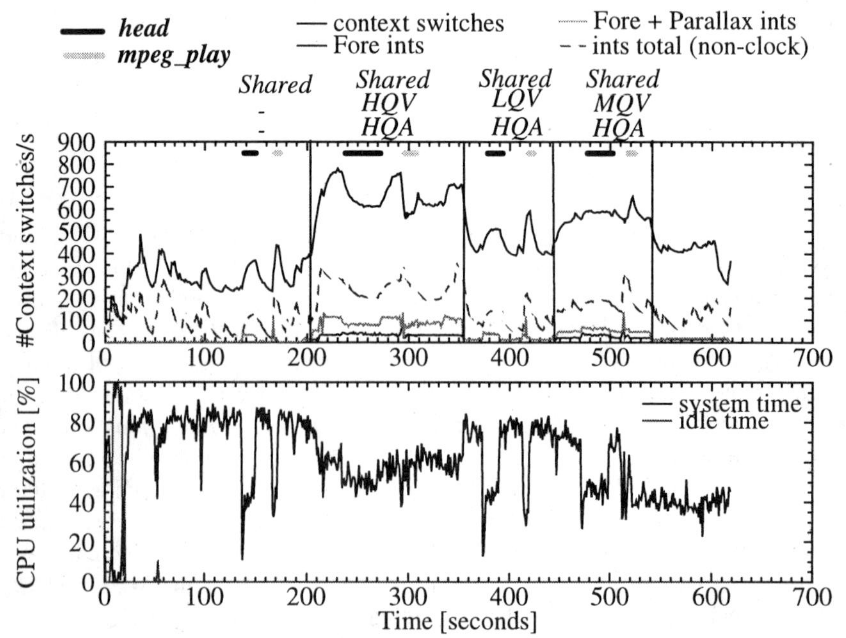

Figure 5: *Sharing head and mpeg_play with PicturePhone enabled*

ules. With the LQV or with the video disabled, the application sharing has time to finish the necessary execution in between video processing, and the system time decreases dramatically while the application executes. The CSR and IR level peaks coincide with the *Framemaker* operations. These peaks are mainly caused by the increased interrupt level due to responses of the distribution of application sharing traffic.

5 Traffic Characterization

5.1 PicturePhone - Audio Component

Packet tracing during JVTOS sessions confirmed that the audio stream is transmitted using UDP and that each audio packet is 4096 bytes long regardless of the audio quality. This is unfortunate since the audio stream hence is very vulnerable to cell loss. The worst possible scenario is of course to lose one cell from each audio packet resulting in a total shutdown of the audio channel. This packet size was allegedly chosen for implementation and portability issues which are relevant to a heterogeneous system. However, over metropolitan and wide area ATM networks this is probably not the best design choice.

The only difference between the two audio qualities is the frequency of the audio packets. A peculiarity of the audio stream is the regular irregularities of the audio packet play out times. The inter packet playout times (IPT) for both the audio qualities have a cyclic pattern. For the highest quality, all packets are evenly spaced (IPT=125ms±4ms) except every fifth packet which is transmitted immediately after the fourth packet (IPT<1ms). The average IPT was measured to 92.9 ms with a COV equal to 0.017. However, the average audio bandwidth is still the expected one. For the lowest quality, the audio packets are evenly spaced with roughly 250 ms IPT. The reason for this erratic behaviour for the high quality is most likely the audio buffering technique used by the audio component of the PicturePhone. The mean bit rate of the 22 khz audio quality is 357 kbps compared to a theoretical 352 kbps. We have not discussed the LQA option in this evaluation due to the fact that we think that that audio quality is not acceptable..

5.2 PicturePhone - Video Component

The video stream is also transmitted using UDP as the transport layer protocol. However, the bandwidth of the video channel depends on the size of each video frame and the frame rate. The size of each frame depends on both the JPEG compression quality and on the information contents of the image. We have during our experiments with JVTOS and the Parallax PowerVideo board discovered that the size of each video frame can vary from roughly 4kbyte for a near single-color image to 25 kbyte for white noise images. Since the maximum UDP segment payload size in SunOS 4.1.3 is 8 kbyte, the 25 kbyte frame must be sent as four video packets. However, we have seldom seen frames larger than 12 kbyte which means two UDP packets (8 kbyte and 4 kbyte) transmitted back to back.

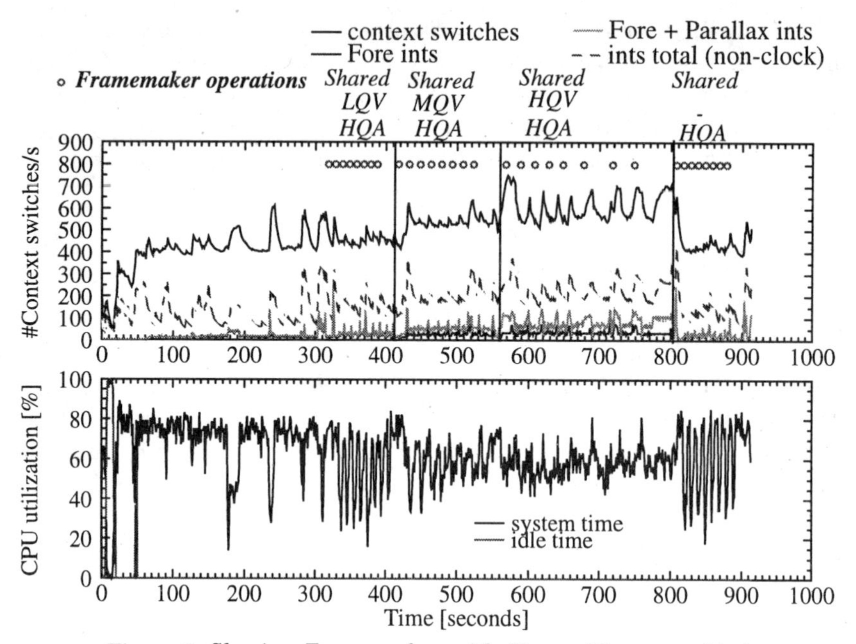

Figure 6: *Sharing Framemaker with PicturePhone enabled*

Table 3 shows some of the important numbers for different configurations of the video channel with and without the audio channel enabled at the same time. We have presented average values for all measured quantities. The COV has been included in parentheses wherever applicable, while the numbers in brackets show the percentage of the frames which are smaller and larger than the maximum UDP segment size respectivly.

PicturePhone setting	Frame rate	Throughput [Mbps]	Frame IPT [ms]	Mean/Min/Max frame size [byte]
HQV/HQA	19.53	1.244	51.2 (0.028)	7936/7420/8748 [96, 4]
MQV/HQA	6.98	0.471	143.6 (0.049)	8405/7960/8860 [25,75]
LQV/HQA	1.00	0.067	1017.45 (0.131)	8387/7944/8564 [19,81]

Table 3: Picturephone statistics

As shown, the measured frame rates are close to the requested frame rates, illustrating that our workstation is powerful enough to provide the HQV/HQA option for the PicturePhone. As illustrated in the previous section, the end-system utilization precludes running other demanding applications concurrently. However, for video conferencing, the JVTOS PicturePhone works nicely.

The measured throughput of the HQV is high compared to existing video conferencing solutions but provides much higher quality. However, the throughput is, as described, propotional to the frame rate. The measured frame inter playout time is close to the requested one with relatively low COV even though the frames usually must be transmitted as two UDP segments. The minimum and maximum frame sizes which occurred during these measurements were 7936 and 8860 bytes respectively. The percentage of frames which were transmitted as two UDP segments varied from measurement to measurement due to the JPEG encoding and the contents of the image.

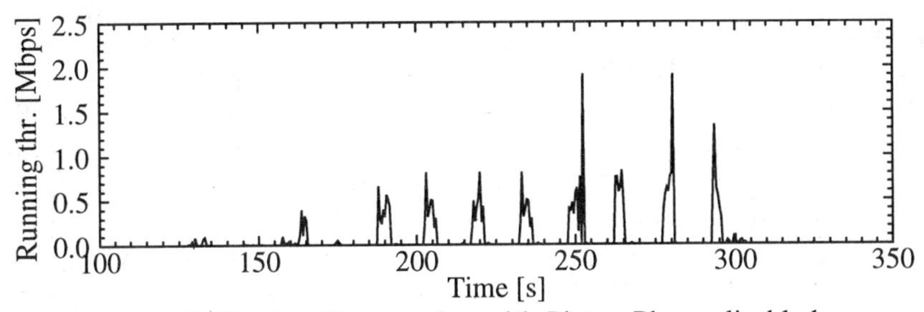

Figure 7: Sharing *Framemaker* with PicturePhone disabled

5.3 Application Sharing

The traffic generated while sharing applications depends heavily on the application itself and the behaviour of the users. Among our test applications, *Framemaker* is the only application with some user interaction. Application sharing traffic will usually be very bursty due to user interaction. This is illustrated in figure 7 which shows the *running throughput* while running the *Framemaker* test sequence. *Running throughput* is in this paper defined as the throughput within 0.5s time slots during the JVTOS session. Figure 7 shows the *running throughput* for the 8 operations done while testing JVTOS. In most of the 0.5 s time slots where there is traffic, the *running throughput* is close to 0.5 Mbps. However, in some time slots the throughput reaches 2 Mbps. Thus, the burstiness of the application sharing traffic is very high. The 8 operations are clearly visible in the figure, and the running throughput is highest when the first page in the document is zoomed. The distribution of the packet lengths of the application sharing traffic were also characteristic. More than 70% of the packets were shorter than 256 bytes, while 20% were exactly 4096 bytes.

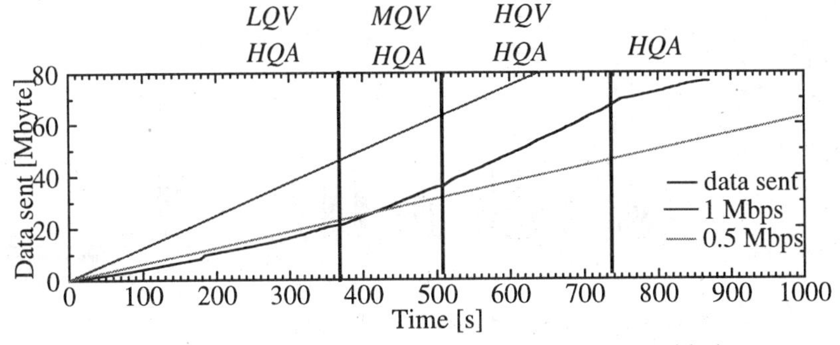

Figure 8: Sharing *Framemaker* with PicturePhone enabled

The throughput during a JVTOS session depends on the components which are enabled and their settings. In addition, the shared applications will also influence the generated traffic pattern. During our experiments the throughput was more than 1 Mbps when all components were active. This is illustrated in figure 8 which shows the amount of data sent during a JVTOS session while sharing the *Framemaker* application. The PicturePhone settings are illustrated in figure 9 a) which shows the *running throughput* of the same JVTOS session. As expected, the throughput is close to 0.5 Mbps when the LQV option is enabled and increases above 1 Mbps when the MQV option is enabled. The running throughput of the HQV illustrates that this option suffers from the concurrent application sharing which is illustrated in figure 9 b). Thus, the overall throughput does not increase as much as the HQV option would suggest. One interesting observation of the results in figure 9 is that the HQA does not seem to be influenced by the application sharing at all. Thus, the audio channel should work properly even when the video quality degrades severely. The drops in the running throughput of the HQV coincides with the operations done with the shared application. In some 0.5s the *running throughput* drops to 0 Mbps which means that no video frames were transmitted within that time slot. Obviously, the video quality is severely reduced compared to the requested 20 frames/s. The *running throughput* of the application sharing of the *Framemaker* application shown in figure 9 b) confirms that the HQV, and the application sharing influence each other. This figure illustrates that the *running throughput* is in general lower than requested especially for the HQV but also for the MQV. However, unlike the LQV/HQA and HQA options, there are less silent periods which means that the mean throughput is higher as illustrated in figure 8.

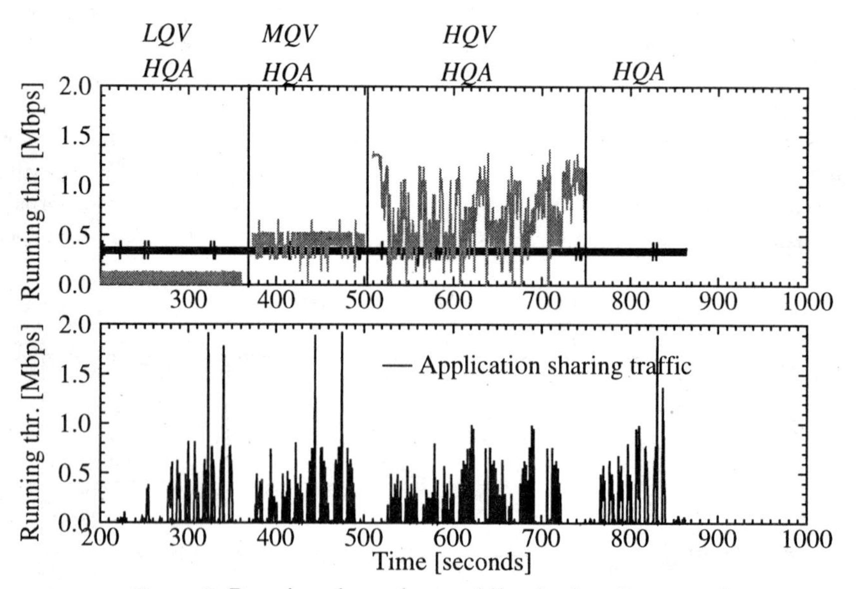

Figure 9: Running throughput while sharing *Framemaker*

5.4 Protocol Issues

JVTOS over TCP/UDP

One of the design goals of JVTOS is to offer a CSCW application which could be used in heterogeneous networks. Thus, JVTOS needs a higher layer protocol with the necessary functionality, e.g. flow control mechanism, acknowledgement policy, routing, etc. Currently, JVTOS is using TCP and UDP as the default transport layer protocols. However, the audio and video channel can be configured at start up time to use the XTPX [16] protocol, and the application sharing is being ported. In these measurements we have only used TCP and UDP as transport layer protocols.

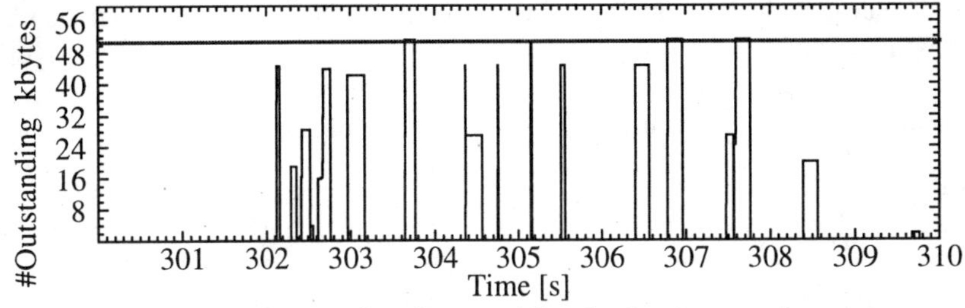

Figure 10: Outstanding bytes while sharing *mpeg_play*

TCP is one of the protocols which are used over ATM networks. Some of the features of the original TCP specification are use of sliding window flow control with a 16 bit sequence space, positive acknowledgements and lack of window size negotiation during connection setup. More recent improvements[10], including congestion control and avoidance, header prediction, fast retransmit, Nagle's algorithm, and window extensions have increased TCP/IP popularity and usability. Unfortunately, the SunOS 4.1.3 implementation has not implemented all these features. For example, lack of the window extensions can degrade the performance of the application sharing. This is pointed out in figure 10 which illustrates that the generated traffic is bursty but can under some circumstances use the entire window size. The maximum window size is indicated with a horizontal line at 52428 bytes. Fortunately, the UDP traffic (i.e. video and audio channels) will not be affected since UDP has no flow control mechanism. The only problem for UDP is an increased probability of cell loss and eventual UDP segment discard. Another problem that JVTOS and other CSCW applications must consider is deadlock situations [12] which might occur if the end-systems in different administrative domains have incompatible TCP configurations. Thus, JVTOS sets the window size on a per connection basis through socket options.

The percentage of small TCP segments (<100 bytes) transmitted as part of sharing an application is high. At the same time the audio channel and the video channel transmit only large UDP segments. Thus, compared to the traffic generated by the PicturePhone, the overhead of the application sharing traffic is high. Usually, Nagle's algorithm [10] is used to reduce the overhead of transmitting small packets. However, in JVTOS it is disabled by the low-level communication sub-module [2] to increase the responsiveness of JVTOS. The unfortunate implication is of course an increase in the number of small packets which in turn increases the CSR and IR levels. Thus, although the responsiveness decreases if Nagle's algorithm is enabled, the end-system utilization might improve. In the wide area, Nagle's algorithm is an efficient way to reduce the number of small packets

TCP/IP over ATM

Even limited cell loss in ATM networks can be a problem for the application sharing in JVTOS as it relies on TCP. As discussed in [13] and [17], TCP relies too much on the retransmission timer, and this timer has a too high granularity in most TCP implementations. In SunOS, the timer granularity is 500 ms, and the minimum value of the timer is 1 s. Thus, the cost of a single expiration of the retransmission timer is high. Some solutions to this problem include the fast retransmit algorithm [10] or the retransmission algorithm used in TCP Vegas [13]. Unfortunately, the implementation of the fast retransmit algorithm is incomplete in SunOS 4.1.3.

Another related issue is the protocol and driver processing of small versus large segments. Network adapters are optimized for high throughput by using DMA in both send and receive path. These adapters have a relatively high per-packet cost in the device driver processing compared to simpler adapters using programmed I/O. We have used the advanced ATM adapter SBA-200 in these measurements. However, the predecessor SBA-100 [18] which uses programmed I/O would most likely perform better for all the small packets.

6 Concluding Remarks

This performance evaluation has confirmed that integrating high quality video, audio and collaborative work into a single application is no simple task with the current technology, i.e. hardware, video boards, monolithic operating systems. It is obvious that the implementation of teleservices should be done with the same care as the implementation of network and transport protocols. Furthermore, the teleservices and the audio and video devices must be closely integrated with the workstation architecture and the operating system. Without suitable integration, the performance will degrade.

In the version of JVTOS we have evaluated, using high and even medium quality video while doing cooperative work will reduce the performance. However, our own experience with JVTOS doing cooperative work is positive. By using only low quality video while sharing documents over the European ATM Pilot [19], the performance was more than adequate. The only problem we experienced was the degradation of the audio quality when cell loss occured.

SunOS 4.1.3 is not the best operating system to implement an advanced teleservice like JVTOS. A better solution would be to use an operating system where JVTOS could be implemented as a single process with multiple threads[1] thereby re-

1. JVTOS is currently being ported to Solaris 2.4

ducing the cost of context switching, and possibly taking advantage of the common address space. Another solution which might be applicable is to use an extensible operating system [20] where for instance the PicturePhone application, although implemented in user space, could be run in the kernel to avoid crossing the protection boundary between user and system space.

Acknowledgements

We would like to thank the JVTOS developers who gave us useful information about the different components of the JVTOS application during our work. This work was partially supported by the European Commission within the framework of the RACE 2060 CIO project.

References

1. Dermler, G et al. Constructing a distributed multimedia joint viewing and teleoperation service for heterogeneous workstation environments. *Proc. of 4th IEEE workshop on future trends of distributed computing system.* Lisbon, Portugal, September 22-24, 1993.
2. Gutekunst. T et al. A distributed and policy-free general-purpose shared window system. *IEEE/ACM transaction on networking*, 3(1), February 1995, pp. 51-62.
3. *Visual Audio Tool.* Audio conferencing tool in the public domain. Available via anonymous ftp and internet retrieval services.
4. ITU Telecommunication Standardization Sector Study Group 13 I.371. Traffic control and congestion control in B-ISDN. 1995. (Frozen issue Paris, March 1995.)
5. Aarstad, E., Venturin, R. Traffic characterization of the JVTOS multimedia application. *Proc. of 12th Nordic teletraffic seminar (NTS)*, Esbo, August 22-24, 1995, pp. 71-81.
6. Helvik, B et al. The synthezied traffic generator; objectives,design and capabilities. *Proc. of Integrated Broadband Communication Networks and Services,* Copenhagen, Denmark, 1993.
7. FORE Systems Inc. *ForeRunner^{TM} ASX-200 ATM switch user's manual.* 1994. (Revision level A.)
8. FORE Systems Inc. *200-series ATM adapter - design and architecture.* 1994.
9. Leffler, S J et al. 4.3 BSD Unix operating system. Reading, Mass., Addison-Wesley, 1989. (ISBN 0-201-06196-1)
10. Stevens, W.R. TCP/IP illustrated, volume 1. Reading, Mass., Addison-Wesley, 1989. (ISBN 0-201-06196-1.)
11. Moldeklev, K., Klovning, E., Kure, Ø. TCP/IP Behavior in a high-speed local ATM network environment. *Proc. of the 19th conference on Local Computer Networks.* Minneapolis, 1994, pp. 176-185.

12. Moldeklev, K., Gunningberg, P. Deadlock situations in TCP over ATM. *Proc. of 4th IFIP workshop on protocols for high speed networks,* Vancouver, British Columbia, Canada. 1994, pp 243-259.
13. Brakmo, L et al. TCP Vegas: new techniques for congestion detection and avoidance. Computer Communication Review. 24(4), 1994, pp. 24-35.
14. Ousterhout, J. Why aren't operating systems getting faster as fast as hardware ?. *Proc. of the summer 1990 USENIX conference.* 1990.
15. FORE Systems Inc, *SPANS signalling protocol.* 1992.
16. Technical University of Berlin. Specification of the Broadband Transport Protocol XTPX. 1993. (RACE 2060, CIO, Deliverable 30.)
17. Bonaventure, O, Danthine, A, Klovning, E, Danthine, O. TCP/IP and the European ATM Pilot. *Proc. of 1995 International Conference on Network Protocols (ICNP95),* Tokyo, Japan. 1995, pp 270-277.
18. FORE Systems Inc. *SBA-100 SBus ATM computer interface - user's manual.* 1992
19. Parker, M et al. The European ATM Pilot. *Proc. of the 15th International Switching Symposium*, Berlin, April, 1995, pp 146-150.
20. Bershad, B et al. Extensibility, safety and performance in the SPIN operating system. To be presented at SOSP'95.

Multimedia Teleservices Modelled with the OSI Application Layer Structure

Erwin van Rijssen
Logica
PO Box 22067
3003 DB Rotterdam
The Netherlands
ery@lbvrtda.logica.com

and Ing Widya
Centre for Telematics and Information Technology
University of Twente
PO Box 217
7500 AE Enschede
The Netherlands
widya@cs.utwente.nl

and Eddy Michiels
Centre for Telematics and Information Technology
University of Twente
PO Box 217
7500 AE Enschede
The Netherlands
michiels@cs.utwente.nl

Abstract. This paper looks into the communications capabilities that are required by distributed multimedia applications to achieve relation preserving information exchange. These capabilities are derived by analyzing the notion of 'information exchange' and are embodied in communications functionalities. To emphasize the importance of the users' view, a top-down approach is applied. The (revised) OSI Application Layer Structure (OSI-ALS) is used to model the communications functionalities and to develop an architecture for composition of multimedia services with these functionalities. This work may therefore be considered an exercise to evaluate the suitability of OSI-ALS for composition of multimedia teleservices.

1 Introduction

With the emergence of cheap computing power and network capacity distributed multimedia applications become feasible. These applications integrate media such as audio, video and text in order to represent information to the users in easily perceptible form [2]. The information is usually geographically dispersed and must be retrieved, processed and transmitted very fast to achieve timely delivery. The main difficulty related to the distribution of multimedia information is the preservation of relations between pieces of information. These relations may have a global perspective (e.g., simultaneous receipt of information at different destinations) and a local perspective (e.g., a time relation between subsequent samples in a video stream). To support a large variety of applications, a framework architecture is required that allows users to compose multimedia teleservices in an efficient and effective way. The architecture must provide generic communications functionalities that enable exchange of information and that preserve the relations mentioned above.

The aims of this work are the identification of some basic functional building blocks and the development of an architecture that describes how the building blocks can be composed to provide multimedia teleservices. The architecture is based on the revised OSI Application Layer Structure (OSI-ALS) [7] and therefore this work also examines the suitability of the OSI-ALS[1] for modelling multimedia teleservices. Before a framework architecture can be defined, it must be clear what communications capabilities are required by multimedia applications. The capabilities are identified by analyzing the notion of *information exchange* and are quantified with QoS parameters. The top-down approach emphasizes the significant role of the users' view. Results found in the literature are used to fit this approach, see for instance [3, 10, 11, 16].

This paper is organized as follows. Section 2 derives communications capabilities that must be provided by multimedia teleservices. In section 3, a teleservice is presented that is meant to support multimedia applications like videophony. A teleservice can be decomposed into an application-oriented protocol layer and an underlying transport service. Section 4 briefly discusses the requirements a transport service must satisfy in order to be suitable for transfer of multimedia information. Section 5 elaborates a framework architecture enabling composition of multimedia teleservices on top of the discussed transport service. This architecture is validated in section 6 by modelling two important mechanisms: synchronization and QoS management. Finally, in section 7 conclusions are drawn.

2 Communications Capabilities of Multimedia Teleservices

A large diversity of communications capabilities is needed to satisfy the communications requirements imposed by all kinds of distributed multimedia applications. As mentioned in the previous section, communications capabilities are identified by analyzing the notion of *information exchange*. The term *information* involves the following capabilities:

- the type of information, i.e. the so-called *media*;
- the relations between various media information.

The term *exchange* involves the following capabilities:

- the communication-structure between the end-users;
- the directions of information flow in the communication structure;
- the distributed synchronism, i.e. the simultaneous submission of information from different sources or the simultaneous receipt of information at different destinations.

In the following sections, each communications capability is analyzed in more detail to identify the necessary functionalities. The functionalities are used to preserve the information characteristics and to guarantee the required relations, mentioned in section 1. How strictly these relations must be maintained can be specified by

[1] At the end of this paper a list with abbreviations is added to improve the readability of this work.

quantifying QoS parameters. These parameters are also derived in the following sections.

2.1 Media

The inter-arrival times of video information units at a destination cleary satisfy a certain regularity. A strong relation exists between the time arrivals at which successive video information-units[2] are received. On the contrary, in the transfer of text information, the inter-arrival times of text information-units usually do not obey to any regularity. These examples illustrate the time relation characteristic of media information. This leads to the following definitions (see also [11]).

- **Continuous Media** information flow is a sequence of information-units of a single type that are related in time. Such an information flow is also called a *stream*. Examples are video and audio information.
- **Still Media** information is a set of information-units that are generally not related in time. Examples are data, text and image information.

The existence of a time relation between (uncoded) continuous media information-units often implies some natural redundancy in the information. The higher the natural redundancy of the exchanged information, the lower the reliability of the communications service can be. The required reliability of a service can be expressed in terms of the QoS *error rate* parameter (see table 1).

To maintain the time relation in a stream, a minimum amount of 'bandwidth' must be continuously available. The amount of bandwidth that is needed during the use of a service can be expressed with the *information-unit rate* and the *information-unit size* parameters (see [17] and table 1).

Due to the time relation, the rate at which continuous media information-units are sent is equal, within a certain tolerance, to the rate at which they are received. In contrast, if still media information-units are transferred, different rate values at the sending and receiving sites of a communication line are conceivable. To preserve the time relation in a continuous media stream between the sending and the receiving SAPs, the QoS *intra-stream jitter* parameter is defined ([6] and table 1).

2.2 Media information Relations

This section identifies the relations that are conceivable between different media information, mentioned earlier. It also describes the functions and QoS parameters that are required to preserve the relations during information transfer.

Synchronization Relations. In a videophony application, someone's lip movements in a video scene must match with the corresponding audio information. In order to obtain 'lip-synchronization' between the individual information-units of a

---

[2]An information-unit is an elementary collection of information of a particular type.

video and an audio stream, the multimedia service must provide so-called *continuous synchronization* [3]. A necessary condition for this synchronization function is that the time between the arrival of corresponding information-units in different streams is restricted to an upperbound. This upperbound is expressed with the *inter-stream jitter* parameter (table 1). The synchronization that is necessary in a video play-out between control information (e.g., 'play' and 'stop') and video and audio information can be achieved by *event-based synchronization* [3]. It is user controllable, which means that a user can explicitly interfere each time that synchronization is required. Since it is used to express causal ordering and it is not applied in situations in which stringent time-constraints exist between information-units, the inter-stream jitter parameter is not relevant for event-based synchronization functions.

Quality Relations. Due to congestion in networks, communications service providers may not be able to maintain the required quality of a service. In this case, the provider may need a user supplied rule to enable the degradation of specific services. for this reason, the *priority* parameter is introduced (table 1). With this parameter, users are able to specify that they want the communications service provider to abort media information under certain circumstances. For example, in a videophony session the users may require the abortion of the video stream, as soon as the audio stream crashes.

2.3 Communication Structures

Th communication structure capability deals with the configuration that is provided by the communications service to enable users to communicate. Four basic communication structures can be distinguished: point-to-point, multicast, multicollect and multipoint structures (see [1] for definitions). This communications aspect does not contribute to the definition of new QoS parameters. However, in order to use these communication structures, functions that allow establishment of multiparty associations are needed.

2.4 Communication Directions

This capability deals with the directions in which information can be exchanged between end-users. In **dialogue communications**, information flows in both directions are allowed. Many distributed applications in which dialogue communications is applied have an interactive character. As a result, information-units must be received within an acceptable time interval, in general. For example, in a videophony application this is 0.25s [4]. The *end-to-end delay* parameter (table 1) is defined to enable users to inform the service provider about this requirement. There are also distributed multimedia applications that apply dialogue communications, but in which the end-to-end delay is not submitted to stringent upper bounds. An example is a multimedia electronic-mail service with confirmation. In some multimedia applications the information that is exchanged in both directions is of different types (e.g., a film-retrieval application), whereas in other applications the information types are equal (e.g., videophony). In the former case, the requirements imposed on the communications service can be quite different and must therefore be specified separately for each direction of transmission. This is also preferable in the latter case, because different end-users can have different QoS requirements.

In **monologue communications** the information flows in only one direction. The inability of a receiver to respond implies, in general, that the end-to-end delay requirements imposed on the communications service are less stringent than for dialogue communications.

2.5 Distributed Synchronism

As mentioned in the introduction of section 2, distributed synchronism is the simultaneous submission of information from different sources or the simultaneous receipt of information at different destinations, within certain human perceptible tolerance. In a 'game', e.g., all players must perceive the destruction of a target at the same moment. This requirement can be expressed with the *replicated-stream jitter* parameter (table 1). This parameter is therefore used to specify another aspect of information distribution in time and space. It can be considered the distributed version of the inter-stream jitter parameter.

2.6 Overview of Quality of Service Parameters

In table 1 an overview is shown of the QoS parameters distinguished in this paper.

End-to-End Delay: the elapsed time between offering an information-unit at the sending Service Access Point (SAP) and its delivery to the receiving SAP.
Error Rate: the estimated probability that an information-unit is subject to loss, or corruption between sending SAP and receiving SAP.
Information-Unit Rate: the rate at which information-units are sent or received at a SAP.
Information-Unit Size: the size of the information-units that are sent or received at a SAP.
Intra-Stream Jitter: the variance in information-unit inter-arrival times within a stream, measured at the receiving SAP.
Inter-Stream Jitter: the difference in inter-arrival times of synchronization-related information-units in different media information, measured at the receiving SAP.
Replicated-Stream Jitter: the difference in submission (respectively, receipt) times of information, measured at the sending (respectively, receiving) SAPs.
Priority: parameter used to distinguish between the relative importance of various media information. High priority media information are served before lower ones. Should the network become congested, lower priority information-units will be dropped before high priority information-units.

Table 1. QoS Parameters

3 A Multimedia Teleservice

This section elaborates a *call* model that provides an abstract view of conceivable communication relations between the users of a multimedia teleservice. The model includes communications capabilities (identified in previous section) that are needed to satisfy the requirements imposed by distributed multimedia applications.

3.1 Multimedia Communications Capabilities

With the communications capabilities mentioned below, it must be possible to support a well accepted class of distributed multimedia applications, e.g., multiparty videophony. Furthermore, users must be able to adjust the functionality of the delivered service. However, the provider of the teleservice must not be too complex to implement.

In many distributed multimedia applications two or more end-users are involved. Therefore, the teleservice must be able to provide *multiparty* calls [8] and support all *basic communication structures* (section 2.3). This service must also support *continuous media* information exchange (section 2.1) in *interactive dialogues* (section 2.4). As continuous media require a guaranteed bandwidth, the service will be *connection-oriented.* Moreover, this service must maintain *synchronization* and *quality relations* between media information (sections 2.2 and 2.5), necessary in applications such as multiparty conferencing.

3.2 A Simple Call Model

The call model describes the communications relation between users and is defined in accordance with the communications capabilities mentioned in section 3.1. The model is inspired by [5] and is discussed in more detail in [14]. This paper only discusses those issues that are relevant with respect to the evaluation of the OSI-ALS in section 6.

First of all, a call definition specifies a set of users that are bound to the call. These users have a *global call view,* which consists of a *global QoS view* and a *global communication structure view.* To maintain the global call view, users are informed about all changes made to the call QoS and call communication structure. A global QoS view is necessary, because all users have to know the quality of the information that other users are able to send or receive. A global communication structure view is needed to know which users are sending information to whom.

The communication structure of a call is described as a set of *service components.* A service component is the basic constituent of a call (see also [5]) and models the information exchange of a single media, such as video or audio. For each service component a set of attached users is specified. The decomposition of a call into service components facilitates the adaptation to changing user needs. A call is decomposed into service components, because the end-users already distinguish between media in practice. Another advantage of treating different media as separate service components instead of combining them in one format is that this avoids the need for a large number of composite signals [9].

Apart from the communication structure, the QoS requirements that users impose on service components, are specified. Furthermore, the synchronization and quality relations that may exist between service components are part of the call model (see, e.g., section 2.2 and [5]). During the establishment of a call the initiating user may specify a so-called success condition. This condition specifies under which circumstances the call establishment can be considered successfully (e.g., 'user *A* must agree with participation in the call').

Figure 1 gives an example of a call with three users. The communication structure consists of two service components: an audio component between A, B and C and a video component between A and C. In the rest of this paper this example is used to illustrate how the developed architecture can be applied.

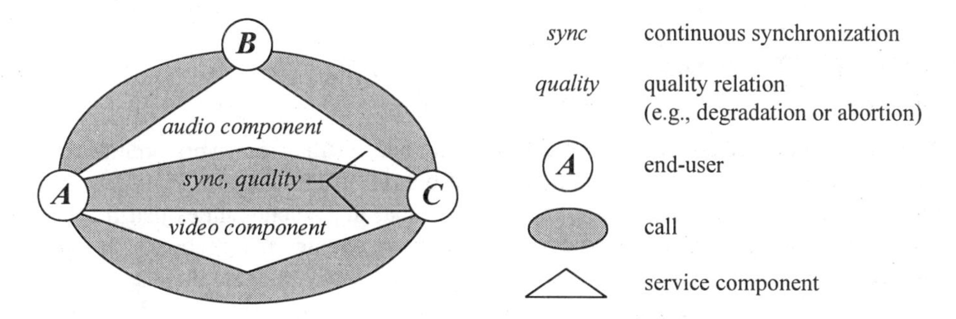

Fig. 1. Example of an Abstract View of a Multimedia Teleservice

4 Transport Service Requirements

Now that the communications capabilities have been defined, the service provider can be decomposed into two parts: an application-oriented protocol layer and an underlying transport service provider. To determine the functionalities of the application-oriented protocol layer, the capabilities of the transport service have to be known. In this paper, the underlying transport service is OSI Transport Service [15] look alike, in order to be compatible with the OSI-ALS. It provides end-user to end-user services. However, this service is extended with new capabilities that are easily provided by todays high speed networks, e.g., ATM-based networks.

The underlying transport service is able to provide multicasting. To support multicasting the establishment of multipoint transport connections is a necessity. Since QoS requirements can be quite different for various directions of information flow, the use of unidirectional transport connections is advocated (see, e.g., ST-II [18]). Moreover, dynamic QoS management is required for such transport connections to adapt to changing user needs (see also [3]). This means that QoS renegotiation must be possible during the information-transfer phase of a connection.

5 A Multimedia Framework Architecture

A functionality gap exists between a transport service that satisfies the requirements described in previous section and the class of teleservices identified in section 3.1. This gap must be filled up with an application-oriented protocol layer. This section does not consider presentation layer functions. It develops an application-oriented architecture which enables the composition of a particular class of multimedia teleservices (see section 3.1). The aim is to develop an architecture that satisfies the following design principles:

- users are capable to tailor the teleservices to their specific needs;
- efficient and effective composition of teleservices by using predefined functional building blocks;
- extensibility of teleservices by adding or removing some building blocks.
- (re)design of teleservices, independent from the underlying transport service.

By using basic functional building blocks that are consistent with OSI-ALS, the suitability of OSI-ALS for modelling multimedia teleservices is evaluated.

First, section 5.1 presents an overview of the multimedia framework architecture. This contributes to a better comprehension of the rest of this chapter. The derivation of the architecture is divided into three steps. First, the functionalities that must be available in an application-oriented protocol layer are enumerated (section 5.2). Then, in section 5.3 three architecture levels are distinguished to facilitate the clustering of functionalities into functional building blocks. The actual clustering of functionalities into OSI-ALS consistent building blocks is performed in section 5.4. Section 5.5 shows the relations between levels, building blocks and functionalities.

5.1 Architecture Overview

Figure 2 shows a framework architecture that enables composition of teleservices. This architecture comprises the functional building blocks that are identified in section 5.4.

In the world of telecommunications, a distinction is made between call control and bearer control (see, e.g., [12]).In the architecture presented in this paper a similar division can be recognized. The upper part of the architecture performs 'call control equivalent' functions, such as call establishment. In the lower part of the architecture 'bearer control equivalent' functions are available, for instance to establish the components needed for exchange of media information (e.g., video information). This separation is illustrated in figure 3.

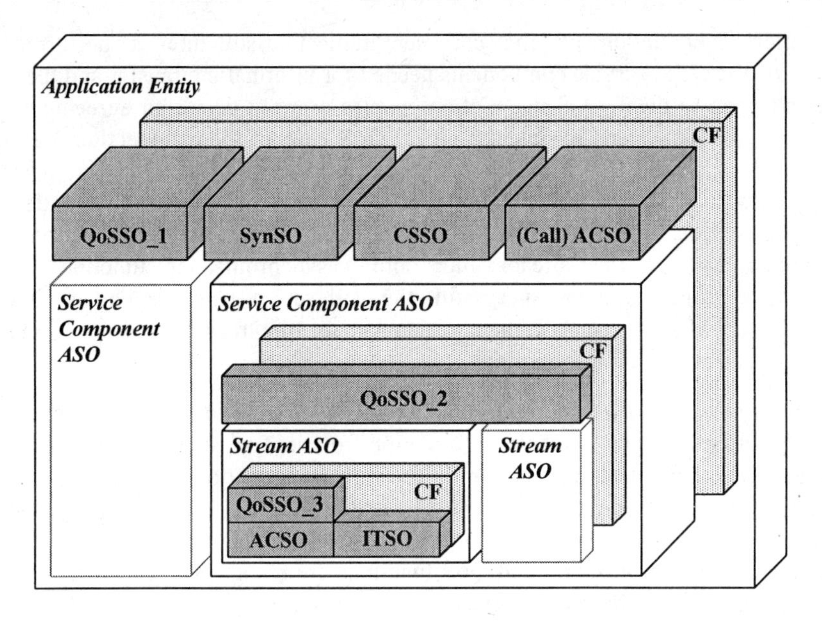

Fig. 2. Overview of the Multimedia Framework Architecture

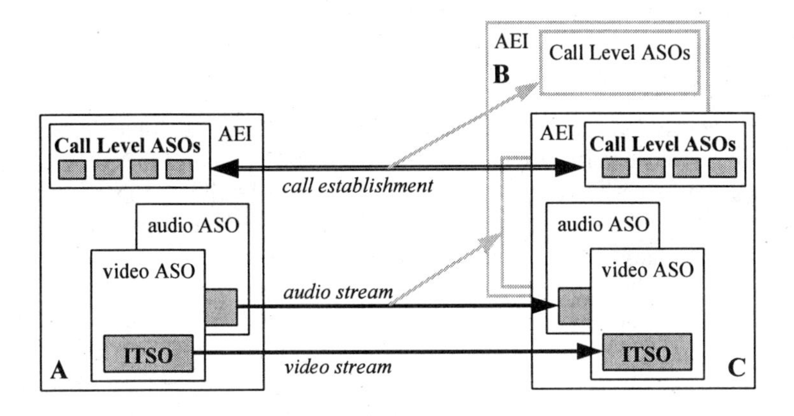

Fig. 3. Separation of 'Call Control' and 'Bearer Control'

5.2 Architecture Functionalities

A first step towards an application layer architecture is to determine the functionalities that are needed to provide multimedia teleservices. Since a connection-oriented teleservice is assumed (section 3.1) three phases can be distinguished during the lifetime of a call: an *establishment*, *information-transfer* and *release* phase.

In the *establishment* phase, the application layer has to perform functions such as negotiation on the call communication-structure, QoS negotiation and QoS parameter translation. Negotiation on the call communication-structure is performed to determine the set of service components needed for information transfer and the set of users attached to them. QoS negotiation enables users to derive an agreement from personal QoS wishes and to obtain a global QoS view of the call (section 3.2). QoS parameter translation [10] is the function that maps some of the QoS parameters of the teleservice (e.g., end-to-end delay and error rate) on those of the underlying transport service. The other QoS parameters are mapped onto application layer functionalities (e.g., inter-stream jitter onto a synchronization function). If the negotiation process appears successful, the call set-up phase is ended with the establishment of the transport connections that enable the transfer of user information, such as video and audio information.

In the *information-transfer* phase functionality is needed to submit information to the transport service and to receipt information from it. To enable users to modify the QoS and the communication-structure of the call, renegotiation functionality must be available. Changes made to the call must be administrated to maintain the global call view. To avoid violation of inter-stream jitter bounds and to satisfy priority requirements, synchronization and QoS functions are recognized. The latter can also be used to maintain abortion relations between service components (see section 2.2).

The *release* phase encompasses all functions that are needed for orderly termination of the call, such as the release of the supporting transport connections.

5.3 Architecture Levels

As said previously, an important requirement imposed on a multimedia framework architecture is that it must provide generic functional building blocks. In this way, teleservices can be easily composed by using predefined building blocks. Moreover, reuse of building blocks results in a cost reduction for the composition of new services. Another advantage is that users can tailor the teleservices to their specific needs by selecting building blocks from a given set and by specifying the way in which these blocks interfere. Independance from the underlying transport service is achieved by nesting of building blocks. Nesting is a new feature in the revised OSI Application Layer Structure (OSI-ALS). The size of the 'gap' between the transport service and the teleservice determines how deep the nesting is.

Functional building blocks are identified by clustering functionalities (see section 5.2) that are strongly related. In this way, each building block is associated with a particular function that is clearly defined. By assigning the clustered functionalities to the building blocks of OSI-ALS, the suitability of OSI-ALS for composition of teleservices is evaluated. In order to find an optimal clustering, it appears useful to examine how a call is mapped onto transport connections.

In section 3.2. a call is defined as a combination of service components, which are monomedia but possible bidirectional. The complexity of a call can be further reduced by dividing the service components into unidirectional, multipoint streams. Each stream corresponds to exactly one multipoint transport connection. In this way the framework architecture is structured into three levels (see also figure 2):

- *Call* level;
- *Service Component* level;
- *Stream* level.

Figure 4 illustrates how the call that is shown in figure 1 is mapped onto transport connections.

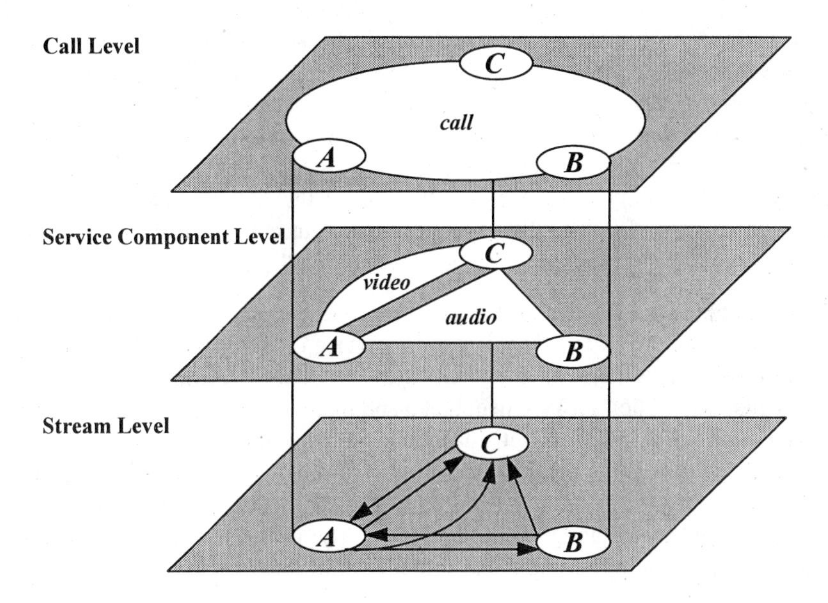

Fig. 4. Mapping a Call onto Transport Connections

5.4 Architecture Building Blocks

In OSI-ALS [7], *Application Service Elements (ASEs)* are defined as elementary building blocks. The ASEs are encompassed by other building blocks, that are called *Application Service Objects (ASOs)*. In this paper, however, ASOs are considered the elementary building blocks and ASEs are not used. This is due to the fact that ASEs only support point-to-point communications, whereas ASOs also allow multipoint communications. In order to benefit from the multicasting capability of the transport service, the building blocks in the framework architecture must support multipoint communications. OSI-ALS allows nesting of ASOs, which satisfies the need to be able to design teleservices independent from the underlying transport service. Besides ASOs, *Control Functions (CFs)* are defined in the OSI-ALS. A CF is part of an ASO. It is able to control local sub-ASOs and to communicate with other (remote) CFs. It perceives all information that enters the ASO and is responsible for a correct distribution of this information over the sub-ASOs. To illustrate that CFs perceive all information, figure 2 shows them in parallel with the other ASOs. Co-ordination between peer CFs in remotely located protocol entities (*Application Entity Invocations (AEIs)*) is performed via the sub-ASOs. This is due to the fact that CFs in different AEIs cannot communicate directly (see [7]). The rest of this section explains for each level which functionalities it comprises and to which building blocks these functions are assigned.

Call Level. Before service components can be established, the users and the provider must a.o. negotiate on the call communication-structure, the call QoS and the synchronization relations between the service components. These functions are placed in a *Communication-Structure ASO (CSSO)*, a *Synchronization ASO (SynSO)* and a *QoS ASO (QoSSO_1)*, respectively (see also figure 2). To enable the negotiation between CSSOs, SynSOs and QoSSO_1s in peer Application Entities, a communication link must be set-up. This link is established by the *Association Control Service Object (ACSO)*, which can be viewed as a multiparty extension of the OSI *Association Control Service Element (ACSE)* [7]. This extension also encompasses the ability to renegotiate the QoS of that link during the information-transfer phase. The result of the negotiation procedure is determined in QoSSO_1, using the user specified success condition (see section 3.2).

During the information-transfer phase, the global communication-structure view is monitored by the CSSO. Synchronization of service components is also a function of the call level and is performed by the SynSO. Quality relations between service components (see section 2.2) are negotiated and maintained by the QoSSO_1. When a user is bound to a call, but not attached to any of the service components, the component level and the stream level in its AEI are empty. Nevertheless, this user must be kept informed about the global call view. Therefore, maintenance of the global QoS view is also assigned to QoSSO_1. The fourth function of the QoSSO_1 is call QoS renegotiation during the information-transfer phase.

The service component ASOs, that are depicted in figure 2 below the ACSO, CSSO, SynSO and QoSSO_1, can perform the actual establishment of the service components.

Service Component Level. At this level, intelligence is needed to decide how service components are decomposed into streams. This function is assigned to a *QoS ASO* (*QoSSO_2* in figure 2). It depends on the QoS requirements of the users whether multipoint or point-to-point streams (and thus transport connections) are used. For example, if one user wants to receive video information of HDTV quality, while for another user normal quality is sufficient, QoSSO_2 must decide whether the video information is multicast over a single multipoint transport connection or transmitted over separate point-to-point transport connections. In the former case the information is submitted once, thereby satisfying the quality requirements of the most demanding user. In the latter case the same information is submitted twice to the transport service, but with different QoSs. QoSSO_2 may also decide to split the video component into, e.g., two streams. Hereby, one stream has a more reliable QoS than the other, to transfer the coded video information-units that are sensitive to errors. The other stream transfers the less error sensitive part of the coded video information.

Stream Level. QoS negotiation which is necessary at this level, is restricted to the transport connections that have to be established. A *QoS ASO* (*QoSSO_3* in figure 2) is used for this purpose. It also maps most application-oriented QoS parameters (e.g., end-to-end delay) onto transport connection parameters (e.g., transit delay). To exchange the media information-units (e.g., video-information units), *Information-Transfer ASOs* (*ITSOs* in figure 2) are used. The stream level ACSO establishes the transport connections that are needed to transfer the media information-units.

5.5 Relations between Levels, Building Blocks and Functionalities

This section presents an overview of the relations between levels in the architecture, the building blocks and the functionalities these building blocks perform.

Level	ASOs	Functionalities
Call	QoSSO_1	QoS (re)negotiation & maintenance
		Maintenance of global QoS view
	SynSO	Negotiation & maintenance of synchr. relations
	CSSO	Comm.-structure (re)negotiation & maintenance
		Maintenance of global comm.-structure view
	(Call) ACSO	Establishing link via which peer QoSSO_1s, SynSOs and CSSOs communicate
	SC ASOs	(zero or more)
SC	QoSSO_2	Deciding (based on QoS) whether multicast or point-to-point transport connections are used
	Stream ASOs	(one or more)
Stream	QoSSO_3	QoS
	ACSO	Establishing & releasing transport connections
	ITSO	Submission/receipt of information to/from transport connections

Table 2. Relations between Levels, Building Blocks and Functionalities

6 Evaluation of the Framework Architecture

In section 5.1 an architecture for composition of multimedia teleservices was presented. In order to evaluate the suitability of the architcture, section 6 selects two mechanisms and shows how they can be modelled with the functionalities defined in the architecture. A more complete evaluation would result in too much detail for this paper. The mechanisms selected are synchronization and QoS management. Synchronisation is discussed because it is often applied to satisfy the requirements that the end-users impose on the distributed multimedia applications (see section 2). QoS management is illustrated bcause a lot of advanced distributed multimedia applications require a flexible QoS that can easily be adapted to changing user needs.

6.1 Synchronization

Many synchronization mechanisms can be distinguished, but in this paper only two, frequently applied mechanisms are discussed. The first mechanism is marker synchronization. Hereby, markers are placed in the user information to obtain synchronization between related user information-units. The second mechanism that is dealt with is channel synchronization. This mechanism uses a control channel alongside the user information channels [16].

- **Marker Synchronization**. When synchronization markers are used, the synchronization information is exchanged over the same transport connections as the user information. Therefore, the SynSO of the call level inserts the markers between the user information-units. This is done via the CFs of the service component level (see figure 5a). As a consequence, the ITSOs at the stream level must submit two types of Protocol Data Units (PDUs): synchronization PDUs and user information PDUs (e.g., video PDUs). A SynSO negotiates beforehand with SynSOs in peer AEIs about the synchronization mechanism to be used. This is indicated by the thin dashed arrow under the SynSO in figure 5a. Notice that for simplicity purposes figure 5 represents the CFs in another way than figure 2.

- **Channel Synchronization**. When channel synchronization is applied, the external synchronization channel transfers tuples of identifiers of information-units that have to be synchronized. In this case, the SynSO does not submit synchronization information to the CFs of the service component level (see figure 5b). Instead, the synchronization information is exchanged between peer AEIs via a special transport connection. To enable the SynSO to establish the needed transport connection, the SynSO is refined with an internal ACSO. This is possible since the OSI-ALS allows the nesting of ASOs.

Figure 5 shows how the functionalities of the framework architecture can be composed to provide synchronization. The architecture allows enough flexibility to model both discussed synchronization mechanisms. Furthermore, a close look at figure 5 teaches that the structure of the AEI hardly changes when a teleservice is extended with new service components that have to be synchronized as well. This indicates that the architrecture allows easy extension of teleservices, even when a complex mechanism as synchronization is involved.

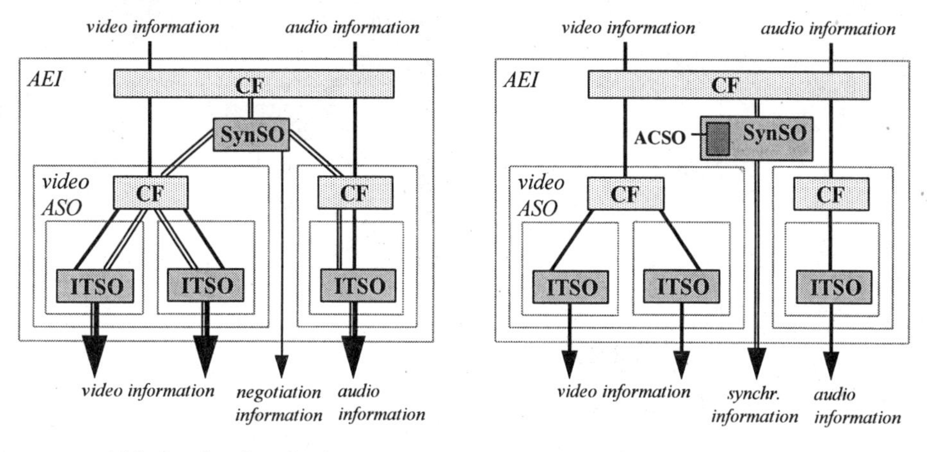

Fig. 5. Application Entity Invocations for different Synchronization Mechanisms

6.2 Quality of Service Management

QoS management comprises several functions, such as QoS (re)negotiation and QoS maintenance.

- **QoS Renegotiation**. Due to changing user needs, QoSSO_1 may be requested to perform QoS renegotiation. The result of the renegotiation procedure is sent by QoSSO_1 to QoSSO_2 (see the dashed line of figure 6 between these ASOs). Then, QoSSO_2 can examine whether the new QoS differs so much from the old one that splitting (or integration) of transport connections is required. Moreover, it can examine the need for QoS renegotiation on existing transport connections. In case this appears necessary, QoSSO_2 asks QoSSO_3 to perform this QoS renegotiation (see the dashed line in figure 6 between these ASOs).
- **QoS Maintenance**. The QoS of a transport connection is monitored by the QoSSO_3 of the stream ASO that corresponds to that connection. When the obtained QoS does not satisfy the requirements, QoSSO_3 can start a QoS renegotation procedure. If this procedure fails, QoSSO_3 may decide to release the transport connection, which is then performed by ITSO. Another kind of QoS management is the surveillance of the quality relations that may exist between different service components. For example, in a videophony session it is conceivable that the users require the abortion of the video component when the audio component has crashed. QoSSO_1 surveys whether this quality relation is still satisfied. Once it notices that the audio service component is crashed, it indicates to all ACSOs in the video component (via QoSSO_2 and QoSSO_3) that the video transport connections must be released as well.

The architecture offers enough flexibility to model QoS management, since changing QoS needs do not result in complicatd changes to the AEI. The modularity that is obtained by using OSI-ALS consistent building blocks makes it possible to compose an AEI with which QoS management can be tailored to specific user needs.

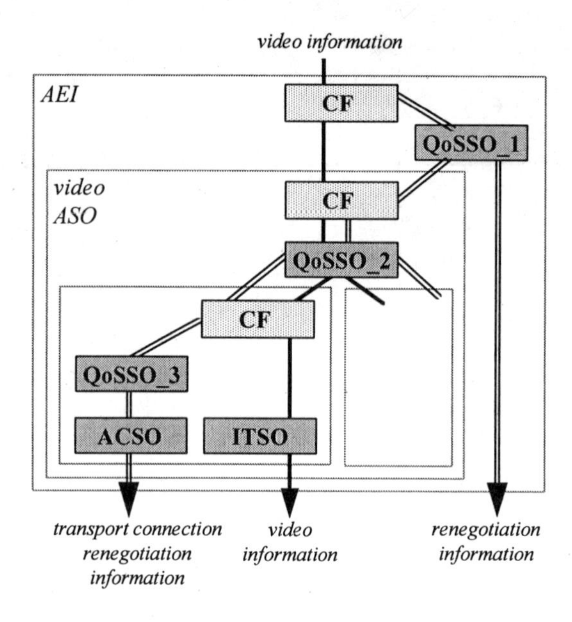

Fig. 6. Application Entity Invocation for QoS Management

Figure 7 shows how the CF at the call level distributes control information over SynSO, QoSSO_1 and CSSO in order to manage a video and an audio ASO. The call control information is sent to peer AEIs over the link that is denoted with the dashed line in the figure. During the call establishment phase, before the negotiation on the QoS and call communication-structure is finished, the video and audio do not yet exist. In this phase, only the call control link is present to enable the negotiation procedures.

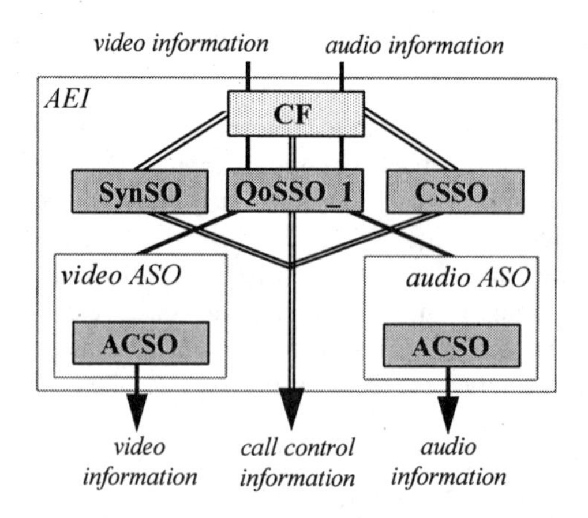

Fig. 7. Application Entity Invocation for Control of Video and Audio ASOs

7 Conclusions

This paper derives in a top-down approach the communications capabilities required by distributed multimedia applications. These capabilities are embodied in functionalities such as synchronization and maintenance of quality relations, that are quantified with application-oriented QoS parameters The functionalities are clustered into functional building blocks, that are defined in an architecture that allows a flexible composition of teleservices. Most application-oriented QoS parameters are mapped onto QoS parameters of an underlying transport service. OSI-ALS is used successfully to elaborate a multimedia framework architecture. This exercise therefore shows to some extent the suitability of OSI-ALS to model multimedia teleservices. However, the composition rules prescribed in OSI-ALS give also rise to some questions. For example, it is not clear why peer CFs in different AEIs have to communicate via underlying ASOs instead of communicating directly.

The elaborated architecture, which consists of three levels, indicates that the nested structure of OSI-ALS provides independence from an underlying transport service. The Application Service Objects (ASOs), defined in OSI-ALS, are appropriate to serve as elementary building blocks of the architecture, since they enable effective and efficient composition of user tailorable teleservices (see the requirements mentioned in section 5). The Association Control Service Object (ACSO), which establishes multiparty associations to exchange information, is expected to be easily realizable with a new transport service like ST-II [18] or the transport service provided by XTP [13]. Notice however that XTP does not support any QoS yet.

Finally, it is worthwhile to remark that the elaborated example shows an architecture which has several similarities with architectures found in the telecommunications area, although the OSI-ALS originates from the OSI datacommunications area.

Acknowledgement

The authors would like to thank Geert Heijenk and Xinli Hou for the intensive discussions on the call model, and Dick Quartel and Robert Huis in 't Veld for their contributions to the first two sections.

Glossary

ACSE	Association Control Service Element
ACSO	Association Control Service Object
AEI	Application Entity Invocation
ASE	Application Service Element
ASO	Application Service Object
ATM	Asynchronous Transfer Mode
CF	Control Function

CSSO	Communication-Structure Service Object
ITSO	Information-Transfer Service Object
OSI	Open Systems Interconnection
OSI-ALS	OSI Application Layer Structure
PDU	Protocol Data Unit
QoS	Quality of Service
QoSSO	QoS Service Object
SAP	Service Access Point
SC	Service Component
SynSO	Synchronization Service Object

References

1. S.R. Ahuja and J.R. Ensor: Coordination and Control of Multimedia Conferencing. *IEEE Communications Magazine*, May 1992, pp. 38-43.

2. S.A. Bly, S.R. Harrison and S. Irwin: Media Space: Bringing People together in a Video, Audio and Computing Environment. *Communications of the ACM* 36(1), January 1993, pp. 28-47.

3. A. Campbell, G. Coulson, F. Garcia and D. Hutchison: A Continuous Media Transport and Orchestration Service. *Sigcomm '92*, 1992, pp. 99-110.

4. D.B. Hehmann, M.G. Salmony and H. Stuttgen: Transport Services for Multimedia Applications on Broadband Networks. *Computer Communications* 13(4), May 1990, pp. 197-203.

5. G.J. Heijenk, X. Hou and I.G. Niemegeers: Service Description of Communications Systems Supporting Multi-media Multi-user Applications. *IEEE Network*, January 1994.

6. ISO: Quality of Service Framework - Working Draft No. 1, November 1992. ISO/IEC JTC1/SC21 N7521.

7. ISO: Final Text of ISO/IEC 9545 Application Layer Structure, May 1993. ISO/IEC JTC1/SC21 N7815.

8. S.E. Minzer: A Signaling Protocol for Complex Multimedia Services. *IEEE Journal on Selected Areas in Communications* 9(9), December 1991, pp. 1383-1394.

9. S.E. Minzer and D.R. Spears: New Directions in Signaling for Broadband ISDN. *IEEE Communications Magazine* 27(2), February 1989, pp. 6-14.

10. K. Nahrstedt and J. Smith: Revision of QoS Guarantees at the Application/Network Interface. *Technical Report MS-CIS-93-34, University of Pennsylvania*, March 1993.

11. C. Nicolaou: An Architecture for Real-Time Multimedia Communications Systems. *IEEE Journal on Selected Areas in Communications* 8(3), April 1990, pp. 391-400.

12. A. Paglialunga and M. Siviero: ISCP (ISDN Signalling Control Part): the Best Candidate for the Target Broadband-ISDN Signalling Protocol. *Communication Networks* 4(2), April 1993, pp. 193-200.

13. Protocol Engines, Inc.: XTP Protocol Definition, Revised 3.4, July 1989.

14. E.H. van Rijssen: A Survey of Distributed Multimedia Requirements for Application Layer Structuring. Master's Thesis, University of Twente, Department of Computer Science, Enschede, The Netherlands, October 1993

15. SIS Standardiseringsgrupp: Information Processing Systems - Open Systems Interconnection - Connection-oriented Transport Service Definition, ISO 8072-1986, December 1986.

16. R. Steinmetz: Synchronization Properties in Multimedia Systems. *IEEE Journal on Selected Areas in Communications* 8(3), April 1990.

17. R. Tokuda, Y. Tobe, S.T.-C. Chou and J.M.F. Moura: Continuous Media Communication with Dynamic QOS Control Using ARTS with an FDDI Network. *Sigcomm '92*, August 1992, pp. 88-99.

18. C. Topolcic: Experimental Internet Stream Protocol, Version 2 (ST-II). Internet Request for Comments No. 1190, RFC-1190, October 1990.

Service Definition of a Multimedia Partial Order Connection

Christophe Chassot, Michel Fournier, Michel Diaz, André Lozes
(chassot, fournier, diaz, lozes@laas.fr)

LAAS du CNRS
7, avenue du Colonel Roche
31077 Toulouse - France

Abstract : Starting from a new concept of connection, the partial order connection (POC), we introduce the design principles of a multimedia partial order Transport connection (MM-POC), that allows one to define a multimedia QoS consistent with multimedia application requirements. A complete definition of the proposed multimedia partial order service (MM-POS) is provided including complete definition of the QoS parameters and the service primitives needed to open/close a connection, negotiate/renegociate the QoS parameters, transmit data and notify service users when loss occurs or the negotiate QoS is not being provided.

Key words : partial order, partial order connection, connection-oriented protocol, connectionless protocol, multimedia and high speed protocol, quality of service, multimedia Transport protocol, multimedia Transport service.

1 Introduction

Present data transfer protocols use either connectionless (CL) or connection-oriented (CO) paradigms. A new concept, the Partial Order Connection (POC), for which CO and CL connections appear to be special cases, has been introduced in [1] and [2] through its reliable (*R-POC*) and unreliable (*U-POC*) aspects. A POC is an end-to-end connection that allows one to define and use for transferring data any partial order protocol and any partial order service between communicating entities. In a POC, *order* and *reliability* appear to be both quality of service parameters : they are specified by the service user during the first partial order service invocation ; once known by both sending and receiving POC entities, 'order' and 'reliability' service parameters are translated into protocol parameters, so as to allow the start-up of partial order and reliability management mechanisms.
A new multimedia architecture, establishing the POC concept, is detailed in this paper. Special attention is focused on gains in time latency generated by using a MultiMedia Partial Order Connection (MM-POC) as a Transport support for multimedia applications. This architecture has been initially presented in [3] ; this preliminary work is extended in this paper with a first definition of the corresponding MultiMedia Partial Order Service (MM-POS).
A brief presentation of the POC concept is given in Section (2). Section (3) presents the main principles of a multimedia Transport connection. Establishment of the POC concept (through its reliable and unreliable aspects) in this kind of Transport connection is presented in Section (4) ; it is specially shown in which way a MM-POC generates interesting gains in term of transit delay, still respecting the 'order' and 'reliability' Quality of Service (QoS) parameters, as they have been specified by the sending user

during the connection set up phase. Finally, Section (5) presents a first definition of the Multimedia Partial Order Service associated with the previously detailed Transport connection.

2 Related work about the POC concept

A POC is an end-to-end connection that allows one to define and use for transferring data any partial order protocol and any partial order service between communicating entities. In a POC, the objects - typically service data units (SDUs) - can be delivered to the receiving user in an order that is different from the sending order, i.e. the order they have been submitted by the sending user. The acceptable difference between the submitted sequence and the different but acceptable delivery sequences to the receiving user precisely results from the definition of the (selected) partial order. [4] [5] show through examples how these different delivery sequences lead to transfer speed-up and save resources at both sending and receiving sides. Finally, the POC concept provides a conceptual link between present unreliable CL protocols and reliable CO protocols, that appear to be two extreme cases of this new general definition of a connection.
[1] and [2] respectively introduce a reliable and an unreliable version of a POC, the latest allowing the service to lose a predefined subset of the exchanged objects. In a U-POC, transfer speed-up and saving resources are enforced, still respecting the user requirements as they are expressed in term of both order and reliability.
A formal specification of POC, written in Estelle [6] [7] indicates how the receiving partial order entity dynamically determines if arriving protocol data units (PDUs) satisfy a given partial order without or with loss. [4] provides a quantification metric of partial order complexity. [5] defines a space of protocols including order, reliability and time and presents two other metrics, based on an analogy with entropy and incertainty notions. Finally, a multimedia Transport architecture including the unreliable POC concept is presented in [3]. So as to understand the main section of this paper, the main points of this architecture are now presented.

3 A multimedia Transport Architecture

From the application point of view, it clearly appears that high speed Transport protocols will be used for multimedia communications. However, as the requirements of multimedia applications are multiple and diversified with regard to the kind of data flows they imply, it is now necessary to define different transfer characteristics for each of these flows. As an example, let us assume an application be composed of partially synchronised text and video flows : the used Transport service has to guaranty both a perfect reliability with respect to the data flow transfer, and a sufficient throughput with respect to the video one ; however, it is clear that a totally reliable Transport connection is not needed for the video flow and it is also true that a high speed transfer is not a major requirement for a data text type communication. Moreover, the use of a reliable service implies transmission latency, often inconsistent with an acceptable high speed video distributed application.
Taking this into account, and as an alternative to the multimedia Transport service detailed in [8] [9] [10] [11], it is proposed in this paper to define a multimedia partial order Transport connection, including and managing monomedia partial order connections. Each of these POCs provides a given quality of service (including order and reliability parameters) according to the user requirements expressed for each of the different flows constituting the multimedia application.

Multimedia Transport connection: set of multiple monomedia Transport connections, each with a given QoS

As far as the multi-flow aspect of a multimedia application is concerned, it clearly appears the need for a new multimedia Transport architecture providing a set of QoSs, each of them being dedicated to one of the different flows composing the multimedia application. In the proposed architecture, a multimedia Transport connection implies the establishment and then a specific co-ordination of several monomedia connections, each of them with a distinct QoS.
As an example, consider the application presented in Figure 1(a), which is a subset of an example initially detailed in [4]. The distributed application is composed of a broadcast bulletin, delivered on two different video windows : the biggest one is dedicated to a classical broadcast presentation (video 1) ; on the other one appears an analogous presentation given in signs language (video 2). Two audio outputs are dedicated to the sound presentation. The associated multimedia Transport architecture is made of three monomedia connections, each of them supporting a given QoS, respectively adapted to the presentation characteristics of audio, video1 and video2 flows (Figure 1(b)).

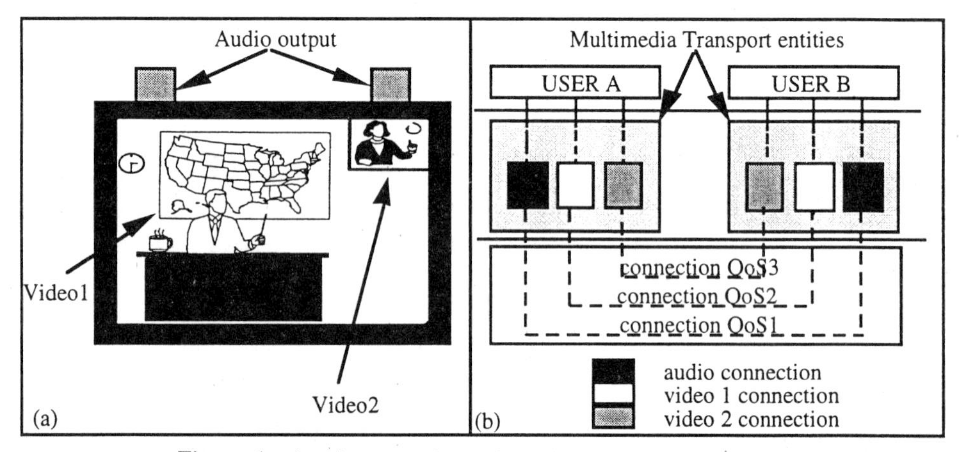

Figure 1 : Architecture of a multimedia Transport connection

Multiplexing of several monomedia POCs on a same underlying connection is not considered in this paper, as many network connections are assumed to be available in future networks.
Let us now consider how the unreliable POC concept (*U-POC*) can be handled in this architecture.

4 Establishment of the U-POC concept in a multimedia Transport connection

As far as new requirements of multimedia applications are concerned, current studies about the Transport layer suggest the addition of new functionality's. As an example, the enhanced Transport service presented in [11] introduces new QoS parameters related to transit delay and transit delay jitter ; moreover, the notion of "threshold" and "compulsory" QoS values define new semantics for a Transport service.

Our approach does not rely on the specification of new classical (monomedia) Transport mechanisms but could use them : it is proposed here to extend a subset of the existing mechanisms, with regard to both conceptual limits of current protocols and requirements of multimedia applications. More precisely, this section shows how the establishment of the POC concept inside a Transport architecture will extend classical Transport functionality's, and seems to be of a great interest for multimedia applications.
Let us note here that, due to implementation and real management difficulties, temporal synchronisation is supposed to be managed at a higher conceptual level than the Transport level.

Among the different models for the specification of multimedia synchronisation scenarios, the TSPN model has been proposed [12] [13]. It extends Time Petri nets to formally describe timed behaviour of multimedia objects and streams, in asynchronous distributed systems.
Let us consider Figure 2 ; the illustrated Petri net is deduced from the application defined TSPN model of the example developed in Section (3). Consider each place of the Petri net as a meaningful SDU (for instance, white and grey places represent full images). It then appears that this Petri net provides a logical representation of the partially ordered synchronisation constraints of a given period, as they are expressed at the user level.
This representation defines what we call a 'multimedia partial order', in which the SDUs of a given period are related together. Note that the first (respectively the last) transition of the net defines a total order between the previous period of related SDUs and the current one (respectively the current period of related SDUs and the next one).
Service users are then allowed to define their most adequate multimedia partial order, and then, should be able to select the associated multimedia partial order Transport service.

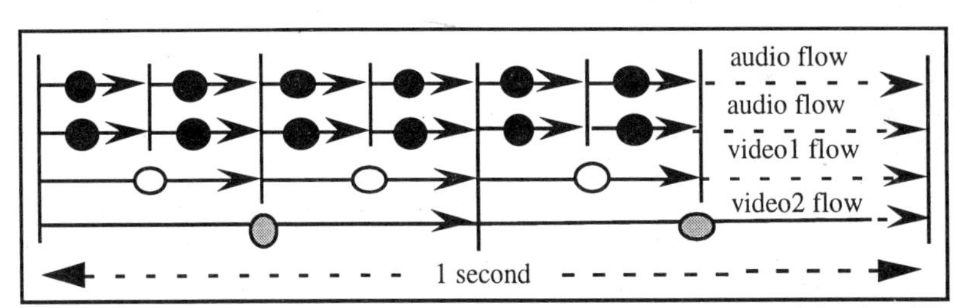

Figure 2 : Multimedia partial order associated with the example of Section (II)

Let us now study how a multimedia partial order such as the one defined by Figure 2 may be handled in a multimedia partial order Transport connection. Reliability aspect is not considered in paragraph (4.1) : multimedia partial order handling is thus presented as *the establishment of the reliable POC concept (R-POC) in a multimedia partial order Transport connection*. In paragraph (4.2), the reliability aspect is then taken into account : handling of both multimedia partial order and reliability is presented as *the establishment of the U-POC concept in a multimedia partial order Transport connection.*

4.1 R-POC concept establishment

So as to clearly understand the establishment of the R-POC concept in a multimedia Transport connection, let us consider Figure 3 the first 16 objects of the Petri net illustrated in the previous paragraph. This Petri net includes two kinds of dependence relationship between these objects (called SDUs in the following) :

- intra-flow dependency relationship ; as an example, 'SDU 3 precedes SDU 6' is an intra-flow dependency relationship ;
- inter-flow dependency relationship ; as an example, 'SDU 2 precedes SDU 8' is an inter-flow dependency relationship.

Note that the initial and last transitions of the Petri net define both intra and inter-flow dependency between SDUs from two consecutive periods.

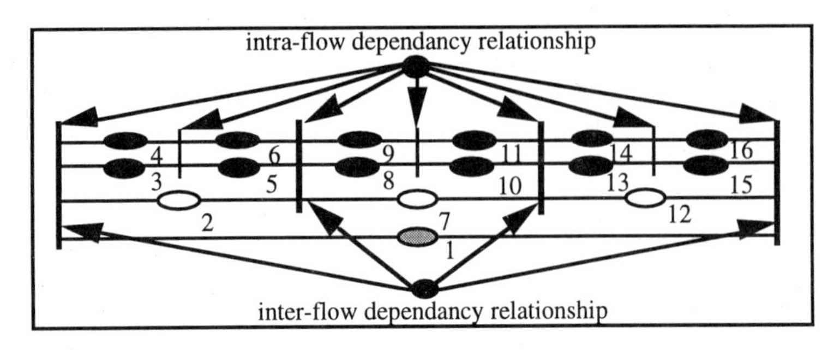

Figure 3 : Intra and inter-flow dependency relationship in a multimedia partial order

The R-POC concept establishment may be viewed at two levels : a *monomedia* establishment, and a *multimedia* establishment, intuitively related to the intra and inter-flow dependency relationship management.

4.1.1 'Monomedia' establishment of the R-POC concept

The R-POC concept is first established in each of the monomedia connections constituting the multimedia connection, so as to handle intra-flow dependency relationship. Such an establishment may be considered as a *monomedia* establishment of the R-POC concept.

Service users are then allowed to select all *monomedia* partial order services for each of their monomedia flows. Respecting a monomedia partial order on each monomedia POC generates the same benefits as those reminded in Section (2) ; particularly, in a POC, user data are delivered sooner than in a classical reliable connection (for instance TCP-like one). Such a transit delay reduction is a major interest with regard to temporal constraints of multimedia applications. Of course, a service user is still allowed to select *total order* or *no order* as order QoS parameter for each of the POCs ensuring the transport of its different flows.

4.1.2 'Multimedia' establishment of the R-POC concept

To synchronise the SDU delivery of the different media according to the inter-flow dependency relationship of the a given multimedia partial order, it is now proposed to define a *multimedia* establishment of the R-POC concept.

Let us consider again Figure 3 and look at the inter-flow dependency relationship, in the multimedia partial order connection presented. Handling of inter-flow dependency relationship is given on Figure 4.

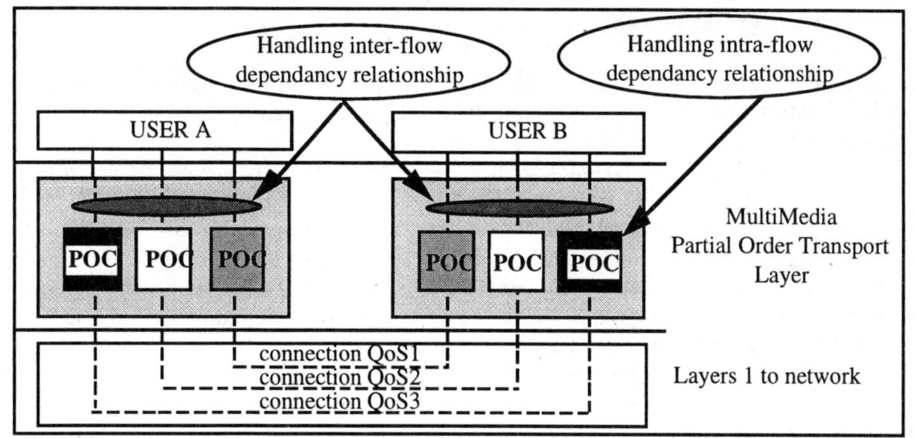

Figure 4 : Establishment of the R-POC concept

With regard to [14], multiplexage of several flows on a same POC is not used in the previous architecture of the multimedia partial order Transport entity. As a result, inter-flow dependency relationship management appears to be an *inter-POC* dependency relationship management.

Let us now study the establishment of the U-POC concept in a multimedia Transport connection.

4.2 U-POC concept establishment

In a classical U-POC, the associated protocol is allowed not to recover the PDU losses, when the lost SDUs do not exceed an unacceptable number deduced from the user requirements. In this section, it is shown how the establishment of the unreliable aspect of a POC in a multimedia partial order connection allows user data to be delivered as soon as possible, issuing gains in transmission latency or/and in buffer usage.

The service provider of an end to end reliable data transfer through an imperfect medium has to manage all PDU retransmissions at the cost of their transit delay. Such a cost may be not acceptable as far as real time aspects of multimedia applications are concerned. In other words, it is now more suitable to decrease transit delays at the expense of reliability losses rather than the opposite. As an example, the loss of one picture every 30 during the transfer of a 30 pictures per second video movie has no impact on the final user visual perception.

Expression of reliability in a Multimedia Partial Order Connection (MM-POC)

To show the impact of U-POC concept establishment in a multimedia Transport connection, the *reliability* QoS parameter will be simply expressed as a vector $k=(k_1, ..., k_n)$, where n defines the number of monomedia POCs composing the MM-POC. Thus defined, the reliability quality of service is guaranteed by the partial order multimedia Transport service when the associated protocol does not exceed k_i consecutive losses on each POC_i. Note that a complete and more realistic reliability expression is given in Section (5).

U-POC concept establishment

Reliability management of a MM-POC is proposed through two different implementations of adequate error control mechanisms. Each of them is based on the following rule :

"delivery of a given SDU makes obsolete all SDUs that are not yet delivered (they can be lost or not) preceding it in the multimedia partial order".

When processing an *out of partial order* SDU (i.e. not deliverable with regard to the multimedia partial order), the protocol will then deliver it if the number of SDUs made obsolete does not exceed the tolerated loss level on each POC. Such a management allows the protocol to deliver SDUs *as early as possible*, depending on the level of reliability the user may tolerate. The corresponding two approaches are given in the next two sections (paragraph 4.2.1 and 4.2.2).

4.2.1 Media per media reliability management

When processing the delivery of an out of partial order SDU (say A) to the user on POC_i, the protocol will deliver the data if the PDUs made obsolete on this i^{th} connection are still fulfilling the reliability requirements. Also, it cannot deliver it if this delivery would need to generate a loss declaration of one or more valid SDUs (not yet delivered) on any of the other POCs ; a *media per media* reliability is thus defined. Let us then note that :

- A is *normally* delivered (i.e. without generating loss(es)) when A is deliverable with regard to the multimedia partial order ;
- A is delivered with intra-POC loss(es) when A is directly deliverable with respect to POC_i monomedia partial order, or when the number of not yet processed (i.e. neither delivered nor declared lost) SDUs preceding A in the corresponding POC_i partial order is smaller than k_i. In this case, SDUs preceding A in POC_i and not yet processed are never delivered.

4.2.2 Per group of media reliability management

Generalising the previous approach, the *per group of media* protocol may now deliver to the user one SDU, say A, even if it generates a tolerable number of losses of one or more valid SDUs on any media different from POC_i. This reliability management is said to be *per group of media*. Taking this into account :

- A is delivered with inter-POCs loss(es) when A is directly deliverable with respect to the inter-flow partial order, or when the number of SDUs preceding A in $POC_{j(j \neq i)}$ partial orders (for all monomedia POCs), neither yet delivered nor declared lost, is smaller than $k_{j(j \neq i)}$.

5 Service definition

As defined in the OSI transport service [15], the QoS concept becomes inadequate to fulfil multimedia application constraints. First, it relies on the best effort service semantic : if the service provider does not succeed in maintaining the level of the negotiated QoS, no specific action is taken ; in other words, the only way for a service user to know the actual value of a QoS parameter is to measure it by itself. Second, the set of associated QoS parameters need to be revised in order to integrate both the features

of new underlying networks (such as the ATM), and of the multimedia application constraints.
Recent studies about new Transport services lead to the definition of a service semantic, different from the best effort one [11] ; when the negotiated level of a QoS parameter is no more maintained, the Transport service has to notify it to the service user. Consequently, the transport layer has to monitor QoS parameter values. Applied to a given QoS parameter (say *P*), this semantic is well suited when the service user may benefit from the notification of *P*'s value degradation.
Among newly introduced QoS parameters, let us note that the *transit delay* and the *transit delay jitter* parameters have been defined so as to allow the service users to specify time constraints for isochronous applications. *Throughput* definition has been extended in order to take into account traffic burstiness. Finally, the definition of new error control parameters takes into account the main features of all media (for instance, audio and video applications could react differently to bit or packet error rates).
Multimedia application requirements are strongly related to temporal aspects. It is important to notice that these temporal constraints have to be respected at both the application and the Transport level. However, implementation of an applicative process in an asynchronous system environment seems to make difficult the respect of temporal constraints, whatever the used Transport service. In such an environment, even if newly defined sets of Transport QoS parameters allow one to express temporal applicative constraints at the Transport level, it is not clear that the application user will be able to benefit of it.
Taking this into account and assumed to be implemented in an asynchronous system environment, our multimedia Transport service is not aimed at providing service user with temporal guarantees. Indeed, it is proposed to consider *order* and *reliability* as meaningful QoS parameters, so as to allow all service users to manage in the best possible way the temporal benefits generated by the establishment of the U-POC concept in a multimedia Transport architecture.

MM-POS : a peer to peer asynchronous MultiMedia Partial Order Transport Service

Our Transport architecture is assumed to be implemented in an asynchronous system environment. Consequently, whatever the features of the used underlying network, the MM-PO service has a best delay service semantic as far as QoS parameters having temporal aspects are concerned, except for a minimal value of the throughput, for which any degradation may be notified to the user. This latest choice is motivated by the following remark : new high speed networks such as the ATM networks are expected in the future to provide a certain quality of service related to throughput. Let us note that presently, this QoS is only expressed by a guaranteed peak rate. In the following, the provided service semantic will be defined for each of the QoS parameters.

5.1 QoS parameters

Assume a generic multimedia partial order connection be made of *n* monomedia POCs, each of them having a given QoS. Let us now define the QoS parameters, which specification occurs during the MM-PO connection set-up phase. In the following, POC_i defines the ith monomedia POC of the MM-POC.

5.1.1 Order

a) Expression

[4] provides a first representation of the *order* parameter in a given partial order protocol : in both sending and receiving POC entities, the partial order is locally represented in $N(N-1)/2$ bits as an $N*N$ upper-triangular matrix, where N is the number of objects in the partial order. In a partial order matrix, say M, $M[i,j] = 1$ means object j is preceded by object i in the given partial order ; else, $M[i,j] = 0$.
As a result, in a MM-POC, the order QoS parameter may be coded as a vector which size is $N(N-1)/2$. An additional vector, which size is N, has to be specified in order to identify the monomedia POC each object is related to.

b) Service semantic

QoS parameter 'order' value is always respected.

c) Negotiation

Partial order is specified to the MM-PO service as a QoS parameter in the first service primitive invoked by the sending user to establish a MM-POC. Receiving users receive it through an indication primitive sent by the MM-PO service at the receiving side.

5.1.2 Reliability

a) Expression

The first component of the *reliability* parameter is expressed as a (**k**, **w**, **l**) structure :

$$\mathbf{k} = (k_1, k_2, \dots k_n)$$
$$\mathbf{w} = (w_1, w_2, \dots, w_n)$$
$$\mathbf{l} = (l_1, l_2, \dots, l_n)$$

where :
- k_i defines the upper bound of consecutive SDU losses the user may tolerate on POC_i and furthermore :
- at any time, no more than l_i SDUs among the last w_i sent SDUs are allowed to be lost on POC_i.

This expression is similar to the one defined in [16]. *Burst Loss Sensitivity*, *Loss Sensitivity* and *Loss Interval* are the defined parameters for k_i, l_i and w_i respectively.
As a second component of the reliability parameter (Figure 6), four additional *reliability classes* have been defined for each POC, in order to allow the service user to select if he wants to be delivered corrupted SDUs or to be notified loss SDU occurrences. If a received corrupted SDU, say *A*, cannot be delivered to the receiving user, *A* may be declared lost by the MM-POC entity, if this loss is acceptable with regard to (k_i, l_i, w_i) parameters.
Loss occurrences may be notified to the user by either of the following expressions : the number of lost SDU(s) or the list of lost SDU(s).

		DELIVERY with CORRUPTION	
		authorized	not authorized
loss	not signaled	1	2
	signaled	3.x	4.x

with { x = 0 when the number of loss is signaled
x = 1 when losses are listed

Figure 6 : Reliability classes

b) Service semantic

Two service semantics can be selected flow per flow using the vector of booleans **v**=$(v_1,...,v_n)$.

- In the first one, reliability is always enforced : this can produce a limited increase of the transit delay.
- In the second one, reliability can be degraded.

A reliability degradation consists in the occurrence of SDU losses, such that one of the (k_i, l_i, w_i) parameters is violated. In response to a service violation : either the MM-POC is released or the QoS degradation is notified to the user. When a QoS violation has to be notified, the MM-PO service indicates to the service user which of the two k_i and l_i parameters is no more respected. QoS parameter 'reliability' is summarised in Figure 7.

Reliability					
flow ident	reliability class	w	l	k	v
1					
2					
n					
Service Violation Semantic					
0			1		

with { 0 = MM-POC released
1 = degradation notified

Figure 7 : 'Reliability' QoS parameter

Note: releasing one degraded POC only (rather than releasing the entire MM-POC) is not envisaged in this definition.

c) Negotiation

Reliability is specified to the MM-PO service as a QoS parameter of the first service primitive invoked by the sending user to require the establishment of an MM-POC. Receiving user is provided with it through an indication primitive invoked by the receiving MM-PO service.

5.1.3 *Throughput*

a) Expression

Two special values allow the service user to characterise the required *throughput* on each POC : a minimal value, Th_{min}, and a maximal one, Th_{max}. Let us note that the throughput may be defined either *in byte per second* when supported media are audio like or video like one, or in *message per second* when supported media need a Transport service message mode.

b) Service semantic

On each POC, service users are ensured to get either the Th_{min} value, or a notification of reaching Th_{min} (degradation).
The Th_{max} value is the submitting throughput upper bound the sending user has to respect. As a result, the receiving service user is ensured to be delivered SDUs at a throughput smaller than or equal to Th_{max}. In response to a service violation : either the MM-POC is released or the QoS degradation is notified to the user. QoS parameter 'throughput' is summarised in Figure 8.

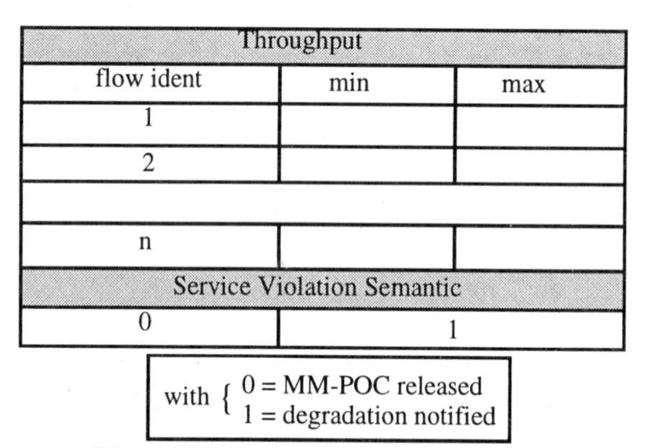

Throughput		
flow ident	min	max
1		
2		
n		
Service Violation Semantic		
0	1	

with { 0 = MM-POC released
1 = degradation notified

Figure 8 : 'Throughput' QoS parameter

Note : releasing one degraded POC only (rather than releasing of the entire MM-POC) is not envisaged in this definition.

c) Negotiation

The MM-POC set up phase may be cancelled when the Transport service has sufficient knowledge of the underlying network capacities to reject the required Th_{max} value. Let us note that the ATM network allows one to specify a peak rate value at the establishment of an ATM connection. Connection set up may then be cancelled when the Transport service is provided with a peak rate value smaller than the Th_{max} value, specified by the service user.

5.1.4 *Transit Delay*

a) Expression

Transit delay is defined as the elapsed time between the submission date of a SDU at the sending side, and the delivery date of the same SDU at the receiving side. Two values allow the service user to characterise the required transit delay on each POC : a minimal value, Td_{min}, and a maximal one, Td_{max}.

b) Service semantic

As far as the Td_{max} value is concerned, a best effort service semantic is applied. At the opposite, the Td_{min} value is always respected ; a sufficient condition to ensure the Td_{min} value is to specify it as half the lower bound of the measured round trip time (*RTT*) value.

c) Negotiation

The Td_{min} value is not initially specified by the service user. Indeed, this value is determined by the service provider during the connection set up phase, and then provided to users, as a service parameter of the indication and confirm connection set up primitives. Connection set up may be cancelled if the Td_{max} value, initially specified by the service user, is greater than the Td_{min} value, measured by the service provider.

5.1.5 *SDU size*

Knowledge of the maximal SDU size may be useful to the service provider ; it is specified by the service user as a service parameter during the connection set-up phase.

5.2 MM-PO Service primitives

MM-PO service primitives are now presented. Let us note that these primitives are related to the MM-POC, and not directly to the monomedia POCs constituting the MM-POC.

5.2.1 *Connection set-up primitives*

Four primitives are required to manage the MM-PO connection set up phase : MM-PO_CONNECT.Req, MM-PO_CONNECT.Ind, MM-PO_CONNECT.Resp and MM-PO_CONNECT.Conf (Figure 9).

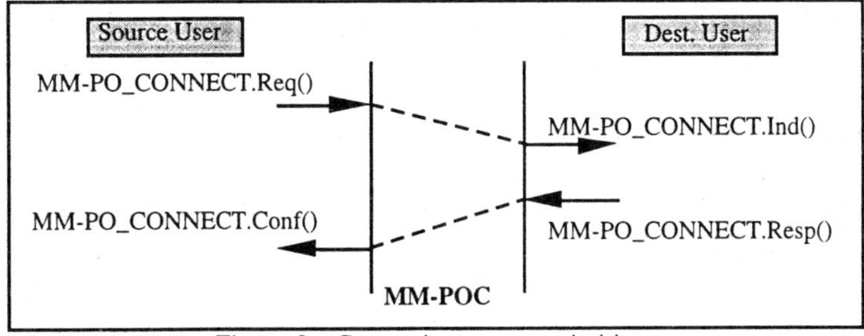

Figure 9 : Connection set-up primitives

Service primitive parameters

• **MM-PO_CONNECT.Req** :
- destination address, source address,
- flow number (say n),
- intra-flow partial order for each of the n flows,
- inter-flow partial order,
- (k_i, l_i, w_i, v_i) for each of the n flows,
- reliability class for each of the n flows,
- reliability service violation semantic,
- Th_{min} and Th_{max} for each of the n flows,
- throughput service violation semantic,
- Td_{max} for each of the n POCs,
- maximum SDU size for each of the n POCs,

• **MM-PO_CONNECT.Ind** :
- destination address, source address,
- *cep_0*, *cep_1*, *cep_2*, ..., *cep_n*,
- flow number (say n),
- intra-flow partial order for each of the n flows,
- inter-flow partial order,
- (k_i, l_i, w_i, v_i) for each of the n flows,
- reliability class for each of the n flows,
- reliability service violation semantic,
- Th_{min} and Th_{max} for each of the n flows,
- throughput service violation semantic,
- Td_{max} for each of the n POCs,
- Td_{min} for each of the n POCs,
- maximum SDU size for each of the n POCs,

where :
- *cep_0* identifies the MM-POC connection end point at the receiving side ;
- *cep_i* (for i from 1 to n) defines the POC_i connection end point at the receiving side.

• **MM-PO_CONNECT.Resp** : - status = accept or reject ;

where :
- reject reason is assumed to be coded in the *status* field

• **MM-PO_CONNECT.Conf** :
- status = accept or reject
- *cep_0*, *cep_1*, *cep_2*, ..., *cep_n*,
- Td_{min} for each of the n POCs ;

where if status = accept, then :
- *cep_0* identifies the MM-POC connection end point at the sending side ;
- *cep_i* (for i from 1 to n) defines the POC_i connection end point at the sending side.

5.2.2 Data transfer primitives

Data transfer is invoked with two service primitives : MM-PO_DATA.Req and MM-PO_DATA.Ind (Figure 10).

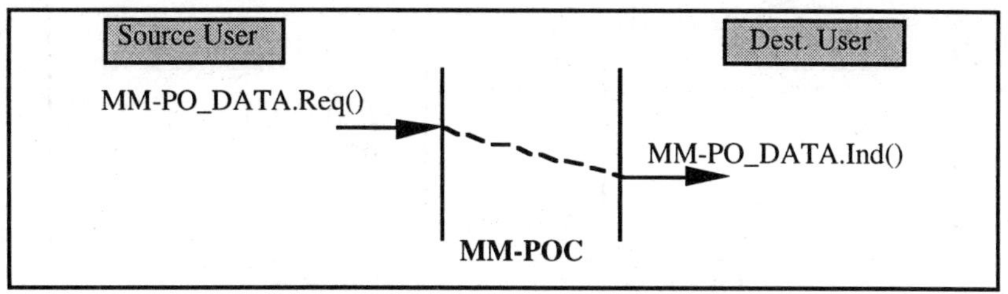

Figure 10 : Data transfer primitives

Service primitive parameters

• **MM-PO_DATA.Req** : - *cep_0*, *cep_i*,
- data,
- length ;

where :
- *cep_0* identifies the MM-POC connection end point at the sending side ;
- *cep_i* identifies the POC connection end point, the submitted 'data' is relating to ;
- *length* defines the data length.

• **MM-PO_DATA.Ind** : - *cep_0*, *cep_i*,
- data,
- length,
- acceptable_loss ;

where :
- *cep_0* identifies the MM-POC connection end point at the receiving side ;
- *cep_i* identifies the POC connection end point, the delivered *data* is relating to ;
- *length* defines the data length ;
- the *acceptable_loss* field is defined further on.

It has been shown in Section (4.2) how a SDU related to a given POC (say POC_i) might be delivered to the user at the cost of SDU loss(es) declaration on POC_i and/or $POC_{j(j \neq i)}$.
Such a delivery may be considered as
- a "delivery with acceptable loss" if the MM-PO QoS is maintained, as far as reliability QoS aspect is concerned ; with regard to the selected reliability class, such acceptable losses might have to be notified to the user.
- a "delivery with unacceptable loss" if the MM-PO QoS is no more maintained. With regard to the selected service semantic, either such unacceptable losses have to be notified to the user, or the MM-PO connection has to be released.

The *acceptable_loss* field provides service users with a notification of acceptable loss(es) occurred on the POC (say POC_i) the delivered 'data' is relating to. When the delivery is done with acceptable loss(es) on POC_j such that $j \neq i$, then these losses are notified (if it is required) using the MM-PO_LOSS.Ind primitive defined below (Figure 11).

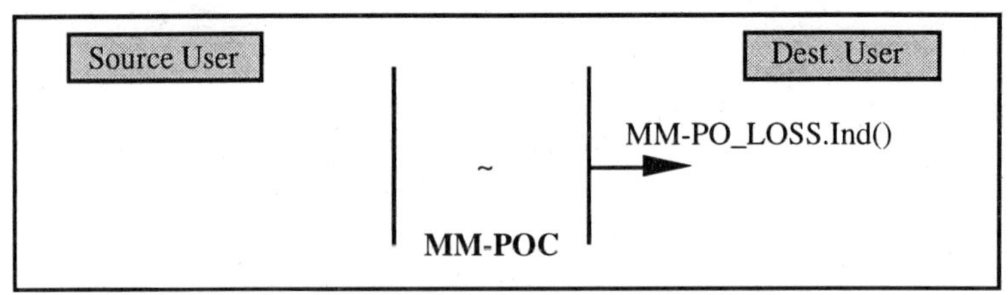

Figure 11 : Loss indication primitive

Service primitive parameters

• **MM-PO_LOSS.ind** : - *cep_*0, *cep_i*,
- acceptable_loss ;

where :

- *cep_*0 identifies the MM-POC connection end point at the receiving side ;
- *cep_i* identifies the POC connection end point, notified losses are relating to ;
- *acceptable_loss* is the same fields as the one defined in the MM-PO_DATA.Ind primitive.

This primitive is not used when acceptable loss(es) have ever been notified to the service user through a MM-PO_DATA.Ind primitive ;

'acceptable_loss' field definition

The acceptable_loss field is composed of two segments : a code and an associated message (Figure 12).

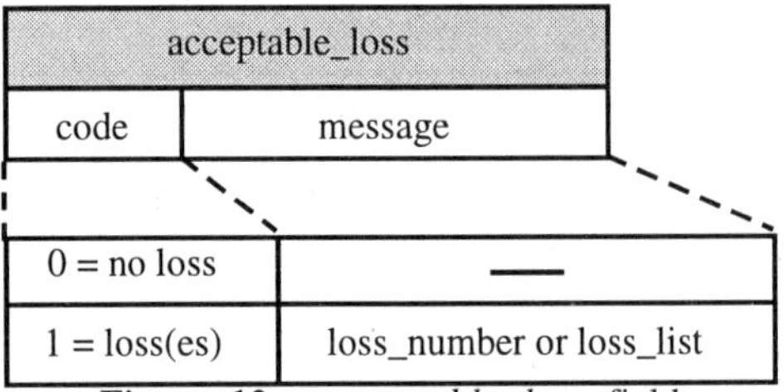

Figure 12 : *acceptable_loss* field

***code* = 0** when :

- either the service has not to notify acceptable loss occurrences, i.e. when the selected reliability class equals either **1** or **2** (see paragraph 5.1.2) ;
- or the delivery is done without loss.

The associated ***message*** field is then not meaningful.

***code* = 1** when :

- the service has to notify acceptable loss(es) occurrences (i.e. when the selected reliability class equals either **3**.x or **4**.x (see paragraph 5.1.2)), and the delivery is done with acceptable loss(es).

The associated ***message*** field has a meaningful value and :

- if the service has to notify the **number** of the observed acceptable loss(es) (i.e. when reliability class equals either 3.**1** or 4.**1**), then the message field contains the number of acceptable loss(es) declared on POC_i ;
- if the service has to notify the **list** of the observed acceptable loss(es) (i.e. when reliability class equals either 3.**2** or 4.**2**), then the message field contains the list of acceptable loss(es) declared on POC_i.

5.2.3 Primitive notifying a QoS degradation

The *Service Violation Semantic* field associated with each of the throughput and reliability QoS parameters allows the service user to select if he wants to be notified any reliability or throughput degradation. If not the service has to release the entire MM-PO connection.

One indication service primitive, MM-PO_QoS-FAILED.Ind, is introduced so as to allow the service to notify either a reliability or a throughput degradation (Figure 13).

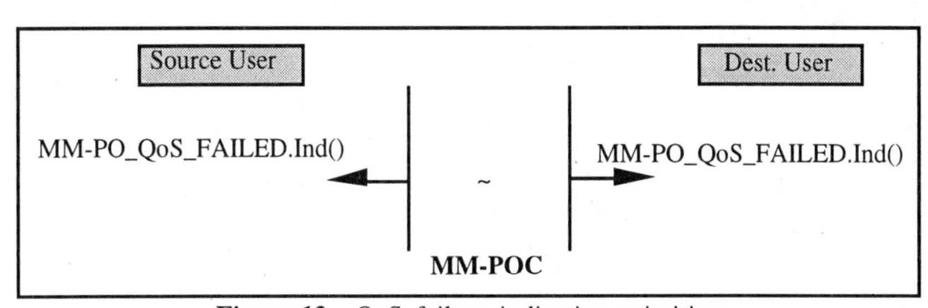

Figure 13 : QoS failure indication primitive

Service primitive parameters

- **MM-PO_QoS-FAILED.Ind** : - *cep*_0, *cep_i*,
  - QoS_degradation ;

where :

- *cep*_0 identifies the MM-POC connection end point.
- *cep_i* identifies the degraded POC connection end point,
- *QoS_degradation* is defined as follows.

The *QoS_degradation* field is composed of two segments : a code and a diagnostic (Figure 14).

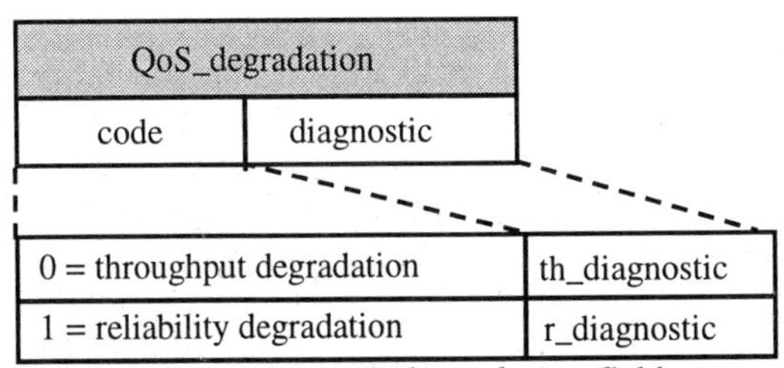

Figure 14 : *QoS_degradation* field

***code* = 0** when a **throughput degradation** has occurred on one of the *n* POCs ;
- the associated diagnostic is given through the *th_diagnostic* field, providing user with the reaching throughput value on degraded POC_i.

***code* = 1** when a **reliability degradation** has occurred on one of the n POCs ;
- the associated diagnostic is given through the *r_diagnostic* field, providing user with the (unacceptable) value of losses number and/or consecutive losses number.

5.2.4 QoS renegociation primitives

Four primitives, MM-PO-RENEGOC.Req, MM-PO-RENEGOC.Ind, MM-PO-RENEGOC.Resp, MM-PO-RENEGOC.Conf are now introduced in order to allow the service user to renegociate QoS without closing the MM-POC (Figure 15). When a renegociation phase is started up, all except one of the initially specified QoS parameters may be modified. Indeed, at the present time, user has to keep the flow number parameter (*n*) unchanged.

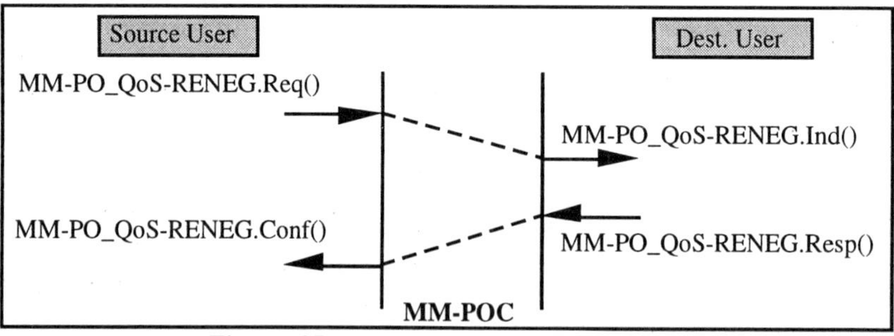

Figure 15 : QoS renegociation primitives

Service primitive parameters

- **MM-PO_QoS-RENEG.Req** :
  - intra-flow partial order for each of the *n* flows,
  - inter-flow partial order,
  - (k_i, l_i, w_i, v_i) for each of the *n* flows,
  - reliability class for each of the *n* flows,
  - reliability service violation semantic,
  - Th_{min} and Th_{max} for each of the *n* flows,
  - throughput service violation semantic,
  - Td_{max} for each of the *n* POCs,
  - maximum SDU size for each of the *n* POCs,

- **MM-PO_QoS-RENEG.Ind** :
  - intra-flow partial order for each of the *n* flows,
  - inter-flow partial order,
  - (k_i, l_i, w_i, v_i) for each of the *n* flows,
  - reliability class for each of the *n* flows,
  - reliability service violation semantic,
  - Th_{min} and Th_{max} for each of the *n* flows,
  - throughput service violation semantic,
  - Td_{max} for each of the *n* POCs,
  - Td_{min} for each of the *n* POCs
  - maximum SDU size for each of the *n* POCs,

• **MM-PO_QoS-RENEG.Resp** : - status = accept or reject ;
where : - reject reason is assumed to be coded in the *status* field.

• **MM-PO_QoS-RENEG.Conf** : - status = accept or reject
- Td_{min} for each of the *n* POCs ;

5.2.5 Connection release primitives

Two primitives are required to manage the MM-PO connection closing phase (Figure 16). Note that a diagnostic may be provided in a close indication, if it follows a QoS failure.

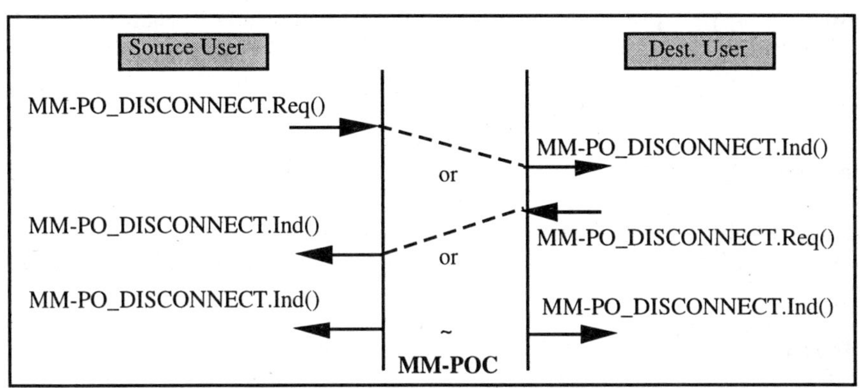

Figure 16 : Connection release primitives

Service primitive parameters

• **MM-PO_DISCONNECT.Req** : - *cep*_0 ;
where : - *cep*_0 identifies the MM-POC connection end point.

• **MM-PO_DISCONNECT.Ind** : - *cep*_0,
- diagnostic ;

where :
- *cep*_0 identifies the MM-POC connection end point ;
- *diagnostic* indicates if the connection release has been required by the peer user or if it follows a QoS failure (Figure 17)

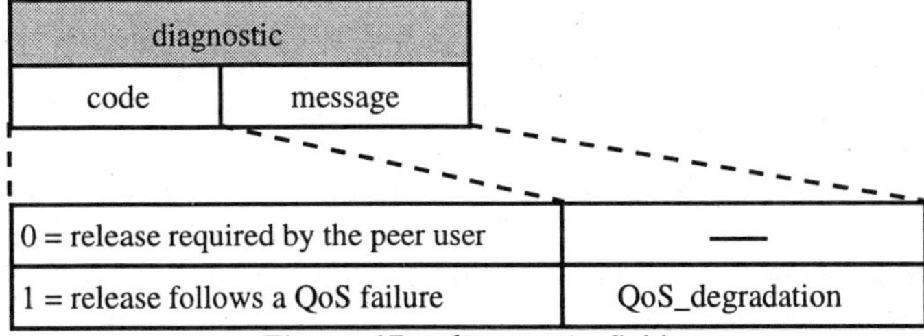

Figure 17 : *diagnostic* field

• ***code* = 0** when the connection release has been required by the peer user ;
- the associated ***message*** is then not meaningful ;

• ***code* = 1** means the connection has been released on a QoS degradation ;
- the associated ***message*** is the QoS_degradation field defined in the previous (5.2.3) paragraph.

Conclusion

A new concept, thc concept of a partial order connection, has been developed that extends the previous concepts of connectionless and connection-oriented protocols.
From this concept, the establishment of a POC concept in a multimedia Transport architecture is first presented in this paper. It is shown how such general POC connections can be defined by the application layer, and how this definition can be transferred to the communication software, in the transport layer. Finally, a first definition of a Multimedia Partial Order Service is given.

Acknowledgements

This work has been supported by CNET-FRANCE TELECOM under Grant 92 1B 178 as part of the CNET-CNRS CESAME project on the design of High Speed Multimedia Cooperative Systems.

References

1 P.D. Amer, C. Chassot, T. Connolly, M. Diaz : Partial Order Transport Service for Multimedia Applications: Reliable Service. 2nd High Performance Distributed Computing Conf. (July 1993) - Spokane, Wash.
2 P.D. Amer, C. Chassot, T. Connolly, M. Diaz : Partial Order Transport Service for Multimedia Applications: Unreliable Service. INET'93 3rd International Conference, San Francisco, CA, (August 17-20 1993).
3 C. Chassot, M. Diaz, A. Lozes : From the Partial Order Concept to Partial Order Multimedia Connections. To appear in JHSN, 1995.
4 P.D. Amer, C. Chassot, T. Connolly, M. Diaz, P. Conrad : Partial Order Transport Service to Support Multimedia Connections. IEEE/ACM Transactions on Networking, Oct. 1994, vol.2, n° 5.
5 M. Diaz, A. Lozes, C. Chassot, P.D. Amer : Partial order connections : a new concept for high speed and multimedia services and protocols. Annals of Telecommunications, tome 49, n° 5-6, may-june 1994.
6 Information Processing Systems - OSI. ISO IS9074 : Estelle - A Formal Description Technique Based on an Extended State Transition Model.
7 S. Budkowski, P. Dembinski : An introduction to Estelle : A specification language for distributed systems. Computer Networks and ISDN Systems, DLB92(1), 3-23, 1987.
8 A. Campbell, G. Coulson, D. Hutchison : A Multimedia Enhanced Transport Service in a Quality of Service Architecture. 4th International workshop on Network and Operating Systems Support for Digital Audio and Video, Nov. (3-5) 1993, Lancaster, UK.

9 G.S. Blair, F. Garcia, D. Hutchison, W.D. Shepherd : Towards New Transport Services to Support Distributed Multimedia Applications. 4th IEEE COMSOC International workshop, Monterey, USA, April 1-4, 1992.

10 H. Leopold, A. Campbell, D. Hutchison, N. Singer : Towards an integrated quality of service architecture (QoS_A) for distributed multimedia communications. 4th IFIP Conference on high performance networking, Liege, Belgium, Dec. (DLB92-18) 1992.

11 A. Danthine, Y. Baguette, G. Leduc, L. Leonard : The OSI95 connection-mode Transport service - the enhanced QoS. 4th IFIP Conference on high performance networking, Liege, Belgium, Dec. (DLB92-18) 1992.

12 M. Diaz, P. Sénac : Time Stream Petri Nets : A Model for timed Multimedia Information. ATPN 95, Int. Conf. on Application and Theory of Petri Nets, Zaragoza, Spain, June 1995.

13 P. Sénac, M. Diaz, P. de Saqui-Sannes : Toward a formal specification of multimedia synchronization scenarios. Annales des télécommunications, tome 49, n° 5-6, mai-juin 1994.

14 D.L. Tennenhouse : Layered Multiplexing Considered Harmful. Protocols for High-Speed Networks, Elsevier Science Publishers B.V. (North-Holland), 1989.

15 ISO International Standart 8072. Information Processing systems - Open Systems Interconnction - Transport service definition, 1986.

16 G. Partridge : "A proposed Flow Specification", Internet Request for Comments n°1363 (RFC 1363), Sept.1992.

Network Support for Multimedia Communications Using Distributed Media Scaling

François Toutain, Laurent Toutain
{ftoutain, toutain}@rennes.enst-bretagne.fr

Ecole Nationale Supérieure des Télécommunications de Bretagne
Antenne de Rennes - Département Réseaux et Services Multimédia

Rue de la Châtaigneraie - B.P. 78
35512 Cesson-Sévigné CEDEX - France

Abstract. This paper focuses on the transmission of continuous multimedia streams in packet switched networks, through the use of distributed media scaling techniques. These techniques make it possible for any receiver of a multicast stream to get best possible quality, while avoiding any waste of network resources. Enabling distributed media scaling requires to design network nodes with media scaling capabilities, as well as to design proper control mechanisms that will ensure optimal filter tuning in any node and at any time. A multicast transport protocol allowing network-level media scaling through a special memory management is proposed, and a distributed algorithm which enables optimal tuning of the network node filters is presented.

1 Introduction

Multimedia continuous streams (audio, video streams) are rarely thought to be transmitted in raw format. Instead, compression and coding schemes are applied in order to reduce the amount of bandwidth required for their transmission [4]. Some of these schemes are scalable, i.e. they allow delivery of various media qualities by tuning some compression factor. Some are hierarchical, i.e. they produce media subbands which can be played independently, each one improving the quality of the presentation.

Media scaling is the activity of modifying the stream structure, in order to modify its transportation requirements, with the drawback of usually also altering the perceived quality of a stream. In [2], media scaling is defined as *subsampling a data stream and only present some fraction of its original content*, and the difference is made between *continuous scaling*, which means degrading the content of a substream, and *discrete scaling*, which is done by adding or removing substreams of a hierarchically encoded stream. It should be noticed that discrete scaling techniques require to dispatch the original stream into substreams before entering the network, as well as to reassemble the remaining substreams into one coherent stream in end-systems [10]. The latter operation could be performed through inter-stream synchronization mechanisms. However it involves expensive computations to take place at each receiver [2]. In contrast, continuous scaling does not require any special processing at the sinks, provided that it is done in a proper way with respect to the original stream structure.

Distributed media scaling

Multimedia networking researchers have been more and more considering media scaling techniques for a few years, as a way to adapt continuous streams to various constraints arising from the users as well as from the underlying, heterogeneous systems (computers and networks). Former approaches were targeted towards "best-effort" adaptive applications, being able to perform stream degradations driven by controlling codecs in feedback loops [1, 17]. Another approach is to combine media scaling and service guarantees [2, 3, 12]. In such a case, media scaling can be done at the source of the data stream, so as to conform to the previously negotiated QoS contract (so-called traffic shaping). However this scheme has the drawback of restricting multicast communications to a single common QoS level. This issue can be solved by performing media scaling at the sinks, therefore allowing each receiver to select its own QoS level, while the source sends the highest level to everyone. In this scheme however, network resources may be wasted as they are required to carry the highest QoS level, whereas receivers filter the data stream in order to get a lower one.

To address these issues, the distribution of the scaling activity inside the network has been proposed [5, 11, 14, 21]. This latter scheme seems well adapted: whereas source-based scaling imposes a single QoS level for all receivers, and while receiver-based scaling overuses the network resources, in contrast *network-based scaling allows to propagate media filters to the point where each receiver enjoys its own best possible QoS, with only the actually useful network resources being involved in the flow transportation* (fig. 1). Obviously, the latter point is of prime importance, considering that network power will always be a limited resource to be shared among users.

Some people think of application-level filters, with the ability to scale multimedia streams based on their semantics, or to perform format conversions [3, 11, 13, 15, 21]. The main drawback of such high-level filtering is that these application-level filters, possibly provided by the user himself, must be run by networks entities (routers, gateways), at the price of added complexity and lack of security. Other people think about network-level scaling, possibly embedded in a transport protocol such as ST-II [7], where "dumb" network entities can filter incoming packets using a priority field [2, 5, 6, 9]. The issue here is to define priority schemes that adequately reflect a graceful degradation of the media streams. Nevertheless, it can be noticed that the network-level scaling activity and priority schemes are independent, thus allowing future improvements such as dedicated scalable coding schemes, whereas high-level filters rely on the streams semantics and structures.

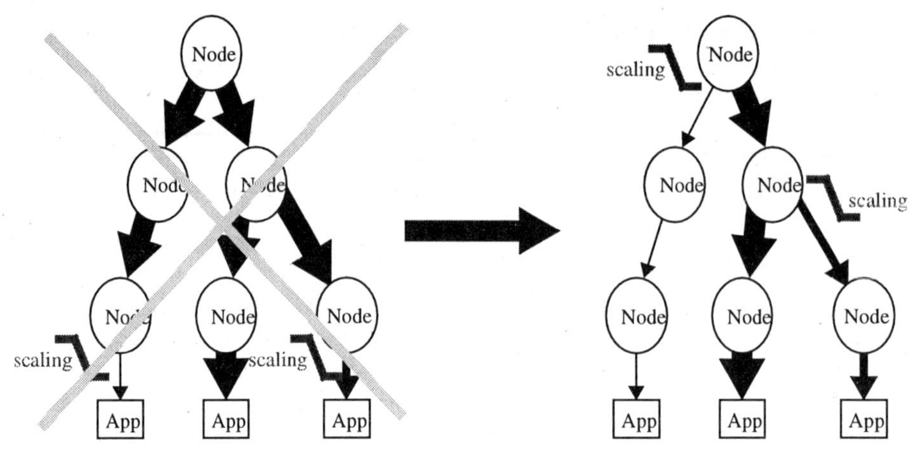

Fig. 1. media scaling inside the network

This paper focuses on such a network-level approach and describes ongoing work in the PRISM[1] project. Basically, enabling distributed media scaling consists of:

- *giving network nodes the ability to filter streams.* We began to investigate this issue in a LAN environment, as presented in section 2, and we currently design solutions for general packet switched networks (these will be the subject of a forthcoming paper).

- *controlling these nodes in order to ensure they deliver the best possible quality but no more than requested, i.e. managing filter tuning inside the network.* Section 3 presents an general algorithm dealing with distributed management of filter tuning inside a packet switched network, which also applies to the restricted LAN case.

A major assumption to be stated relates to the *network service model*, as the choice of a service model obviously impacts the design of nodes being able to perform media scaling: the PRISM project does not rely on guaranteed services, i.e. it does not require deterministic or statistical guarantees about throughput, delay, and jitter, in constrast to most current approaches [14, 18, 19]. However it does not expect straightforward *best-effort* service either, but requires that losses be managed such that:

- *lost data are always the oldest in the nodes.* Losses are due to some nodes being congested: our service model requires that incoming data be stored, possibly causing already stored data to be thrown away. In such a case, the latter is selected based on the time it spent inside the node.

- *end-point applications do not have to deal with these losses, i.e. streams consistency is preserved.* To ensure this property, streams must be structured into independent presentation blocks, and a packet loss must induce the loss of the entire block it belongs to.

Discussing the network service model is beyond the scope of this paper.

1. a french acronym meaning "distributed platform for integration of multimedia services"

2 A Multicast Transport Protocol with Media Scaling Capabilities

As a first step towards integrating media scaling in multimedia networking, we have developped a transport protocol whose goal is to allow low-level media degradation, i.e. independent of any particular coding format. For this purpose a generic, abstract stream structure had to be designed. The scope of this protocol was restricted to local area networks, with the idea to make use of low-level broadcasting capabilities, thus allowing efficient multicast communications to take place. This allowed us to define an addressing scheme whose purpose is to differentiate between multimedia streams rather than hosts. We called this protocol PRISM Transport Protocol (PTP). PTP is fully described in [8], and an overview of PRISM can be found in [16].

2.1 Generic Stream Structure

An efficient, unexpensive media scaling technique is to drop packets based on some tag in their headers. So packets are a good data unit to start with. However multimedia streams usually have a semantical structure, i.e. they are made of a number of *atoms*, or data blocks that may be independently presented, and cannot be further split without breaking dependencies (for instance frame dependencies inside an MPEG stream). Hence we use the *information unit* concept to capture these semantics, with the thought that such units will be atomically delivered to the receiving applications. For instance, information units of a MJPEG video stream would be frames, whereas those for a MPEG stream would be groups of pictures (GoPs). The usefulness of information units will be explained in section 2.2.

To do the link between the two views, we state that information units are constituted of packets. Hence sending applications (or possibly libraries that are aware of the streams semantics) will have to submit packets with appropriate tagging and indications of information units boundaries. On the other side, the protocol memory management will be in charge of reassembling information units before delivering them to the receiving applications.

2.2 Memory Management

We found that classical packet queues that are used in almost any protocol may not be well suited for multimedia services. The reason is, that packet losses occur in an uncontrolled manner once the queue is full. Worse, packets that are rejected usually are the most recent, whereas packets residing in the queue may be out of date. Hence we designed circular buffers being able to store information units. As recent information units reach the memory manager, they possibly erase the oldest units still lying in it. Every packet shall then indicate the information unit it belongs to, as well as an offset indicating its location inside the unit.

Circular buffers may be seen as the combination of two mechanisms:

- a mechanism to reassemble incoming packets into information units,
- a FIFO queue handling information units, with a congestion resolution policy which erases the queue head rather than the incoming unit.

This allows to have transmission and presentation rates being partially independent. Thus receiving applications are permitted, but not forced to consume data units as long as they are available, i.e. stored in a circular buffer and not already read. This indeed is possible since information units are atoms of the presentation activity: *no information unit depends upon others, thus any received unit can be processed.*

Slow applications may possibly not read every unit, if the transmission rate is greater than their own presentation rate. In such a case, using circular buffers strongly reduces the delay between optimal and actual presentation times, or *skew*, as they decouple the transmission and presentation rates. Figure 2 illustrates this benefit: in case 1, where a FIFO packet queue is used, a presentation rate lower than the data arrival rate induces some buffering (until no more memory is available) along with an always increasing shift. On the other hand (case 2) the PTP scheme allows a nearly optimal presentation timing, only degraded in case of unresolved losses[1]. The choice of the most recent available data unit implicitly reduces the skew, thus smoothing the long-term traffic behaviour.

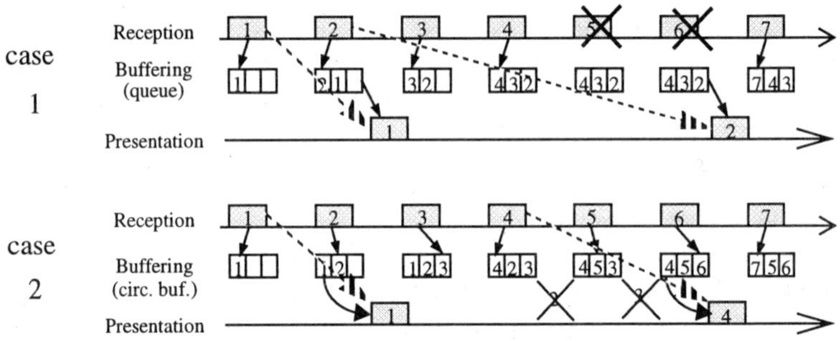

Fig. 2. PTP skew smoothing effect

This scheme is also useful for intra-stream re-synchronisation issues: in case 1, when the buffers are full, the synchronisation of the multimedia stream is lost. Thus, the upper-level mechanisms must take some actions in order to resynchronise: free buffers (which implies loosing data), then wait for a synchronisation point before restarting the presentation. With PTP, these actions are done in an automatic manner: instead of letting a congestion of the buffers occur, the data discarding rate dynamically adapts to the transmission and presentation rates, thus making it easier for synchronization mechanisms to perform their tasks.

We also allowed several applications to be bound to the same circular buffer,

1. IUs incur *unresolved losses* when they are waiting for retransmission of lost packets. Packet loss handling in PTP is fully described in [8].

and thus to listen to the same data stream. This was done by designing *virtual circular buffers*, mapped onto physical ones, which give to the applications a local view of the available data units. The actual number of buffers determines the delay that the data will spend before its presentation, along with the overall quality of this presentation. Indeed, the longer a block is buffered, the more time is available to retransmit possibly lost packets, thus improving the perceived quality. However, long delays obviously reduce the degree of interactivity. It was decided to let applications specify the number of the buffers to be used, while avoiding resource wastes through virtual buffers management. Such a tradeoff between presentational quality and interactivity is actually an application-level one.

2.3 Media scaling

Although circular buffers provide some kind of best-effort stream sampling, it was possible to add some filtering based on hierarchical streams, in order to improve the perceived quality.

The reason is depicted by figure 3: applications having a presentation rate lower than the transmission rate will jump over a number of information units while processing one of them. The resulting quality may be poor, depending on the width of the gap. Filtering information units, so as to make them faster to decode, allows the slow receivers to reduce this gap (to increase the stream evenness), and thus to present a larger number of units, all being of the same quality. Hence the proper filtering threshold is a function of the receiver presentation rate.

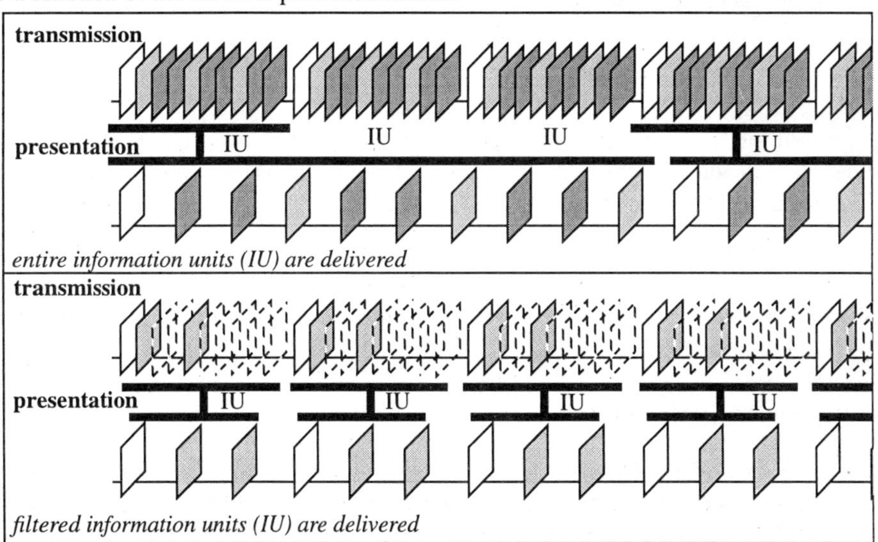

Fig. 3. filtering information units enhances the stream evenness

We implemented a fast, efficient scaling mechanism in two steps. Assuming that receiving applications may specify a *quality level* (actually a priority threshold) they want, incoming data packets are first filtered before they enter the circular buffer they belong to, based on the highest quality level that was requested by any application bound to this buffer. The assignation of priorities is performed, on a per-packet basis, by the sending application itself (or possibly a streaming layer). Second, by replacing the kernel-to-user-memory copying with a conditional one, packets (reassembled into information units) are again filtered, this time with the quality level asked by the application that performs the "receive" system call. Strategies for choosing quality levels are beyond the scope of a transport protocol, as they relate to quality of service issues. This will be discussed in section 2.4. Similarly, PTP allows receivers only to perform media scaling, thus potentially wasting network resources, as depicted in section 1. A mechanism that allows filtering to propagate inside the network will be presented in section 3.

2.4 Filter Tuning Strategies

Since the filtering threshold is a function of the receiver's presentation rate, we decided to investigate applications ability to tune this threshold. This section just presents envisionned solutions, as the issue is still under study.

A first solution is to monitor the IU transmission rate as well as the decoding process: knowing the time it takes to decode each subband of the stream, it is possible to select a filtering level, for which the decoding rate is as close as possible to the transmission rate.

A second approach is to calculate an estimator of the incoming stream evenness. By studying a function such as

$$f(x) = \frac{\alpha}{1 + k \cdot \sigma}$$

(with α being the percentage of displayed subbands and σ being the standard deviation of the units interarrival) for each filtering level, it is possible to find the level which maximizes f(x).

A third solution is for the protocol to deliver IU numbers and a flag denoting if the *receive* system-call was blocked. A discontinuity in IU numbers causes the filter level to decrease, whereas a *receive* call being blocked causes the level to increase. Of course, an implementation will have to avoid oscillation troubles.

2.5 Implementation and future work

PTP has been implemented inside the kernel of the SunOS 4.1. Unix operating system, and an implementation inside Solaris is underway. Both directly interface with Ethernet MAC layer. Experimentations of best-effort video communications are conducted using hierarchised MPEG streams.

Next step is to extend this approach to packet switching networks, by designing routers with the ability to handle information units and to perform stream filtering. Such a research is currently underway in our team.

3 Media Scaling Propagation

Assuming protocol entities with media scaling capabilities, we must design a mechanism that allows to propagate effective scaling activities inside the network, in order to avoid unnecessary resource commitment. A first approach could be considered with respect to PTP, by enabling an application-level feedback from the receivers to the sources of multimedia streams. But such an approach is unable to propagate filter tuning outside the local network, or does so without taking into account external nodes with resource constraints and filtering capabilities. Thus a more general mechanism must be thought of.

Most current approaches (eg. [3, 9, 10, 14, 21]) rely on *QoS monitors* to track the communication performances, and enable feedback control of *QoS filters*, located at the edge or within the network, to shape the streams appropriately. This scheme actually performs indirect measurement of the aggregated constraints which are located between the monitor and the filter. This is consistent with end-to-end QoS management: end-systems deal with high-level QoS parameters such as the level of filtering applied to the streams (QoS negotiation). These parameters are mapped into low-level ones (bandwidth, delay, ...), which are then monitored to enforce the QoS of the communications. However, optimal placement and tuning of QoS filters is still an open issue in these approaches (the most elaborate being [21], which introduces a protocol allowing to propagate some kinds of "nomadic" filters, but fails to address network constraints as well as dynamicity).

In contrast to these approaches, we chose to use high-level parameters (actually the scaling level, or *quality level*) as an abstraction layer on which dynamic, optimal tuning of media filters can be performed. This allows to move QoS mapping from end-systems into network nodes, thus defining part of a network service interface: the nodes must be able to understand high-level parameters and to turn them into locally sensible low-level parameters, for instance using previously delivered *flowspecs*[1]. In other words, a node, filtering a given stream up to a given quality level, must determine if it can handle the resulting stream, regarding its resource requirements. We believe that letting each node performing local QoS mapping will allow heterogeneous nodes to coexist inside networks.

Let us assume that network-level media scaling capabilities are available at each node in a multicast tree, allowing to selectively discard both incoming and outgoing data packets of a multimedia stream[2], as depicted by figure 4 (the _ symbol stands for a scaling activity):

This indeed is quite close to PTP filters layout, and could also easily be implemented inside *mrouted*-like routers[3]. Our goal is to allow nodes inside the tree to

---

1. connection setup is outside the scope of this paper
2. filters are instantiated for each stream crossing the node, since quality levels of different streams are unrelated.
3. mrouted are the multicast routing daemons of the Internet MBONE.

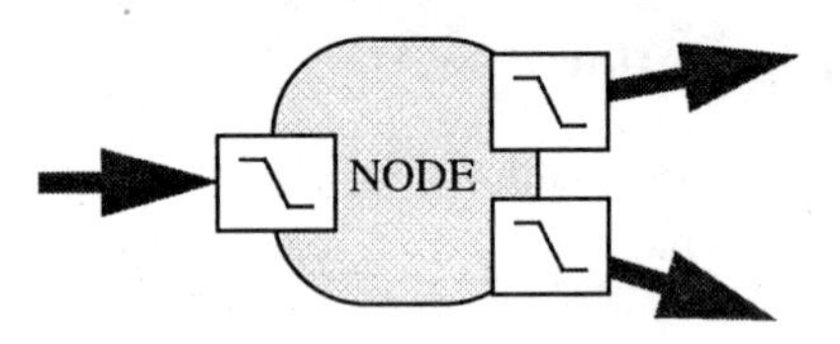

Fig. 4. network level media scaling

exchange control messages, in order to propagate scaling levels as close to the source as possible, and to react to user actions as well as network events. Any equilibrium in the control (i.e. without any message remaining inside the network) will have to ensure that any node receives the best possible quality level and that no level higher than required is transmitted, that is no network resource is wasted. Such a behaviour will be embedded inside a distributed algorithm, which will rely on:

- some mechanisms for local resource control (possibly reservation, admission control, scheduling, and/or dynamic monitoring[1]...), available at each node inside the network ;
- some mechanisms for multicast tree establishment (connection setup, multicast routing), by which a node can know who its peers are;
- reliable message passing between nodes (no message is corrupted, lost, duplicated, or out of sequence)[2].

Two assumptions are made about these mechanisms: first, local resource control is in charge of determining the achievable service level, in terms of node (cpu, buffers, ...) and network resources (for outgoing links), for each incoming stream. Stream concurrency is managed by the local resource control too, through resource allocation policies. Its interface shall offer a REQUEST call, used by the scaling control to ask for a certain quality level for a stream, and answered with the actually achievable level. In addition the local resource control can issue an asynchronous ALERT trap, indicating a change in the achievable level (the local resource control embeds the QoS mapping functionalities described before). Second, a static routing scheme is assumed, i.e. once established a path in the multicast tree will remain unchanged. Dynamic routing schemes with rerouting capabilities require further research.

Under these assumptions, the following subsections detail an algorithm for dynamic scaling control, whose goals are to deliver the best achievable quality level to each receiver, while minimizing the underlying-system resources consumption.

3.1 Algorithm

The local media scaling control must maintain a context for each stream cros-

---

1. dynamic monitoring seems more adapted to the PRISM service model. However the algorithm needs to make no assumption about the underlying management policy.

2. message passing reliability could indeed be obtained by the algorithm itself. The assumption here is made for the sake of clarity.

sing the node. Such a context consists of the following variables:

- I is the incoming quality level, reflecting the level at which the incoming stream is filtered by the upper nodes,
- P is the quality to ask to the node's predecessor (parent),
- S_i is the quality requested by the *ith* node's successor (child),
- T_i is the quality actually transmitted to the *ith* node's child,
- R_i is the locally achievable quality under resource control constraints, for the *ith* child. R_i is obtained by issuing a REQUEST to the local resource control, or is carried by an ALERT trap from it.

In addition, let $C_i = \min(S_i, R_i)$ for each child i (contracted quality level, i.e. the level that could be delivered to the child with respect to local constraints only)

Then T_i is given by :

$\forall$ child i, $T_i = \min(I, C_i)$,

(taking into account the upper nodes constraints), and P can be determined by:

$$P = \underset{i}{MAX}(C_i)$$

Given this context, incoming scaling can be tuned to P and outgoing scaling to the *ith* child can be tuned to C_i. In addition, some events shall trigger the context (re)evaluation:

- the local resource control issues an ALERT trap, carrying a new value for R_i,
- or the *ith* child modifies its needs, thus modifying the S_i value,
- or the actually received quality changes, thus modifying the I value.

When a change in the P value occurs, the node sends a PROPAGATE message to its parent, carrying the new P value. Similarly a change in a T_i value triggers an ALARM message to be sent to the *ith* child, with the new T_i value in it. Thus incoming PROPAGATE messages carry one child's wills, whereas incoming ALARM messages denote changes in the quality level received from the parent node.

The nodes behaviour may then be depicted by the distributed algorithm given below[1]. Propagate and Alarm messages carry a sequence number in order to prevent the troubles associated with several simultaneous propagations (this implements a classical *incremental approach*: propagation "waves" are tagged, hence each node is able to detect late waves, which must not be processed).

---

1. the algorithm presented here only deals with a single stream crossing the node, for the sake of clarity; the multi-stream version is straightforward.

```
var counter init 0;

upon reception of Propagate(q , n) from child j
   Sj ← q;
   QoS_eval(n);
end

upon reception of Alarm(q , n) from parent
   if n ≥ counter then
      I ← q;
      QoS_eval(+∞);
   endif
end

upon reception of ALERT(q) from resource control for child j
   Rj ← q;
   QoS_eval(+∞);
end

procedure QoS_eval(n)

   Pold ← P;
   P ← 0;

   for each child j
      Told ← Tj;
      Rj ← REQUEST(resource control for child j);
      Cj ← min(Sj , Rj);
      Tj ← min(I , Cj);

      if Tj ≠ Told then
         send Alarm(Tj , n) to child j;

      P ← max(P , Cj);

      update outgoing link j scaling level: Cj

   end foreach;

   if P ≠ Pold then
      counter ← counter + 1;
      send Propagate(P , counter) to parent;
   endif

   update incoming link scaling level: P

end
```

Clearly, it can be seen that users' wishes propagate towards the media source, whereas resource usage constraints propagate towards the sinks and are aggregated along the paths. Furthermore, it must be noticed that the algorithm behaviour is asymmetric. In case of an increased quality level query from a receiver, the algorithm follows a two-phase scheme (figure 5) if previous level is the same on each node (and smaller than 9). If however an upper node has a previous level which differs from the client's one, the behaviour is that depicted by figure 6, where an early alarm depicts the fact that the upper node immediately starts to increase the quality of the stream (N1 on the figure had previous level 7, while N2 and client had a lower one).

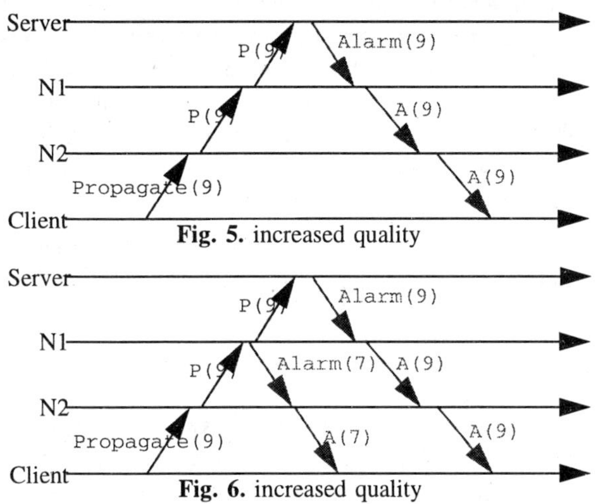

Fig. 5. increased quality

Fig. 6. increased quality

Last, a decreased quality level query induces the behaviour depicted by figure 7 (previous level greater than 5)

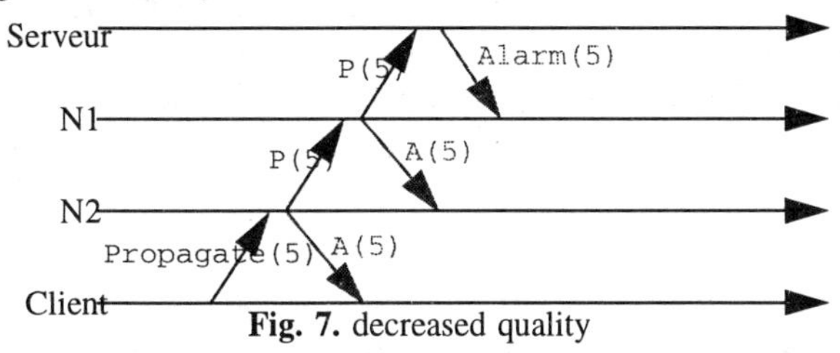

Fig. 7. decreased quality

Such a behaviour is in accordance with the following (good-sense) property: a node always has the capability to deliver a decreased quality level, and the new level will not be further degradated by the previous nodes in the path, since they were actually able to deliver an even greater one. Thus an *Alarm* can immediately be issued to the child. The reciprocy is false: a node may be able to increase the quality level, provided that its predecessors also have the capability. That is why the *Alarm* message is delayed until the scaling propagation reaches a node where the quality is actually available.

So far, nodes were thought to be some kinds of routers, with media filtering capabilities. In fact, the root and leaves of the multicast tree (source and receivers) could also apply the algorithm: for the source, I would denote the quality level of the media flow leaving a codec, and *Propagate* messages would be used to appropriately tune the codec output. Inside receivers, S_i would denote an application's requirements, while R_i would characterize the operating system resource usage (cpu, memory, I/O). Thus several applications could receive a common media flow with different quality levels[1].

As already said, each node implements a scaling activity for the incoming packets of each multimedia stream. Since the incoming scaling level is propagated to the node's parent, such an activity may seem useless. However it allows the node not to receive a quality greater than requested for the time its parent establishes the new scaling level. This indeed saves the node's resources (buffer occupancy, cpu). Moreover, it allows several receivers to share a same physical link: in such a case, we indeed come back to PTP filtering mechanisms, with the control algorithm being used both between the source and sinks, and also *inside* each receiver, to deliver the multimedia streams to multiple applications.

Appendix A presents a proof of the algorithm correctness, and appendix B demonstrates its behaviour on an example network.

4 Conclusion

Distributed media scaling inside a packet switched network raises two issues: network nodes must be enhanced in order to allow hierarchised multimedia streams to be filtered, and efficient distributed control of these nodes must be designed. Assuming PRISM network service model, which does not require real-time service guarantees, we presented a number of concepts allowing to answer these two issues. The first one was addressed in a restricted case (LAN communications), providing helpful background to answer the general, switched network case. The second one was given a general solution, through a distributed algorithm which interfaces both with media filters (and thus tunes the filtering activity), and with a minimally-designed resource controller, which assesses local constraints with respect to resource availability.

Distributed media scaling also raises a number of issues worth investigating. Among them, hierarchical preemption of communications is promising, as a mean to better share network resources, possibly cooperating with Class Based Queuing methods proposed in [20], and could ultimately induce novel QoS-based routing strategies.

The design of routing nodes which conform to our service model is part of ongoing work in the PRISM project, and will be the subject of a forthcoming paper.

---

1. for instance, consider a stereo audio stream: application A could redirect it to a monaural loudspeacker with reduced quality, while application B could store the full quality stream on a local disk.

The authors wish to thank A. Léger for his support, and are indebted to O. Elloumi and O. Huber for their help and fruitful comments.

References

1. J. C. Bolot, T. Turletti, I. Wakeman. *Scalable Feedback Control for Multicast Video Distribution in the Internet.* In Procs. SIGCOMM'94, pp 58-67, 1994.

2. L. Delgrossi et al. *Media Scaling for Audiovisual Communications with the Heidelberg Transport System*, Proc. ACM Multimedia'93, 1993.

3. N. J. Yeadon. *Supporting Quality of Service in Multimedia Communications Via the Use of Filters.* Research Report MPG-94-10, Lancaster University.

4. W. Tawbi, F. Horn, E. Horlait, and J.-B. Stefani. Video compression standards and quality of service. *The Computer Journal*, 36 (1), 1993.

5. D. Hoffman, M. Speer, G. Fernando. *Network Support for Dynamically Scaled Multimedia Data Streams.* In Procs. NOSSDAV'93, pp. 240-251, 1993.

6. L. Zhang, S. Deering, D. Estrin, S. Shenker, D. Zappala. *RSVP: A New Resource ReSerVation Protocol.* IEEE Network Magazine 7(5), pp 8-18, Spetember 1993.

7. CIP Working Group. *Experimental Internet Stream Protocol, Version 2 (ST-II).* RFC 1190, IETF, 1990.

8. L. Toutain, F. Toutain. *PTP : A Reliable Multicast Protocol for Multimedia Communications on LANs.* Procs. International Switching Symposium'95 (Berlin), Vol. 1 pp 185-189, and Research Report no RR-94022-RSM, Télécom Bretagne, 1995 (http://bloodmoney.enst-bretagne.fr/prism_paper.html)

9. J. Sandvoss, J. Winckler, H. Wittig. *Network Layer Scaling: Congestion Control in Multimedia Communication with Heterogeneous Networks.* Technical Report 43.9401, IBM European Networking Center, Heidelberg, 1994.

10. S. Gumbrich, B. Kownatzki, J. Sandvoss. *Media Scaling for Real-time Packet Video Transport.* in 4th Open Workshop on High Speed Networks Proceedings, September 1994.

11. J. Pasquale, G. Polyzos, E. Anderson, V. Kompella. *The Multimedia Multicast Channel*, Proc. NOSSDAV'92, pp. 185-196, 1992.

12. A. Krishnamurthy and T.D.C. Little, *Connection-Oriented Service Renegotiation for Scalable Video Delivery* , Proc. 1st IEEE International Conference on Multimedia Computing and Systems (ICMCS'94), Boston, May 1994, pp. 502-507.

13. H. Schulzrinne, S. Casner, R. Frederick, V. Jacobson. *RTP: A Transport Protocol for Real-Time Applications*. Internet draft draft-ietf-avt-rtp-07, march 1995. Work in progress.

14. A. Campbell, D. Hutchison, C. Aurrecoechea. *Dynamic QoS Management for Scalable Video Flows*. Proc. NOSSDAV'95, Durham, april 1995 (http://spiderman.bu.edu/nossdav95/NOSSDAV95.html)

15. A. Eleftheriadis, D. Anastassiou. *Meeting Arbitrary QoS Constraints Using Dynamic Rate Shapping of Coded Digital Video*. Proc. NOSSDAV'95, Durham, april 1995.

16. L. Toutain, H. Afifi, J.P. Le Narzul. *The PRISM distributed multimedia platform*. Proc. IEEE ISCC'95 (Alexandria), and Research Report no RR-94026-RSM, Télécom Bretagne, 1994 (http://bloodmoney.enst-bretagne.fr/prism_paper.html)

17. H. Kanakia, P.P. Mishra, A. Reibman. *An Adaptive Congestion Control Scheme for Real-Time Packet Video Transport*. In Proc. SIGCOMM'93, september 1993.

18. D. Ferrari, A. Banerjea, H. Zhang. *Network Support for Multimedia - A Discussion of the Tenet Approach*. in Computer Networks ans ISDN systems, vol.26, pp. 1267-1280, july 1994.

19. D. Clark, S. Shenker, L. Zhang. *Supporting Real-Time Applications in an Integrated Services Packet Network: Architecture and Mechanism*. Proc. ACM SIGCOMM'92, august 1992.

20. I. Wakeman, A. Gosh, J. Crowcroft, V. Jacobson, S. Floyd. *Implementing Real-Time Packet Forwarding Policies using Streams*. In Proc. USENIX 1995 Technical Conference , New Orleans, january 1995.

21. J. Pasquale, G. Polyzos, G. Anderson, V. Kompella. *Filter Propagation in Dissemination Trees: Trading off Bandwidth and Processing in Continuous Media Networks*. Proc. NOSSDAV'93, Lancaster, november 1993, pp. 259-268.

Appendix A

This appendix presents a proof of the propagation algorithm correctness. This is done by verifying that each node always sends the best possible quality level and no more than the requested level (otherwise network resources would be wasted), unless there are messages in transit, and then prooving that an event will trigger a finite amount of messages.

(I) Node context

(from the algorithm)

- **I** is the incoming quality level, reflecting the level at which the incoming stream is filtered by the upper nodes,
- **P** is the quality to ask to the node's predecessor (parent),
- $\mathbf{S_i}$ is the quality requested by the *ith* node's successor (child),
- $\mathbf{T_i}$ is the quality actually transmitted to the *ith* node's child,
- $\mathbf{R_i}$ is the locally achievable quality under resource control constraints, for the *ith* child..
- $\forall$ **child i,** $\mathbf{C_i = min(S_i, R_i)}$,
- $\forall$ **child i,** $\mathbf{T_i = min(I, C_i)}$,
- $P = MAX(C_i)$

(II) A node never sends a level higher than requested

(*by definition*)

By definition, $\mathbf{T_i = min(I, S_i, R_i)}$

The level would be such that $\exists \mathbf{i} (\mathbf{T_i > S_i})$ which is impossible

(III) A node always sends the best possible quality

(*by contradiction*)

Let us assume that there exists a better level **q**

This is depicted by $\exists \mathbf{i} [\exists (\mathbf{q \le I}) (\mathbf{T_i < q \le S_i})]$ (using (II))

By definition, $\mathbf{T_i = min(I, S_i, R_i)}$

Thus either $\mathbf{T_i = I}$ so $\mathbf{T_i < q \le T_i}$, **q** does not exist

or $\mathbf{T_i = S_i}$ so $\mathbf{T_i < q \le T_i}$, **q** does not exist,

or $\mathbf{T_i = R_i}$ so $\mathbf{R_i < q}$, which is impossible (**q** cannot be higher than permitted by the resource control).

(IV) Between two adjacent nodes, the level is not higher than requested, unless there are messages in transit

(*by contradiction*)

Let **A** be the parent node and **B** be the child node

By definition, $\mathbf{T_B(A) = min[I(A), S_B(A), R_B(A)]}$ (1)

No message in transit is equivalent to $\mathbf{T_B(A) = I(B)}$ (2)

and $\mathbf{S_B(A) = P(B)}$ (3)

If the incoming level on **B** is greater than requested, then $\mathbf{I(B) > P(B)}$,

which is equivalent to $\mathbf{T_B(A) > S_B(A)}$ from (2) and (3) (4)

But (1) implies $\mathbf{T_B(A) \leq S_B(A)}$ (5)

(4) and (5) contradict each other

(V) Between two adjacent nodes, the level is the best possible, unless there are messages in transit

(by contradiction)

Let **A** be the parent node and B be the child node

By definition, $\mathbf{T_B(A) = min[I(A), S_B(A), R_B(A)]}$ (1)

No message in transit is equivalent to $\mathbf{T_B(A) = I(B)}$ (2)

and $\mathbf{S_B(A) = P(B)}$ (3)

If the level is not optimal, then $\exists \mathbf{q}\ [\mathbf{I(B) < q \leq P(B)}]$ (using (IV)) (6)

For **q** to be consistent, we have $\mathbf{q \leq min(I(A), R_B(A), P(B))}$,

which is equivalent to $\mathbf{q \leq min(I(A), R_B(A), S_B(A))}$, applying (3)

thus $\mathbf{q \leq T_B(A)}$ applying (1) (7)

But (6) gives $\mathbf{q > I(B)}$, so $\mathbf{q > T_B(A)}$ applying (2) (8)

(7) and (8) contradict each other

(VI) Inside the tree, all delivered levels are the best possible and not higher than requested, unless there are messages in transit

Any node sends to each of its children the best possible level and no more than what this child requested (from II, III), and these levels are consistent with the children ones, unless there are messages between them (from IV, V).

(VII) Any event triggers a finite amount of messages

Possible events include a receiver asking for a level with a Propagate message, the source indicating a level change by sending an Alarm message, and any node sending Alarm and possibly Propagate messages due to an Alert from its resource control.

A Propagate message can only be sent upstream. (9)

An Alarm message can only be sent downstream. (10)

By construction, a Propagate message is sent if the value of P is modified,

and $\mathbf{P} = \mathbf{MAX_i}\left[\mathbf{min}(\mathbf{R_i},\mathbf{S_i})\right]$

Since an Alarm message does not modify R_i neither S_i,

A node receiving an Alarm message will not send a Propagate message (11)

from (9), any event triggers a finite amount of Propagate messages.

from (10), any event triggers a finite amount of Alarm messages.

from (11), any incoming Propagate message triggers a finite amount of Alarm messages.

Thus the overall number of messages created by any event is finite.

(VIII) Conclusion

From VII, the algorithm eventually reaches an equilibrium, where the delivered and received quality levels are optimal for any node (from VI).

Appendix B

To better figure out how the algorithm performs, this subsection demonstrates its behaviour on an example network.

The multimedia stream to be transmitted will be a MPEG coded video sequence, with fixed group of pictures (GoP) structure containing 16 frames, as depicted in figure 10. Five hierarchy levels are defined, allowing to scale the stream down with five different qualities (the higher the level, the better the quality). Such a scaling is a temporal one, as entire frames are discarded, thus reducing the frame rate[1].

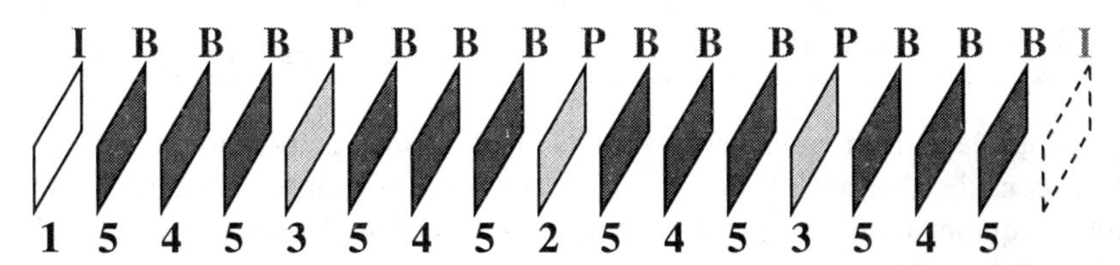

Fig. 8. GoP structure and hierarchy

At a first step, let us consider that node A is a video server, and that node I is willing to receive the MPEG stream with level quality set to 3 (i.e. I and P frames). The routing mechanisms may select the A-E-F-I path, as illustrated in fig 9a.

1. this very simple scheme was chosen to illustrate the algorithm behaviour; in practice more powerful and somewhat more complex schemes would have to be considered.

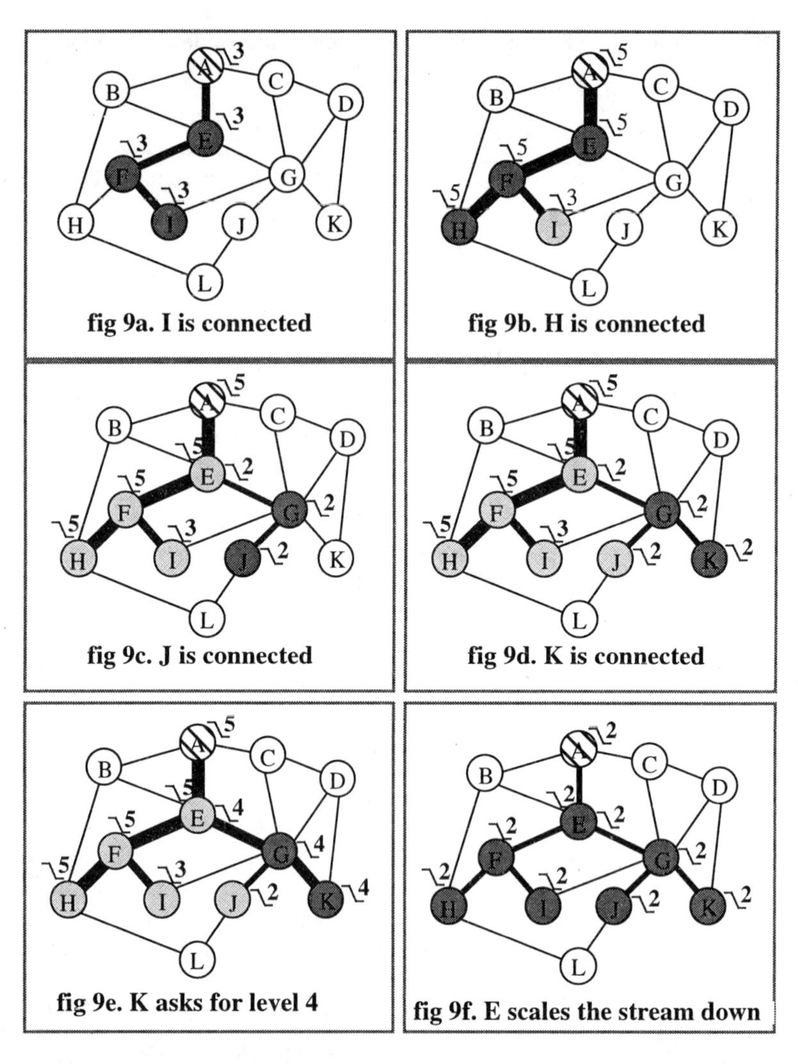

fig 9a. I is connected

fig 9b. H is connected

fig 9c. J is connected

fig 9d. K is connected

fig 9e. K asks for level 4

fig 9f. E scales the stream down

Thus each node in this path tunes its media scaling activity to the quality level 3. Now consider that node H wants to connect to the stream with quality set to 5 (H could be equipped with some hardware decompression tool, and so, could be able to process a full quality stream). The routing algorithm may indicate that F is a good router to connect to. By sending a Propagate message to F, H lets it know that it needs quality level 5. In turn, F informs E that it needs level 5, and E also sends a Propagate to A. I does not see any change in the quality it receives, since F indeed scales the stream down on its link to I (fig 9b.).

Consider now that nodes J then K wish to receive the video stream with level 2 (fig 9c.). J connects to the router E via G. Since E is able to deliver the level 2, A will not be informed of any change. K just connects to G, and no propagation takes place (fig 9d.).

Now, consider that K is willing to improve the quality it receives up to level 4. It sends a Propagate message to G, which in turn informs E of its wills. Again E is able to deliver such a quality level (fig 9e.).

Finally, let us imagine that router E receives an ALERT trap from its local resource controller, due to some resource preemption. Thus E is now unable to serve the quality level 5, and can only handle a stream with level 2. Therefore E sends a Propagate message to A, asking for this level, along with Alarm messages to F and G, making them aware of the degradation (fig 9f.). These Alarm messages propagate until they reach the sinks I, H and K, but J does not receive any Alarm since the qualities it gets remain unchanged.

The ACCOPI Multimedia Transport Service over ATM[1]

Laurent Mathy and Olivier Bonaventure

Institut d'Electricité Montefiore B28, University of Liège, B-4000 Liège, Belgium
{mathy, bonavent}@montefiore.ulg.ac.be

Abstract. This paper discusses some of the issues in mapping the ACCOPI Multimedia Transport Service (AMTS) over the ATM (Asynchronous Transfer Mode) network technology. The AMTS is a transport service mainly designed to support multimedia applications. ATM offers two important characteristics required by multimedia applications (namely QoS and multicast). We first discuss general architectural issues such as multiplexing and QoS negotiation. Then we outline how the main functionalities of the AMTS (call control, data transfer, call join, etc.) can be supported in an ATM environment.

1 Introduction

As we are entering the "telecommunication century", people wish to communicate as easily and naturally as possible and therefore multimedia is emerging as a powerful communication tool. Indeed, multimedia allows to communicate by using concurrently different information types (i.e. different media) such as video, audio, text, still images, and so on.

Moreover, beside the simultaneous use of several media, "natural interpersonal" communications also benefit from the possibilities offered by group communications, that is communications involving more than two peers. Such multimedia group communications among computers allow geographically distant people to communicate as if they were in the same meeting room.

1.1 Heterogeneity in Multimedia Communications

The different media making up a multimedia message have their own characteristics, consequently they also have differing requirements. For instance, some media are

[1]This work has been partially supported by the Commission of the European Communities under the RACE M1005 project ACCOPI: "Access Control and Copyright Protection for Images".

said to be continuous[2] (e.g. video, audio) while others are said to be discrete (e.g. text, still images). While continuous media are characterised by stringent communication performance (e.g. time constraints), discrete media are often characterised by the need to avoid any error during communication.

From such a diversity comes the need, for a multimedia application, to be able to describe the requirements of each of the media it is using to communicate. This is done through the use of Quality of Service (QoS) parameters. Those QoS parameters are used to describe the traffic characteristics required by a given medium.

1.2 The Technology Push

For some years, many researches have been done to drastically improve the communication networks. The most obvious of those improvements is surely the gain in the achievable data rates. Bandwidth in the range of hundreds of megabits to several gigabits per second is now a reality (e.g. FDDI, B-ISDN).

But performance improvement has not been the only one. We have also seen a clear trend to the design of networks providing performance guarantees [1][7][9][18][19]. Although there exist different types of guarantees (e.g. statistical or deterministic) and even though not all the networks provide them, performance guarantees provide support for real-time communications on a wide-area basis.

Besides performance improvement and guarantees, another feature that is becoming ubiquitous [1][18][6] in the communication networks is multicasting, that is the conveying of a same data to several destinations in one sending. Multicasting obviously allows a better utilisation of network resources (bandwidth, buffer space and processing time in network nodes) than when sending out a separate copy of the data to each destination. It is, of course, the favourite transmission mode for distributing data within the networks.

The key point with most of those features is that they are now leaving the LAN/MAN environment where they have been available for quite a long time to reach the WAN environment. This is pushing for the introduction in the WAN environment of applications that were formerly only possible in the local environment.

The networks offering those features are said to be integrated and are able to deal with different types of communications (e.g. multimedia communications, bursty data communications) at the same time.

---

[2]Here, the term "continuous" refers to the user's impression of the data, not necessarily to its internal representation.

However, in order that the applications benefit from those integrated networks, enhanced transport layers are needed, which are well suited to both new application and network environments.

In the sequel, after a brief overview, in section 2, of the ACCOPI Multimedia Transport Service (AMTS) [15], we describe, in section 3, architectural issues related to the direct support of a transport service over ATM. Section 4 then describes specific issues in the support of the AMTS over ATM. Section 5 concludes the paper.

2 An Overview of the AMTS

The AMTS is a transport service designed to support multimedia applications. It provides support for both QoS and group communications.

The QoS semantics provided is the enhanced QoS [5] defined in the OSI 95 peer-to-peer Transport Service [17]. With this semantics, a QoS parameter is seen as a structure of three values, respectively called "compulsory", "threshold" and "maximal quality". Each value has its own well-defined meaning and is the result of a contract between the service users and the service provider.

As regards group communications, the AMTS provides the transport service users with centralised transport calls [15].

A call is a set of multicast (i.e. 1→N) connections [10]. It is said to be centralised when there are only one sender and several receivers. A centralised call as provided by the AMTS is depicted on figure 2.1. In this figure, there are 2 (1→3) and 1 (1→2) connections.

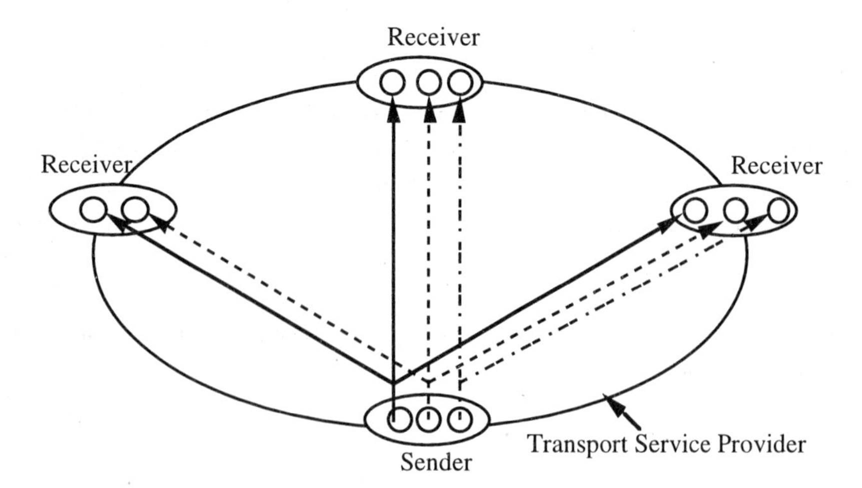

Fig 2.1 Centralised transport call.

The reasons that led to such topologies in the AMTS are the following:

- The idea of a call (e.g. a set of multicast connections) comes from the fact that a multimedia communication is made up of different information types (i.e. the different media). Each of these media (or even each "part" of a given medium) having its own characteristics, also has different performance requirements. Those media will often be transferred on different multicast connections, each one being tailored to offer adequate performance. Therefore, "grouping" multicast connections together into calls, allows to handle, in the transport layer, multimedia communications as a whole. Indeed, the multicast connections carrying the different components media of a multimedia communication are no longer treated independently from one another, as it is the case with most of the current transport services.
- The centralised topologies allow the sender to control the entire call. This gives to the sender the possibility to express QoS dependency relations among multicast connections of a call. A QoS dependency relation simply expresses the equality of the selected values for a given QoS parameter on all the multicast connections involved in that relation. There can only be one QoS dependency relation per parameter. It should be noted that the set of multicast connections involved may be different from one dependency relation to another. Such QoS dependency relations on the transit delay and delay jitter can, for instance, ensure near-synchronisation, at the transport service interface, between audio and video connections.
- In order not to jeopardize performance at the transport layer, we naturally selected multicast (i.e. $1 \rightarrow N$) connections as the basic communication scheme since such connections are the simplest for group communications. Moreover, our previous work on QoS support in group communications [14] has shown that it is really difficult to deal with QoS on connections other than multicast connections.

Moreover, conditions on the existence of a call can be expressed. Such conditions are called integrity conditions [16] and are of two types:

- A condition on the active group of the call (i.e. the set of service users using the call). This condition is called Active Group Integrity (AGI) and specifies requirements on the set of receivers in the call.
- A condition on the topology of a call. This condition is called Association Topology Integrity (ATI) for historical reasons. The ATI is actually composed of two kinds of conditions:
    - Conditions on the existence of each multicast connection of the call, expressed in terms of multicast connection active group integrity (MC-AGI).
    - Conditions on the call as a whole in terms of existing multicast connections (i.e. the connections for which the MC-AGI is verified).

  The ATI condition is considered as verified if the conditions on the call as a whole are verified.

From this description of the AMTS, it is clear that the relationship among the connections of a call is expressed in terms of negotiated QoS values and integrity conditions (ATI). Therefore, data concurrency between the transport connections (i.e. the ability to apply the same function to multiple pieces of data concurrently) is in no way reduced, which is a condition to achieve high-performance [8].

Therefore, we see that the AMTS provides the transport service users with a service interface well suited to the transport of multimedia communications. For more details on the AMTS, see [15].

3 Architectural Issues

Providing a multipeer transport service over ATM networks poses several interesting design problems. In the following sections, we will look at some of them.

3.1 The protocol Stack

In an homogeneous ATM network, the network layer is not an essential part of the protocol stack as all the destinations may be reached directly through the ATM network. Thus, in such an ATM network, the internetworking functions, that are traditionally provided by the network layer are not necessary. Therefore, it is acceptable to use the protocol stack shown in figure 3.1. This figure also shows that the transport protocol providing the AMTS will interact directly with the ATM Adaptation Layer of the User Plane of the B-ISDN Reference Model [11], but also with the lower layers of the Control Plane of the B-ISDN Reference Model (UNI 3.1 [1]).

Another, complementary, reason to bypass the network layer is that the "service" provided by the combination of the lower layers of the User [12] and the Control Plane [1] of the B-ISDN Reference Model shown in figure 3.1 is comparable to an unreliable OSI Connection-Oriented Network Service [13]. In this paper, the service provided by the combination of the lower layers of the User and Control Plane of the B-ISDN Reference Model will be referred to as the ATM-CONS.

Both the OSI and the ATM-CONS offer a connection-oriented service with distinct connection establishment and release phases. Both use the addresses encoded in the NSAP format (ATM-CONS may optionally also use E.164 addresses). The connections offered by the ATM-CONS are called Switched Virtual Circuits (SVCs). They both support QoS parameters, but the definition and the semantics of each QoS parameter is not exactly the same. Both the ATM-CONS and the OSI CONS provide in-sequence delivery of the submitted Service Data Units (SDUs). However, the OSI CONS supports unlimited NSDUs while there is an upper limit on the size of the SDUs supported by the ATM-CONS due to the AAL protocols. The ATM-CONS, with the point-to-multipoints extensions to the signalling supports point-to-multipoints simplex connections while the OSI CONS only supports point-to-point connections. A detailed analysis of the differences and similarities between the OSI CONS and the ATM-CONS is outside the scope of this paper, but the service provided by the ATM-CONS is deemed to be powerful enough to build a transport layer providing the AMTS directly over it.

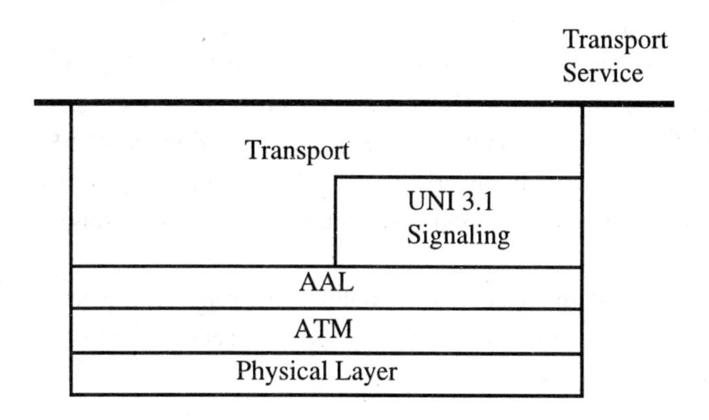

Fig 3.1 Proposed protocol stack.

3.2 Multiplexing

Multiplexing is a protocol function allowing to support multiple connections of a given communication layer onto a single connection of the next lower layer.

It has been shown that using multiplexing simultaneously at different layers of the protocol stack leads to inefficient communications systems implementations [8]. The reasons are manifold [8]:

- It has been shown that demultiplexing (the reverse function of multiplexing) in only one layer of the protocol stack is less expensive in processing costs than multi-layered demultiplexing.
- Multi-layered multiplexing introduces a replication, in the protocol stack, of some functionalities. This is, for instance, the case for flow control which is necessary in any layer where multiplexing takes place. This is mainly due to the fact that flow control applied to a given connection affects all (upper layer) connections multiplexed into it. Therefore, if an independent flow control is required for each connection, flow control has to be applied before multiplexing.

- The more independent connections, the greater parallel processing of data. Therefore, since multiplexing, and all the more so multi-layered multiplexing, reduces the number of connections, it also reduces parallelism and consequently performance.

Moreover, from a QoS point of view, multiplexing can be devastating. Indeed, when several connections are multiplexed, they all share the same QoS associated with the lower layer connection they are multiplexed into. In our architecture (figure 3.1), multiplexing in the transport layer would mean that it is impossible to use the QoS provided by the ATM network to support the QoS requested by the transport service users as all transport connections with various QoS requirements share a single SVC with a fixed QoS.

From this discussion, we can conclude that connection multiplexing should only occur in the lowest layer where QoS are provided [8]. That is the reason why we will, most of the time, only rely on the multiplexing provided in the ATM layer which is commonly supported in hardware in the ATM adapter. This means that for a multimedia transport call, each transport connection should be mapped on (at least) a single SVC.

The only case where transport layer multiplexing can be accepted is when no QoS support is required on the multiplexed transport connections and the SVCs are rare and expensive.

3.3 QoS Negotiation on Top of the ATM-CONS

In this section, we will examine issues of QoS negotiation raised by the simultaneous use of connection-oriented services (e.g. ATM-CONS and AMTS).

QoS Negotiation on Point-to-Point Connections

Seen from the service user's point of view, a QoS negotiation[2] on a point-to-point connection requires three steps (figure 3.2):

1. The calling user transmits its QoS requirements to the service provider
2. Based on the QoS requirements, the negotiation rules and its available resources, the service provider will propose a modified set of QoS requirements to the called user.
3. The called service user will conclude the negotiation and select the QoS for the connection. The service provider and the calling user are then informed about the QoS selected for the connection.

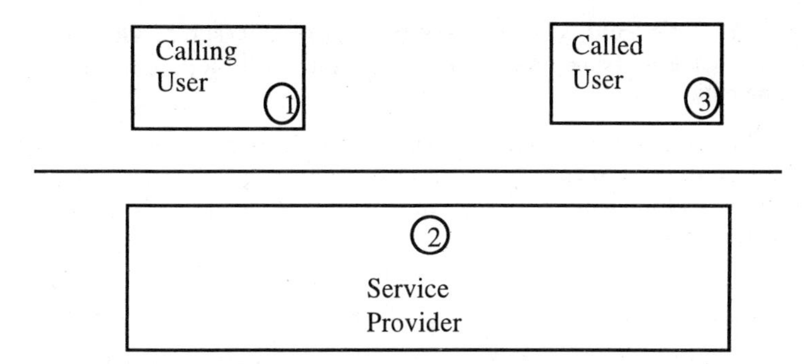

Fig 3.2 QoS negotiation seen from the calling user's point of view.

However, seen from the service provider's point of view, the QoS negotiation is more complex. Indeed, it requires five distinct steps (figure 3.3).

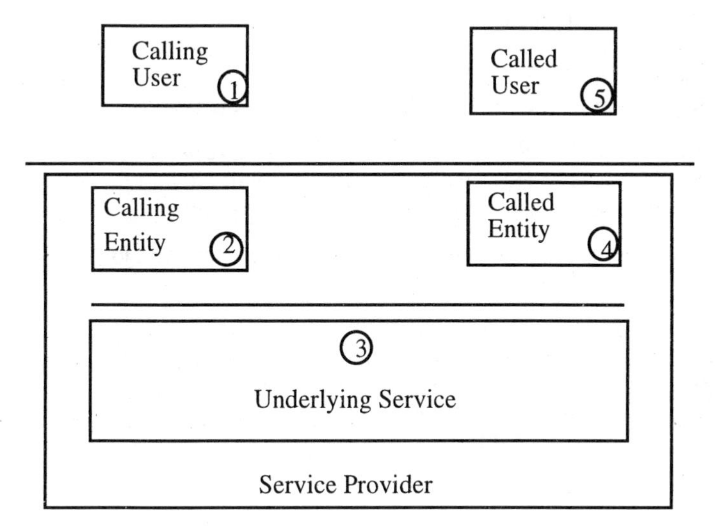

Fig 3.3 QoS negotiation seen from the service provider's point of view.

1. The calling user transmits its QoS requirements to the service provider
2. The calling entity, based on the QoS requirements, the negotiation rules and its available resources, will propose a modified set of QoS requirements. This modified set of QoS requirements will be mapped onto QoS requirements for the underlying service
3. The underlying service, based on the QoS requirements of the calling entity, its negotiation rules and available resources, will propose a modified set of QoS requirements to the called entity.

4. The called entity will possibly modify this set of QoS requirements according to the negotiation rules, and then will propose the modified set to the called user.

5. The called service user will conclude the negotiation and select the QoS for the connection. The called entity, the underlying service provider, the calling entity and the calling user are then informed about the QoS selected for the connection.

When the underlying service is connection-oriented, there is an additional problem, which is linked to the QoS negotiation. The users' QoS requirements will have to be mapped on QoS requirements for a connection established through the underlying service (an example of QoS mapping may be found in [3]). For the establishment of this underlying connection, there are two possibilities:

a. The underlying connection is established during step 3

However, the QoS requested during step 3 will usually be stronger that the QoS that will result from the complete negotiation. With this solution, unnecessary resources will be reserved during the establishment of the connection. They will have to be released later on (e.g. by using a re-negotiation facility if the underlying service supports it).

As the ATM-CONS [1] does not provide a re-negotiation facility, if the ATM SVC is established during step 3, it will have to be released and then re-established with the appropriate QoS during step 5.

b. The underlying connection is established after the QoS negotiation (i.e. during the second part of step 5).

With this solution, the ATM SVC will be established with the agreed QoS requirements, and thus this solution causes less signalling overhead than the previous one. However, the main drawback of this solution is that, if there are not enough resources in the underlying service to establish the requested connection, the establishment of the transport connection may fail even if the QoS negotiation has been successful.

QoS Negotiation on Point-to-Multipoint Connections

On multicast connections, the QoS negotiation is more complex. In [14], we have shown that an additional step is required or, in other words, that the negotiation is achieved through a three-way handshake (figure 3.4). Seen from the service provider's point of view, the QoS negotiation on point-to-multipoints connections (figure 3.5) requires the following steps :

1. The calling user transmits its QoS requirements to the service provider

2. The calling entity, based on the QoS requirements, the negotiation rules and its available resources, will propose a modified set of QoS requirements. This set of QoS requirements will be mapped on QoS requirements on the underlying service

3, 3', 3", ... The underlying service, based on the QoS requirements of the calling entity, its negotiation rules and available resources, will independently propose a modified set of QoS requirements to each called entity.

4, 4', 4", ... Each called entity will possibly modify its set of QoS requirements according to the negotiation rules, and then will propose the modified set to its called user.

5, 5', 5", ... Each called service user will select its QoS requirements for the connection. The corresponding called entity, the underlying service provider and the calling entity will then be informed about the QoS selected for the connection by each user.

6 Based on the QoS selected independently by all the called users, the calling entity will determine which users have compatible QoS requirements, and will choose the QoS for the connection based on these compatible QoS requirements. The called users which proposed incompatible QoS requirements will not participate in the point-to-multipoints connection.

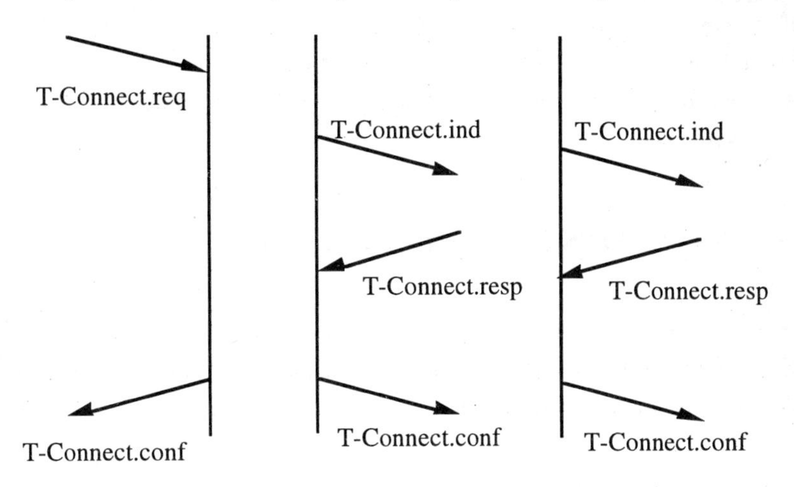

Fig 3.4 Three way handshake.

On multicast connections, some QoS incompatibilities may appear among the values proposed by the called users which have accepted the establishment of the connection. Such a phenomenon, due to the negotiation rules, is studied in [14]. When this QoS incompatibility phenomenon occurs, the transport service provider selects, based on some cost function, one of the conflicting value for the concerned QoS parameter and then "disconnects" the called users which proposed values incompatible with the selected one.

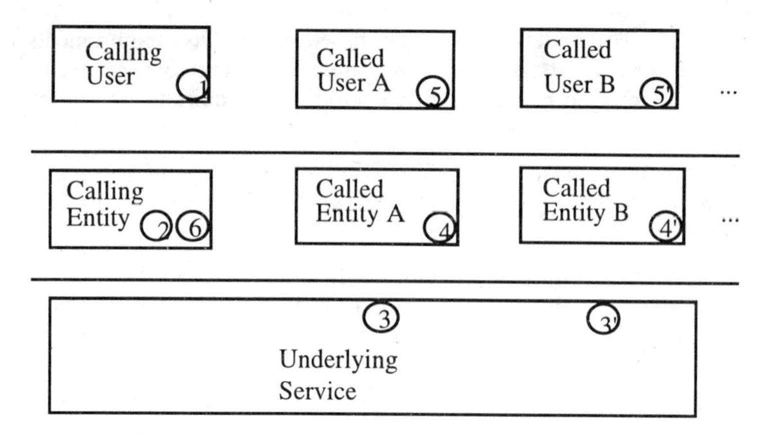

Fig 3.5 Necessary steps for the QoS negotiation on point-to-multipoints connections.

We therefore see that in the multicast case, a sequence of primitives where a *T-Connect.response* is followed by a *T-Disconnect.indication* is acceptable[3]. Thus, using solution b for the establishment of the ATM-CONS connection does not cause the same problems as in point-to-point connections. In the sequel, solution b will therefore be adopted.

4 Principles of AMTS Call Control over ATM

In this section, we will examine how the AMTS can be supported over the ATM-CONS. We will not only focus on call and connection establishment but also outline data transfer, call join and verification of integrity conditions.

4.1 Establishment of a Single Connection Point-to-Point Call

In this section, we will examine how a point-to-point call containing a single connection can be established (figure 4.1). We will use this simple example to show all the necessary steps of the call establishment phase, and the interactions between the transport layer and the ATM-CONS.

To support such a simple call, the transport layer will request the establishment of three SVCs. The first one (which we will name the forward control SVC) is established immediately after the acceptance of the requested call by the local

---

[3] The corresponding sequence on point-to-point connections is valid, but should not be common. On point-to-multipoints connections, this sequence will probably be common.

transport service provider. This forward control SVC will be a simplex SVC (sender→receiver), and it will require a small amount of bandwidth (a few cells per second). This SVC will only be used to transmit Transport Protocol Data Units (TPDUs) which are directly related to the control of the whole call. It is established as a potential point-to-multipoints SVC (i.e. the point-to-multipoints option is selected in the Broadband Bearer Capability Information Element of the SETUP message), and that is the reason why the forward control SVC is not a duplex SVC.

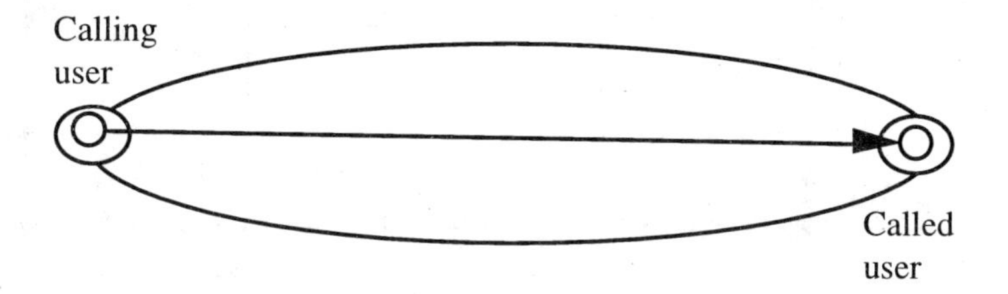

Fig 4.1 A simple call.

Once the forward control SVC has been established, the source transport entity can use it to transmit a TPDU containing all the information related to the call. As in our example there is a single connection within the call, the TPDU will mainly contain the QoS requested for the connection and the transport address of the destination transport service user. The destination transport entity will use this information to decide whether the call and the connection can be accepted and it will open a new simplex SVC (the backward control SVC) towards the source transport entity. This backward control SVC will be used to convey the TPDUs related to the control of the whole call back to the sender. If the destination transport entity accepts the call, it will adjust the QoS requirements for the connection according to the negotiation rules and its available resources and it will inform its local user about the call establishment attempt.

If the called user accepts the call establishment request, the called transport entity will use the backward control SVC to transmit a TPDU containing the acceptance and the negotiated QoS parameters to the source protocol entity. When the calling transport protocol entity receives this TPDU, it will conclude the QoS negotiation and open a new SVC (which will be called the data SVC) towards the destination transport entity. The negotiated QoS are mapped onto ATM-CONS QoS requirements that are used to establish the data SVC. Once the data SVC has been established, the calling transport entity will confirm the establishment of the call to its local user, and it will transmit a TPDU containing the confirmation and the negotiated values for the QoS parameters towards the destination transport entity on the data SVC. A call containing a single simplex connection uses thus three SVCs as shown in figure 4.2.

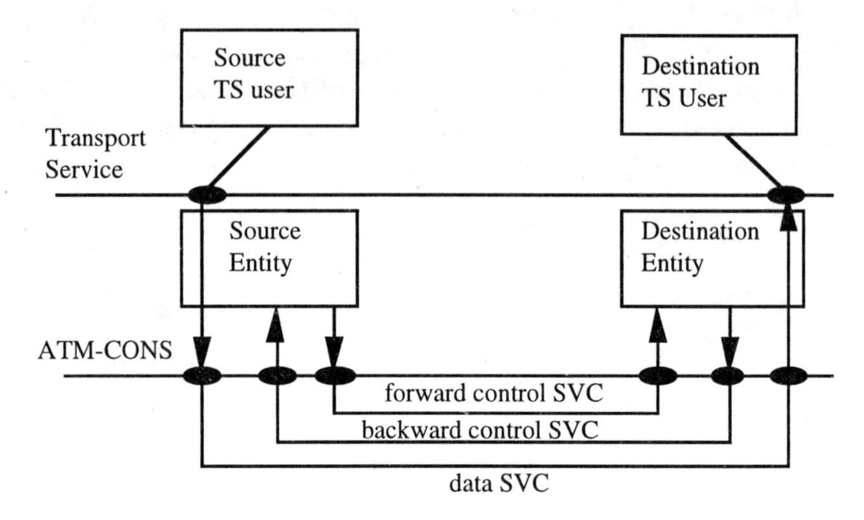

Fig 4.2 Point-to-point call with a single connection.

If either the called user or the called transport entity refuse the call establishment request, the called transport entity will transmit a TPDU containing this refusal and possibly the reason for the refusal on the backward control SVC. The forward and the backward control SVCs will be released when both Transport Users have been informed of the failure of the call establishment attempt.

4.2 A More Complex Topology

In the previous section, we have shown all the steps that are necessary to establish a simple call containing a single point-to-point connection. In order to have a better idea of the role of the control and data SVCs, it is useful to look at the following example. As shown in figure 4.3, we consider a call with 2 point-to-multipoints (1→2) connections and 3 receivers.

The establishment of such a call will result in the establishment of 1 (1→3) forward control SVC, 3 point-to-point backward control SVCs and 2 (1→2) data SVCs. The data SVCs have the same topology as the transport connections shown in figure 4.3. Figure 4.4 shows the four control SVCs.

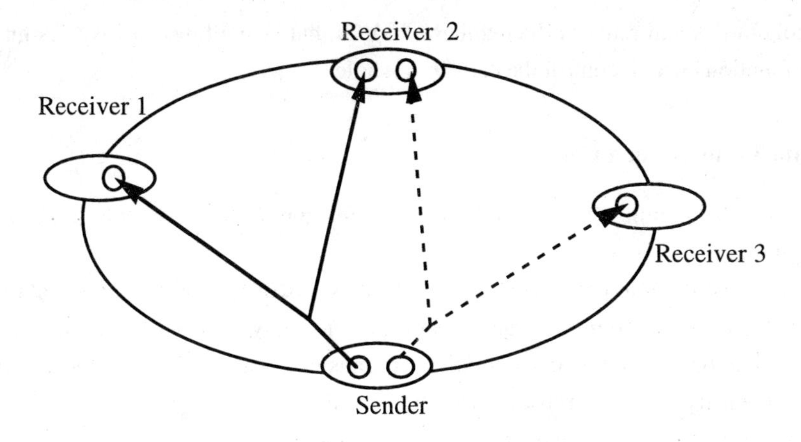

Fig 4.3 A more complex example.

In general, the establishment of a call with 1 sender and n receivers will result in the establishment of :

- 1 (1→n) forward control SVC
- n point-to-point backward control SVCs
- As many data SVCs as there are transport connections

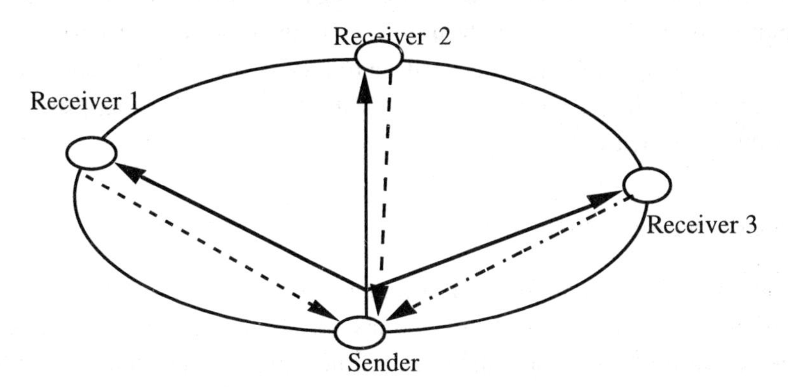

Fig 4.4 Control SVCs .

Another possible choice would have been to use n duplex point-to-point control SVCs instead of 1 point-to-multipoints simplex and n point-to-point simplex control SVCs. However, as we will see later, most of the control TPDUs sent by the sender are targeted to all the receivers, and by thus using a point-to-multipoints forward control SVC will avoid sending several copies of the same control TPDU. Moreover, using only one forward control SVC not only simplifies the design of the transport protocol entities and reduces the number of SVCs, but also allows an easy design of all the functions which control the call "as a whole".

4.3 Data Transfers in a Call

On certain connections of a call, the data transfer may be fully reliable (no error), while on others it may not.

To achieve reliable data transfers, retransmissions of lost or corrupted data are needed. Here, two different retransmission policies may be used by the transport protocol. The first option is to use multicast retransmissions. In this case, the sending transport entity simply retransmits the missing TPDUs on the data SVC corresponding to the concerned transport connection.

On the other hand, the protocol entity may use unicast retransmissions, in order not to generate useless traffic for the receivers which got the TPDU correctly. There are here several possibilities to support unicast retransmissions. An obvious one would be to use "unicast retransmission SVCs", but such a technique would drastically increase the number of SVCs used to support a transport call. That is why, in case unicast retransmission mechanisms are required, we prefer to use duplex backward control SVCs (instead of simplex ones), the direction of transfer from the sending transport entity to the receiving one being used for the retransmissions.

It must however be clear that the usefulness of the forward control SVC is in no way jeopardized by the duplex backward control SVCs, since all the control TPDUs will still be transmitted on it, for simplicity in the design of control mechanisms and bandwidth saving reasons.

4.4 Other Operations on the Call

The control SVCs are used for other operations such as join, leave, late connection establishment... As an example, let us consider what happens when a second receiver asks to be added to the simple call[4] discussed in section 4.1, and how to establish a transport connection within a call.

Transport Call Join

To join the call and the connection, the second receiver will first establish a backward control SVC towards the sender and use this control SVC to transmit a TPDU containing the request to be added to the call and the connection. This TPDU will

[4] In the AMTS, when a user joins a call, it has to join at least one connection of this call. Thus, in our example the new user will join both the call and the connection.

contain the *Callid* of the call, which is a globally unique identification of the call and the identification of the connection[5] to be joined.

If the join is successful, the sender will add the second receiver to the data and the forward control SVCs, and it will transmit a TPDU announcing the successful join on the forward control SVC. We therefore see that any transport entity participating in a call is attached to the forward control SVC of that call.

If the join fails, the sender will establish a point-to-point simplex control SVC to the second receiver, and will use this SVC to transmit a control TPDU containing the reason for the failure to the second receiver. The backward control SVC and the point-to-point control SVC will be released once the receiving user has been notified of the failure of its join request.

Transport Connection Establishment

When the sending transport service user requires the establishment of a new transport connection within the call, this connection is established exactly as described in section 4.1, except that there is no need to establish the control SVCs which already exist .

Thus, the sending transport entity directly transmits a control TPDU on the forward control SVC. That TPDU is therefore targeted towards the receivers of the call only (recall that each receiver in the call is attached to the forward control SVC).

In the same idea, the receivers will use their backward control SVCs to send their responses back to the sender.

4.5 Verification of the AGI and ATI Conditions

The AGI and ATI conditions must be verified by the transport service provider throughout the lifetime of each call. This verification may either be distributed or centralised. In a centralised verification scheme, a single entity has a complete view of the whole call (i.e. the list of established connections, the active participants on each connection, ...) and can use this knowledge to verify that the AGI and ATI conditions are still fulfilled. As the AMTS supports centralised calls, it is natural to use a centralised verification of the AGI and ATI conditions.

In the AMTS, the sender has a central role in the call, and furthermore, it participates in all the connections composing the call. This means that the sender's transport entity has already a good knowledge of the whole call and can use this knowledge to verify the AGI and ATI conditions. For example, let us consider the

---

[5]This connection-id is unique within a single call.

example call of section 4.2. After the establishment of the call, the sender's transport entity has the information provided in table 4.1 about the topology of the call.

	Receiver 1	Receiver 2	Receiver 3
Connection 1	participant	participant	
Connection 2		participant	participant

Table 4.1 Topology information for the call shown in figure 4.2.

However, the sender's transport entity will have to update this information throughout the whole lifetime of the call. For this, it will use several complementary mechanisms :

- The ATM-CONS will indicate any problem with the data and control SVCs (e.g. failure of a point-to-point SVC, failure of a branch in a point-to-multipoints SVC, ...)
- All the explicit operations which are performed on the call (e.g. join, leave, ...) involve the sender's transport entity (e.g. this entity will add participants to the forward control SVC or to the data SVCs).
- The sender's transport entity may periodically exchange control TPDUs with the receivers of a call, in order to cope with situations undetected by the ATM-CONS. Here, the control SVCs are good candidates for the transmission of those control TPDUs. This mechanism could be combined with a possible acknowledgement mechanism on certain (reliable) connections.

4.6 QoS Dependency Relations

As we have seen in section 2, for a given call, the AMTS only allows one QoS dependency relation per parameter, which means that all the connections involved in a QoS dependency relation will have the same selected values for the corresponding QoS parameter.

The involvement of a connection in QoS dependency relations is specified by the calling user at the connection establishment. The very first connection established specifying a QoS dependency relation creates that relation.

Of course, all the data SVCs established to support transport connections involved in a QoS dependency relation are opened, by the sending transport protocol entity, with the same QoS value for corresponding parameter.

5 Conclusion

In this paper, we have examined how the AMTS could be provided over a pure ATM network.

The AMTS provides for centralised calls composed of several connections. As these connections exhibit the same characteristics as the point-to-multipoint SVCs in ATM (simplex 1→N), each of them is mapped on a different ATM SVC in order to exploit the QoS support provided on these SVCs.

However, at the ATM-CONS level, there is an additional cost of providing calls at the transport level: we need to use one control SVC per participant of the call (one 1→N SVC for the sender and one point-to-point SVC per receiver). Those additional SVCs are used to transmit control messages between the sender's and the receivers' transport protocol entities. Those control SVCs are therefore expected to be used optimally but further work is still needed to know their required performance. Moreover, the overhead represented by those control SVCs is not specific to our transport service since such SVCs are likely to be required as soon as transport control is to be sent back to a sender (since the point-to-multipoint SVCs used for data are simplex). Finally, using the same control SVCs to control several transport connections at the same time is also deemed as a factor minimising the number of such SVCs.

References

1. ATM Forum: User-Network Interface (UNI) Specification, Version 3.1, September 1994
2. A. Danthine, O. Bonaventure: From Best Effort to Enhanced QoS, *Architectures and Protocols for High-Speed Networks*, O. Spaniol, A. Danthine, W. Effelsberg, Eds., Kluwer Academic Publishers, 1994
3. S. Damaskos, A. Gavras, "A simplified QoS Model for Multimedia Protocols over ATM", *Participant's Proc.eedings5th IFIP Conference on High Performance Networking*, Grenoble, June 27-July 1, 1994, S. Fdida, ed., pp 239-256.
4. A. Danthine, Y. Baguette, G. Leduc, L. Léonard: The OSI 95 Connection-mode Transport Service - The Enhanced QoS, *High Performance Networking*, IV , IFIP Transactions C-14, A. Danthine, O. Spaniol, eds., Elsevier (North-Holland), pp 235-252.
5. A. Danthine, O. Bonaventure, G. Leduc: The QoS Enhancements in OSI95, *The OSI95 Transport Service with Multimedia Support*, A. Danthine, ed., Springer-Verlag, pp 124-149.

6. S. Deering, D. Estrin, D. Farinacci, V. Jacobson, C-G. Liu, L. Wei: An Architecture for Wide-Area Multicast Routing, *Proc. SIGCOMM '94*, London, Augusts 31- September 2, 1994, Computer Communication Review, Vol. 24, No. 4, October 1994.
7. D. Ferrari, A. Banerjea, H. Zhang: Network Support for Multimedia - A Discussion of the Tenet Approach, Technical Report TR-92-072, International Computer Science Institute, Berkeley, November 1992.
8. D. Feldmeier: A Framework of Architectural Concepts for High-Speed Communication Systems, *IEEE Journal on Selected Areas in Communications,* May 1993, vol.11, No.4, pp 480-488.
9. D. Ferrari, D. Verma: A Scheme for Real-Time Channel Establishment in Wide-Area Networks, *IEEE Journal on Selected Areas in Communications*, April 1990, Vol.8, No.3, pp 368-379.
10. L. Henckel: Multipeer Transport Services for Multimedia Applications, *Participant's Proc. 5th IFIP Conference on High Performance Networking*, Grenoble, June 27-July 1, 1994, S. Fdida, ed., pp 165-183.
11. ITU-T, B-ISDN Protocol Reference Model and its Applications, ITU-T Recommendation I.321
12. ITU-T, B-ISDN ATM Adaptation Layer (AAL) Specification, ITU-T Recommendation I.363
13. ISO/IEC JTC1: Network Service Definition for Open Systems Interconnection, ISO/IEC JTC1/SC6 N7558, 21 Sept. 1992.
14. L. Mathy, O. Bonaventure: QoS Negotiation for Multicast Communications, *International COST 237 Workshop on Multimedia Transport and Teleservices*, Vienna, November 13-15, 1994, Lecture Notes in Computer Science no. 882, D. Hutchinson, A. Danthine, H. Leopold, G. Coulson, eds., Springer-Verlag, pp 199-218.
15. L. Mathy: On the Design of a Transport Service to Support Multimedia Applications, *Participant's Proc. 4th Conference on Broadband Islands*, Dublin, September 4-6, 1995, F. Williams, H. Bräke, J. Nolan, eds., pp 67-75.
16. L. Mathy, G. Leduc, O. Bonaventure, A. Danthine: A Group Communication Framework, *3rd International Broadband Islands Conference*, Hamburg, June 7-9, 1994, O. Spaniol, W. Bauerfeld and F. Williams, eds., Elsevier Science Publishers (North-Holland), 1994, pp 167-178.
17. The OSI95 Transport Service with Multimedia Support, A. Danthine, ed., Research Reports ESPRIT, Project 5341, OSI95, Vol.1, 1994, Springer-Verlag, 515 pp.
18. C. Topolcic: Experimental Internet STream Protocol, Version 2 (ST-II), Internet RFC 1190, 1990.
19. L. Zhang, S. Deering, D. Estrin, S. Shenker, D. Zappala: RSVP: A New Resource ReSerVation Protocol, IEEE Network, September 1993, Vol.7, No.5, pp 8-18.

Towards A Hybrid Scheme for Application Adaptivity

H. S. Cho, M. R. Fry, A. Seneviratne, V. Witana

University of Technology, Sydney
PO Box 123, Broadway,
NSW, AUSTRALIA
{hscho@ee, mike@socs, aps@ee, varuni@dstc}.uts.edu.au

Abstract. This paper presents a framework for providing the infrastructure necessary to support adaptive applications. The suitability of the proposed framework is demonstrated by first analysing some advantages and disadvantages of methods reported in the literature for the development of adaptive applications. We report on some experiments that support our analysis. We define some requirements for adaptive applications that properly decompose the task between the system level and the applications themselves. We present an overall framework for adaptivity based on application level definitions of Quality of Service, user specified preferences, and system level resource management. This scheme is the basis for prototype, adaptive application development at the University of Technology, Sydney.

1 Introduction

Future distributed applications will need to operate on computing systems that have widely varying capabilities, interconnected via networks that provide variable levels of service. For example, a person will have access to a powerful workstation at the office that is connected to a high speed network. At home s/he will have access to a personal computer and a medium speed network. Finally, while traveling s/he will have access to a mobile device connected via a wireless communication link.

Because of this variability in the computing devices and networks, if distributed applications are to realise their full potential, it will be necessary to provide the same applications on all systems, but with functionality that is compatible with the available resources. For example, the same application may provide high resolution full colour video in the office environment, while it may provide only low resolution, slow scan grey scale video on the mobile device. Secondly, the users will want more control of their operating environment to adjust the quality of the applications they are using. They may want to run multiple applications which will enable them to monitor the progress of an activity while working on another, or when at home they may want to customise the operation to minimise cost.

There seems to be two schools of thought as to how to cater for these requirements. One is to have a scheme in which system resources are reserved to guarantee a fixed level of service (Quality of Service) [7]. The other is to have a dynamic model whereby applications adapt to the changes in the operating environment [13]. We contend that a hybrid scheme is more appropriate, where the level of service can be guaranteed and system level support is provided for the development of adaptive applications.

In this paper we first present some experimental results which will provide some justification of our assertion. Then we analyse the support required for implementing adaptive applications, and where this support should be provided. After that we provide a framework that will provide the required support, and finally provide a brief description of a prototype that is being developed at the University of Technology, Sydney.

2 Adaptive Applications

A number of adaptive applications have been reported in the literature. These applications implement the required infrastructure either within the application entirely [1], or within the application with some form of rudimentary support from the system [3, 6]. Both of these approaches can result in anomalous behaviour.

2.1 Application Level Adaptivity - IVS

We consider the adaptivity of the INRIA Video Conferencing System (IVS) reported in [1]. The purpose of this scheme is congestion control. IVS includes a feedback control mechanism in which the parameters of the coder are adjusted according to the observed operational conditions. The feedback control mechanism relies on feedback from the receiver. The receiver provides this information (in the form of a NACK) when it receives an out of order packet. As with slow start [15] the assumption is that missing packets indicate network congestion. The transmitter must be throttled to reduce congestion. This feedback is used by the adaptation mechanism at the transmitter to adjust its output rate by changing either the video frame rate(in Privilege Quality mode) or the quantiser value or the movement detection threshold(in Privilege Rate mode) [1].

To investigate the operation of this scheme we used the Privilege Quality mode [1] , and observed the maximum output rate under varying conditions. The experimental set up consisted of i486 machines(66Mhz) connected via Ethernet running IVS version 3.2. The operating system is Real Time Mach.

Figure 1 shows the unloaded case. IVS transmission speed is initially set to 300kbps(the maximum speed in adaptation mode), and other system and network loads are negligible. The observable oscillations are due in this case to the operation of the feedback scheme. At each peak a NACK is being returned, causing the output rate to drop.

The receiver is observing lost packets. Since there is negligible load on the Ethernet, packet loss must be occurring at the receiver. An examination of the

[1] we believe our results will also hold for the Privilege Rate mode

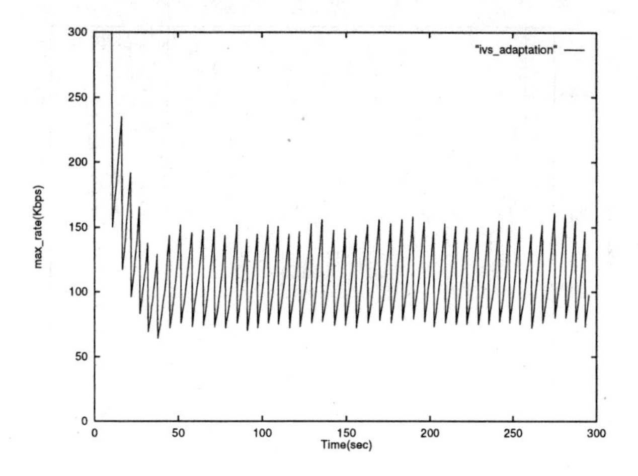

Fig. 1. Throughput of IVS with no Network or Processor Loads

trace file maintained by IVS confirms that NACKs are being generated on a fairly regular basis. This hypothesis is also supported by Figure 2, which shows CPU utilisation of the decoder thread. At an average of about 60% utilisation, the receiver is well loaded.

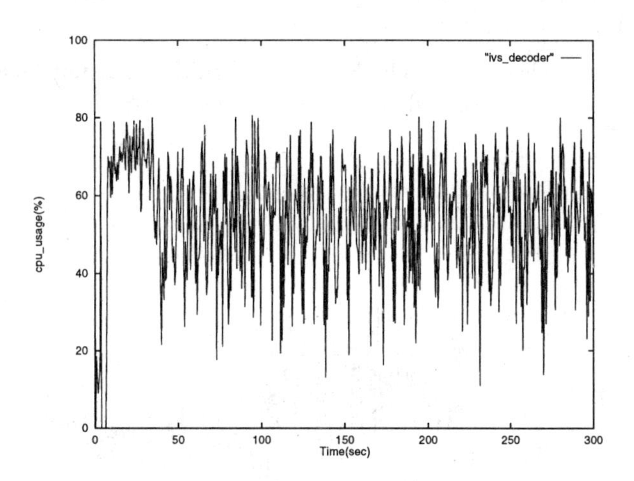

Fig. 2. CPU Utilisation of IVS Receiver

We now introduce additional load at the receiving machine. We concurrently execute a computationally intensive, periodic process which contends with the IVS receiver for CPU resources. This process requires about 30% of CPU cycles (as measured). Figure 3 shows this additional load being applied at the receiver after 3 minutes. As can be seen, this causes the transmitter to actually increase its max_rate.

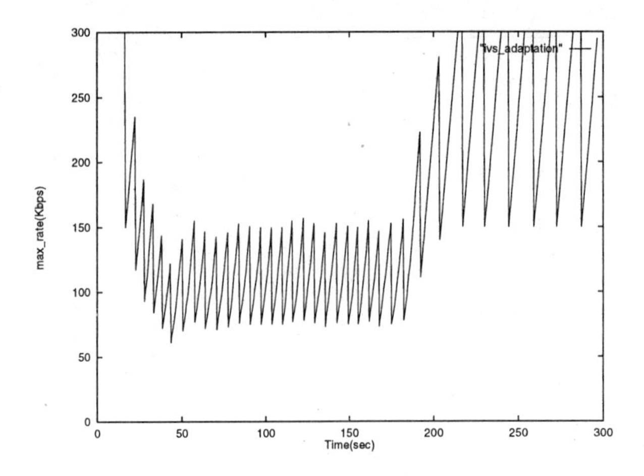

Fig. 3. IVS Throughput with Receiver Load

It is obvious that the frequency of NACK packets from the receiver is reduced. This is not due to any reduction in packet loss, but rather to starvation of the receiving process. It is thus detecting lost packets less frequently, and feedback information is being delayed. In the mean time a substantial amount of the transmitted data is being lost.

This explanation is reinforced by Figure 4. With the additional load the CPU utilisation of the decoder thread is reduced to about 40%.

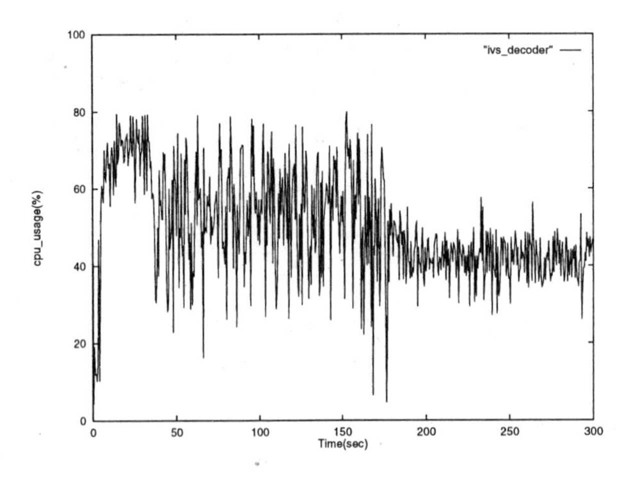

Fig. 4. IVS Decoder Utilisation with Receiver Load

2.2 Adaptivity with System Assistance - An Example

Is it possible for an application to adapt its behaviour through its observation of its access to system resources? To explore this question we developed a simple application which modifies its priority according to the observed system load. The program executes the same function repeatedly, and measures the interval of time for each execution of the function.

The program compares the *i*th execution of the function (E_i) with the *i-1*th execution, (E_{i-1}). As some other programs can run concurrently to ours on the system, the execution time of the function will vary. If another program uses more resources, the execution time of our function increases $E_i > E_{i-1}$. If this occurs, the model assumes that there is a decrease in available system resources and reduces its priority, thus in effect lowering its QoS. In our simulations, we consider execution time modifications as significant if they vary by 10 percent.

The Figure 5. shows the behaviour when we concurrently run two of these applications. It shows that even if both applications are started at the same time and in the same environment, one instantiation of the application achieves a globally higher priority than the other. This is similar to the findings reported in [3]. The applications do not degrade uniformly.

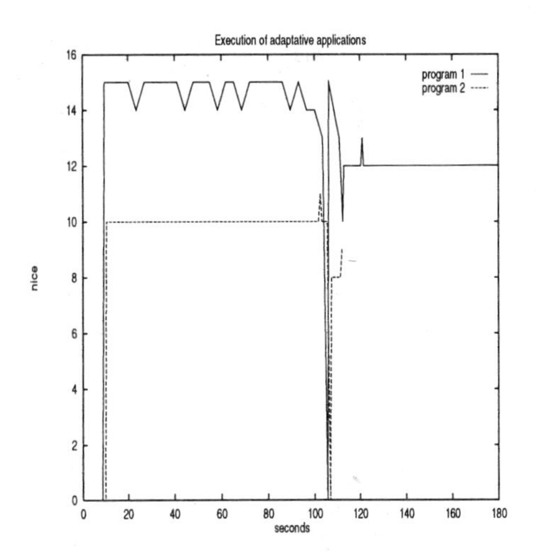

Fig. 5. Adaptive Behaviour of Concurrent Instantiations of an Adaptive Application

2.3 Discussion

Our experiments show that the NACK-based feedback mechanism of IVS is insufficient under certain circumstances. It overlooks the problem of denial of resources at the receiver. It may be possible to enhance the feedback information to throttle the transmitter when the receiver is starved. However, this will not solve the problem of sufficiency of resource allocation. Our second scenario exposed the pitfalls of applications making "blind" adaptive decisions regarding system resource allocations. This can lead to unpredictable and unfair outcomes. The notion that adaptivity cannot simply be based on the observation of one component of the end-to-end system is supported by results reported by others [3, 6, 21].

We argue that some form of system support for resource reservation is necessary for effective adaptivity. To test this contention we used the reserve mechanism of RT Mach [17] to reserve periodic CPU cycles for the decoding thread. Figure 6 shows the throughput when the additional system load is applied after 3 minutes.

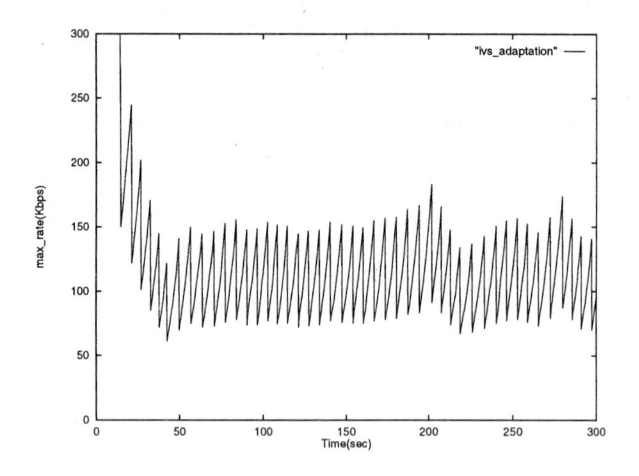

Fig. 6. IVS Throughput with Receiver Load and CPU Reserve

It can be seen that the reserved capacity for the decoder is enabling it to keep up. The throughput and feedback behaviour is close to identical to the case where the receiver is unloaded.

However a general purpose system cannot by itself determine the resource needs of all present and future applications. Some application level specification (or policy) is required, which can then be mapped to system level resource allocation mechanisms. Furthermore, the system cannot make decisions in matching application needs to resource availability, or decide on priorities if conditions change. Such preferences need to be expressed through the application, e.g. as in the Privileged Rate or Privileged Quality options of IVS. However we have shown the need for the system to support such user-expressed policies. It is for these reasons that we propose a hybrid scheme for adaptivity and resource allocation.

3 An Adaptive Framework

For adaptive applications to function properly, a holistic approach to resource management must be adopted. We have derived two requirements for such a scheme. Firstly, the applications should have the autonomy to adapt to the prevailing conditions, thus enabling optimum adaptations while relieving the system of the burden of having to know the semantics of all current and future applications. Secondly, resource management should be centralised to ensure that applications do not take "blind" decisions, so that degradation or enhancement is carried out in a controlled fashion.

Thus the adaptation at the application level will be built into the application by the application designer, and will be specific to each application. The resource management facility on the other hand has to be universal, and offer support to ease the implementation of adaptivity within the applications. The model that can be used for this type of resource management can be based on the notion of Quality of Service (QoS).

The logical structure of such a scheme is shown in Figure 7.

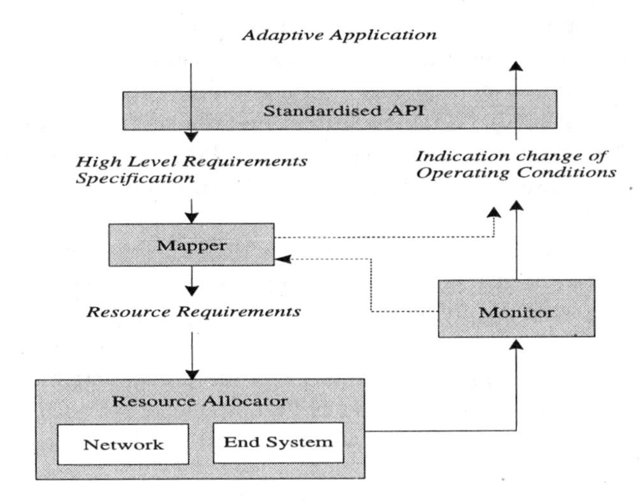

Fig. 7. Logical Structure of a QoS Management Scheme

This model encompasses a number of proposals reported in the literature for managing end-to-end QoS [2, 23, 26, 27]. It requires methods for specifying application level requirements, determination of system resource requirements from this specification, allocation of system resources, and monitoring the availability of these resources.

3.1 High Level Application Specification

A number of methods for specifying application requirements have been reported in the literature. Some schemes have advocated very detailed specification at lower levels, such as the transport layer [19, 2], or have attempted to classify applications into a number of classes such as real time reliable and unreliable, and non real time [29, 12]. Other methods have attempted to formally specify QoS requirements [5, 24]. We believe that lower level specification methods are unsuitable for specification at the application level. In addition, in our opinion the methods based on classification are too coarse grain and inflexible, while those based on formal methods may suffer from excess complexity.

In our proposal we firstly pre-define the application requirements and store them in a system level information base (say, a QoS Management Information Base = QoS MIB). The application can then request its preferred level of service through a well defined API. We believe that this scheme has benefits of operational simplicity and flexibility, while not precluding the use of formal methods.

3.2 Determination of System Resource Requirements

Several studies have proposed methods for determining system requirements from application specifications. At the network level, Moran [19] describes how transport layer parameters can be translated into network parameters. Vogt [28] has shown the mapping between a QoS specification and a token ring adaptor.

At the end system level (which this paper is concerned with) Gopalakrishnan [12] provides a detailed methodology for determining the resource requirements for protocol processing. In [4, 10] the authors describe schemes based on objective functions to determine scheduling requirements. Finally, the QoS Broker model [20] advocates the use of a tuning facility that will enable users to adjust their QoS levels.

However, none of these proposals address the issue of determining resource requirements for a given level of service. We propose a scheme which uses pre-run tests as suggested in [17], and encompasses the ideas proposed in the above studies.

In our scheme, the application QoS levels to be supported are determined by the application designer. The resource requirements for each QoS level are then measured. These requirements will then be stored in a QoS MIB when an application is installed.

We illustrate this idea with a real example. Table 1 shows the measured performance of IVS for DEC 5000/240 workstations over Ethernet and FDDI. In this case we are measuring average bandwidth requirements, but we could equally well measure CPU resource consumption. We have added an uncompressed (raw) format to our version of IVS (IVS++). Obviously this format requires significantly more bandwidth than H261 compressed video. However, it also delivers better image quality.

This shows us the achievable QoS levels for a given configuration, and for variations in application requirements such as picture size and colour. The next step is to design a set of user-oriented service levels. These service levels will be predicated on the notion of *preferences* for different application parameters such as picture size, speed (frame rate), image quality and colour. We then measure the system resources required to support each service level. Each user may have

Format	Size pixels	F/s	Kb/s	Bottle-neck
Uncomp-CIF-Color	352x288	2.3	8200	Network
Uncomp-CIF-Grey	352x288	10.5	8490	Network
Uncomp-QCIF-Color	176x144	11.8	9440	Network
Uncomp-QCIF-Grey	176x144	35.4	7130	Grabbing
H261-CIF-Color	352x288	0.5	135	Encoding
H261-CIF-Grey	352x288	0.9	150	Encoding
H261-QCIF-Color	176x144	3.7	140	Encoding
H261-QCIF-Grey	176x144	5.5	200	Encoding

Performance of IVS++ over Ethernet on a 5000/240

Format	Size pixels	F/s	Kb/s	Bottle-neck
Uncomp-CIF-Color	352x288	3.2	10120	Grabbing
Uncomp-CIF-Grey	352x288	11.5	9280	Grabbing
Uncomp-QCIF-Color	176x144	11.8	10070	Grabbing
Uncomp-QCIF-Grey	176x144	35.4	7720	Grabbing
H261-CIF-Color	352x288	0.6	100	Encoding
H261-CIF-Grey	352x288	0.9	200	Encoding
H261-QCIF-Color	176x144	3.7	150	Encoding
H261-QCIF-Grey	176x144	5.4	200	Encoding

Performance of IVS++ over FDDI on a 5000/240

Table 1. Measured Performance of IVS

their own preferences, and the preference for one quality (say, image resolution) may be different depending on the application, e.g. between a video editor and a video conference.

3.3 Preferences

The user can specify preferences for application level QoS through the application interface. The preferences will in effect provide a user profile. The QoS management system will use these preferences to adjust QoS levels of applications competing for resources. An appropriate user interface for the definition of QoS needs to be developed. Here we focus on the representation of preferences within the system, and the interaction between application and system adaptivity.

To illustrate our scheme, an example of user preferences and adaptation is shown in Table 2.

In our experimental video conferencing system we have defined frame rate, color and picture size as quantities that affect the application QoS level. The possible levels of QoS and resource requirements of each level are shown in Table 2.(a). This table is fixed in the QoS MIB at application installation time.

Table 2.(b) indicates the transition preferences and the transition costs in terms of bandwidth. This tells the system how to degrade or upgrade service levels depending on resource availability. That is, it controls adaptation. A higher preference means higher priority in upgrading, and lower priority in degrading. These preferences may initially be assigned by the application designers. Thereafter, these preferences can be changed dynamically by the user.

(NB Unlike Table 1, the values in Table 2.(b) are not measured quantities. They are used purely for illustration. Likewise we could equally well have chosen CPU utilisation as the dependent resource. In the general case a vector of resources will be associated with each service level.)

As a further suggestion, we adopt a weighing factor that reflects the user's attention. When the user runs two video conferencing applications, s/he can indicate his/her preference by moving the mouse pointer to the more preferring window. In our example, the weighing factor is '1' when the application receives user's attention(Focused), and '0.5' when it does not receive user's attention(Unfocused). Table 2.(b) shows the preferences for each case.

The QoS Manager is notified of changes in resource availability by the Resource Manager. It determines the most appropriate form of adaptation using the information recorded.

Table 2.(c) shows that two instantiations of the video conferencing application are running at QoS level P6 initially. When network bandwidth availability decreases, the QoS manager gets the preferred degradation paths of each application(focused and unfocused), and selects the degradation path with the higher degrading preference(smaller value). At the same time the application is signaled to degrade its service level. This is necessary, for example, to signal a change in encoding method. Table 2.(c) shows that applications degrade(adapt) as directed by the user preferences.

3.4 Allocation of System Resources and Enforcement

As shown in Figure 7, both end system and network resources need to be allocated to each application. Allocation can be guaranteed through the use of a reservation scheme or purely on a "as available basis". A number of bandwidth reservation methods have been proposed [8, 22, 30]. Emerging networks such as B-ISDN promise to provide facilities for resource reservation. Despite this, there is currently a debate as to whether reservation is necessary or feasible on a network wide scale [14]. In this paper, we only consider end-system resource allocation issues.

Although, a number of the end-to-end QoS management schemes advocate reservation of resources [26, 23], to our knowledge only a few reservation schemes have been proposed [17, 18]. The necessity or the viability of reserving more than one resource has still not been addressed.

Despite this, if no reservations can be made, satisfactory performance bounds cannot be provided. As we have shown with our IVS example, reservation at least provides for more predictable delay bounds, which improves the effectiveness of

QoS level	P6	P5	P4	P3	P2	P1
Resolution	480x384	480x384	480x384	320x256	320x256	160x128
FPS	30	20	15	15	10	10
Color	color	color	color	bw	bw	bw
Bandwidth	3,024	2,016	1,511	672	448	110

(a) QoS levels and Resource Requirements

Transition of QoS level	P56	P45	P34	P23	P12	P01
Transition cost in Bandwidth	1,008	505	839	224	338	110
Preference(focused)	3	5	9	13	14	16
Preference(unfocused)	1.5	2.5	4.5	6.5	7	8

(b) Preferences of the QoS transition paths

Change of	QoS level	
Bandwidth	Focused	Unfocused
Initial QoS	P6	P6
-1Mbps	P6	P5
-500Kbps	P6	P4
-800Kbps	P5	P4
-800Kbps	P5	P3

(c) Adaptation Results to the Changing Bandwidth

Table 2. Adaptive Behaviour of Concurrent Instantiations of the Experimental Video Conferencing System

feedback based adaptations. Perhaps reserves can be used to support the minimum level of performance at each service level. In recent discussions on the "mbone" mailing list and other places, it has been suggested that any multimedia stream requires some minimum level of resource guarantee in order to be useful (anyone using real time audio, e.g. vat, would immediately agree with this intuitively).

Therefore, we have initially adopted a resource allocation scheme based on reservations.

This scheme also needs a facility to ensure that the applications only use the allocated resource quotas. This needs to be done by the different resource managers. Schemes for enforcing the use of CPU have been proposed and implemented in RT Mach [25], and on top of Unix [11]. We believe these facilities can be effectively used for enforcing the orderly usage of CPU cycles. However similar schemes will need to be implemented for other resources.

4 Overall System Design

The logical structure of the proposed system is shown in Figure 8. It clearly has similarities with other proposed schemes. Our main focus here is on adaptation through resource allocation and expression of preferences.

When an application is invoked, the QoS Manager is informed of the invocation via a simple API.

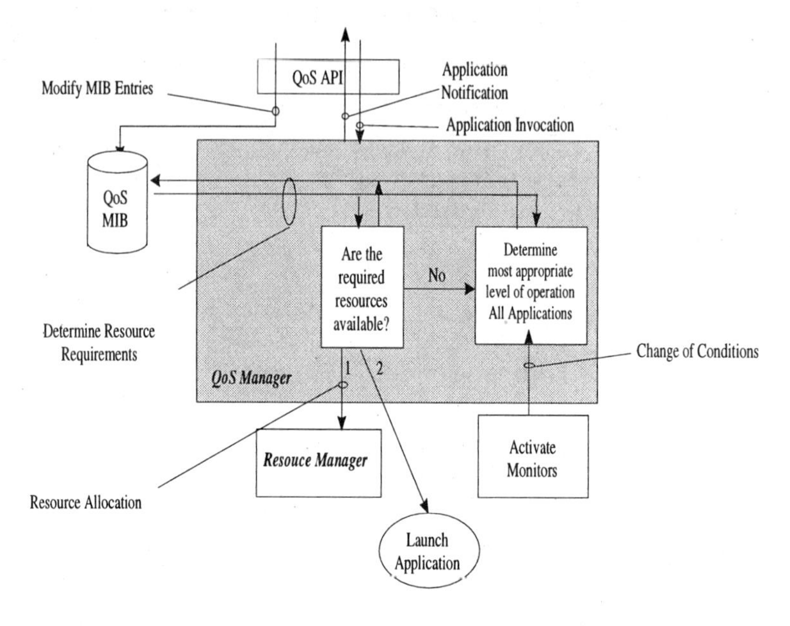

Fig. 8. Logical Structure of the overall system

The QoS Manager (QoSM) in turn obtains the resource requirements for the user's most preferred service level from the MIB [2]. Then the QoSM, determines whether sufficient resources are available to offer this level of service. If sufficient resources are not available, the QoSM checks the preference of the new invocation against the preference of the applications already running. Then the QoS levels of all the applications are adjusted according to their preferences iteratively, until it can be supported with the available resources. Our scheme does not preclude the allocation of preferences across different applications an/or users via some priority policy. The same procedure is carried out on remote machines involved in the association via a negotiation protocol.

If a change in resource availability is subsequently indicated, the QoSM repeats the iterative process as at application invocation to determine the appropriate level of operation.

5 Conclusions and Future Work

We have demonstrated that adaptive applications require system level support to function in a controlled and predictable manner. We have outlined a frame-

[2] The highest preference may not correspond to the highest QoS level

work for providing this support. The framework is based around a QoS Manager which carries out simple admission control, and negotiation of the level of QoS. In addition, the scheme enables the customisation of adaptation through the setting of preferences. We believe our scheme properly separates system and application level functions. The system provides, via the QoS Manager and Resource Manager, generic support for adaptation. Applications apply their own semantics to adaptation via preferences.

There are many outstanding issues that need to be addressed. Some of these issues are as follows. Firstly, methods to accurately determine changes in the operating environment need to be developed. Secondly, the viability of preference-based QoS level adaptation needs to be demonstrated. Finally, effective methods of end system resource allocation and control need to be developed.

Acknowledgments

The work reported in this paper has been funded in part by the Cooperative Research Centres Program through the Department of the Prime Minister and the Cabinet of Australia.

Aruna Seneviratne and Hyun-soo Cho would like to acknowledge the partial financial support for this work from the University of Technology, Sydney, and the Australian Telecommunication Research Board.

Anne Fladdenmuller developed the experimental adaptive application.

References

1. J. C. Bolot and T. Turletti, "A rate Control Mechanism for Packet Video in the Internet," *INFOCOM'94*, June, 1994.
2. A. Campbell and G. Coulson and F. Garcia and D. Hutchinson and H. Leopold, "Integrated Quality of Service for Multimedia Communications," *IEEE INFOCOM'93*,1993.
3. C. Compton and D. Tennenhouse, "Collaborative Load Shedding for Multimedia -Based Applications," *Proc. International Conference on Multimedia Computing Systems*, May, 1994.
4. M. Davies and A. Downing, "Adaptable System Resource Management for Soft Real-Time Systems," *Proc. Symposium on Command and Control Research and Decision Aids,*, California, 1994.
5. M. Diaz and K. Drira and A. Lozes and C. Chassot, "Definition and Representation of the Quality of Service for Multimedia Systems," *Proc. High Performance Networking Conference*, September, 1995.
6. K. Fall and J. Pascale and S. McCanne, "Workstation Video Playback Performance with Competitive Process Load," *Proc. 5th International NOSSDAV Workshop*, Boston, 1995.
7. D. Ferrari, "Reservation or No Reservation," *INFOCOM'95 Panel Discussion*, April, 1995.
8. D. Ferrari, "Client Requirements for Real-Time Communication services," *IEEE Transactions on Communications*, November, 1990.

9. A. Hyden, "Operating System Support for Quality of Service, " *Ph.D Thesis*, University of Cambridge, 1994.
10. M. Jones, "Adaptive Real Time Resource Management Supporting Modular Composition of Digital Multimedia Services," *Proc. 4th International NOSSDAV Workshop*, Lancaster, 1993.
11. R. Gopalakrishnan and G. M. Parulkar, "Real Time Signals: A Mechanism to Provide Real Time Processing Guarantees," *Technical Report*, WUCS-95-06, Washington University, St Louis, 1995.
12. R. Gopalakrishnan and G. M. Parulkar, "Efficient Quality of Service Support in Multimedia Computer Operating Systems," *Technical Report*, WUCS-94-26, Washington University, St Louis, 1994.
13. C. Huitema, "To Share Rather Than Pay," *INFOCOM'95 Panel Discussion*, April, 1995.
14. D. Clark and S. Deering and C. Huitema and S. Shenker, "Panel Session," *INFOCOM'95* , April, 1995.
15. V. Jacobson, "Congestion Avoidance and Control", *ACM SIGCOMM'88*, August, 1988.
16. C. Maeda and B. Bershad, "Protocol Service Decomposition for High Performance Internetworking," *14th ACM Operating Systems Symposium*, December, 1993.
17. C. Mercer and S. Savage and H. Yokuda, "Processor Capacity Reserves for Multimedia Operating Systems," *Technical Report* - CMU-CS-93-157, Carnegie Mellon University, 1993.
18. G. Coulson, A. Campbell, P. Robin, G. Blair, M. Papathomas, and D. Hutchison, "The Design of a QoS Controlled ATM Based Communications System in Chorus." *IEEE Journal on Selected Areas in Communications, Special Issue on ATM LANs*, Vol. 13, No. 4, pp 686-699, May 1995.
19. M. Moran and B. Wolfinger, "Design of a Continuous Media Data Transport Service and Protocol," *Technical Report* TR-92-019, University California, Berkeley, 1992.
20. K. Nahrstedt and J. Smith, "QoS Broker," *IEEE Multimedia Magazine*, Pring 1995.
21. B. Noble and M. Price and M Satyanarayanan, "A Programming Interface for Application-Aware Adaptation Mobile Computing," *Proc. 2nd USENIX Symposium on Mobile and Location Independent Computing*, Michigan, 1995.
22. C. Partridge and S Pink, "An Implementation of the Revised Internet Stream Protocol (ST-II)," *Proc. 2nd International NOSSDAV Workshop*, Heidelberg, 1992.
23. A. Seneviratne and M. Fry and V. Witana and V. Saparamadu and A. Richards and E. Horlait, "Quality of Service Management for Distributed Multi media Applications," *IEEE Communication Society Phoenix Conference on Computers and Communications*, Phoenix-Arizona, 1994.
24. R. Staehli and J. Walpole and D. Maier, "Quality of Service Specification for Multimedia Presentations," *Multimedia System Journal*, August, 1995.
25. H. Takuda and T. Nakajima and P. Rao, "Real Time Mach: Towards a Predictable Real Time System," *Proc. USENIX MACH Workshop*, October 1990.
26. W. Tawbi and L. Fedaoui and E. Horlait, "Dynamic QoS Issues in Distributed Multimedia Systems," *2nd International Conference on Broadband Islands*, Elsevier Science Publishers B.V., 1993.
27. A. Vogel and G. v. Bochmann and R. Dssouli and J. Gecsei and A. Hafid and B. Kerherve, "On QOS Negotiation in Distributed Multimedia Applications," *Technical Report*, University of Montreal, Canada, 1993

28. C. Vogt and R. Herrtwich and R. Nagarajan, "HeiRAT: The Heidelberg Resource Administration Technique Design Philosophy and Goals," *Technical Report* No. 43.9213, IBM ENC, Heidelberg, 1987.
29. M. Zitterbart and B. Stiller and A. Tantawy, "A Model for Flexible High-Performance Communication Subsystems," *IEEE Journal on Selected Areas in Communication*, May, 1993
30. L. Zhang and S. Deering and D. Estrin and S. Shenker and D. Zappala, "RSVP: A New Resource Reservation Protocol," *IEEE Network Magazine*, September, 1993.

Admission Control for End-to-End Distributed Bindings

Laurent Leboucher, Jean-Bernard Stefani
Centre National d'Etudes des Télécommunications (CNET)
38-40 rue du Général Leclerc
92131 Issy-les-Moulineaux – FRANCE
E-mail : { lebouche, stefani } @issy.cnet.fr

France Télécom

Abstract. The paper presents an admission control scheme for distributed bindings with guaranteed end-to-end transfer delay in an object-based architecture for real-time heterogeneous distributed sytems. The scheme is based on recent results on fixed priority scheduling by Tindell [15, 16].

1 Introduction

We are interested in the design and construction of predictable, real-time, open distributed systems. Predictability means that users are allowed to specify *quality of service* (QoS) requirements, and to obtain guarantees about the fulfillment of those requirements. Quality of service measures how well a system or system component performs its functions, and there exists a wide range of relevant quality of service properties such as those related to timeliness, volume, or dependability. Openness characterizes the context of use, which is expected to be typical of future telecommunications information networks. Openness implies adaptability to heterogeneous environments and to varying loads and application requirements (the latter being in contrast to embedded systems which support only a fixed and known set of applications). Heterogeneity can manifest itself at several levels, e.g. hardware, operating systems, networks, communication protocols, programming languages.

The architectural basis of our work is the Advanced Networked System Architecture (ANSA) [6] and the Reference Model for Open Distributed Processing (ODP) [1, 2]. These provide foundations for the construction of heterogeneous distributed systems. We are interested in completing these works with a comprehensive set of functions for QoS support, covering declaration (i.e. means to specify QoS requirements in a rigorous and quantitative form, for measurement and verification), establishment (i.e. means to establish a contractual level of QoS in response to application requirements), monitoring (i.e. means to inspect the QoS achieved in a system), control (i.e. means to ensure given levels of QoS are maintained over a period of time, possibly involving adaptations to changes in the environment).

In order to tackle the complexity inherent in the design of heterogeneous distributed systems, ANSA and ODP identify several viewpoints. Each viewpoint constitutes a complete perspective on distributed systems, together with its specific set of concerns. Our work is essentially concerned with the computational and the engineering viewpoints. The computational viewpoint provides an abstract, language-independent programming model for distributed applications. The engineering viewpoint provides a specification of a (distributed) virtual machine, at operating system level, for supporting the computational model. Engineering issues include for instance those of naming, binding, communications, etc, and in general issues associated with the realization of distribution transparencies.

From the computational viewpoint, a real-time distributed application appears as a collection of interacting objects subject to timeliness constraints. Supporting the computational model implies the ability to instantiate objects that can meet their associated constraints. In the context of an open system, where the set of objects cannot be determined in advance, instantiating a time-constrained object implies some form of admission control to provide the required guarantees.

Constraints and guarantees may take several forms [4]. Constraints can be a combination of, e.g. deterministic bounds (specified as $var \leq bound$ or $var \geq bound$, where var is some QoS parameter such as a delay), fractional bounds (specified as some percentile of var that satisfies $var \leq bound$), probabilistic bounds (specified as $Prob(var \leq bound) \geq prob$), or statistical bounds (specified as bounds on statistical moments such as mean or variance).

In this paper, we detail an engineering-level admission control test for the establishment of distributed object bindings with guaranteed end-to-end communication delay bounds. The constraints we consider are deterministic, i.e. if the admission test is successful, an established binding is guaranteed to meet the constraint throughout its lifetime.

The paper is organized as follows:

- Section 2 provides an overview of our architectural approach and motivates the interest on admission control;
- Section 3 details an admission control test for distributed bindings with guaranteed end-to-end delay bounds;
- Section 4 concludes the paper with some considerations on our current prototyping activities and some indications for further research.

2 Architectural approach

In this section we briefly present the main elements of our architecture, namely the computational and the engineering model. The notion of admission control for distributed bindings is then discussed.

2.1 Computational model

The abstract programming model we use is that of the ODP Reference Model [2], extended with QoS declarations. In this model, distributed applications consist in sets of interacting objects. As usual, an object encapsulates some state and has entry points, called signals, for interaction with its environment. Signals can be grouped in interfaces. An object may have several interfaces. A signal consists in a name (the name of the signal) together with a set of parameters (which can be of basic types or reference to interfaces) an a direction (input or output).

Interactions between given objects is only possible if a binding (i.e. some communication path) has been established between some of their interfaces. A binding between interfaces can be established automatically by the supporting infrastructure (binding is then said to be "implicit"), or be created upon request by a computational object (binding is then said to be "explicit"). Explicit binding results in the creation of a distributed object, called a *binding object*, that supports the binding. A binding object supports interactions between the interfaces it binds. Behaviors of binding objects reflect the communication semantics they support. Implicit binding is possible only between so-called operation interfaces, i.e. which support only standard client/server interactions.

Objects, and binding objects in particular, can be qualified by quality of service assertions that further constrain their behaviour. QoS assertions describe the quality of service that an object provides to its environment, as well as the requirements of that object on its environment to perform correctly. QoS assertions take the form of real-time temporal logic formulas expressing simple safety properties. Examples include delay and delay-jitter bounds. For instance, a server maximum reaction time can be specified as a bound on the delay separating the occurrence of an operation invocation (an input signal) and the occurrence of the corresponding termination (an output signal). A full description of QoS assertions is beyond the scope of this paper but can be found in [12].

The notion of QoS-constrained binding objects has been found useful e.g. in the case of multimedia applications which require explicit control over multiple parallel flows of continuous media data. Quality of service assertions, in this case, typically cover end-to-end communication delay bounds, delay-jitter bounds on the reception of consecutive media samples, and minimum throughput bounds. For a discussion of the computational model in the context of multimedia applications see e.g. [3].

2.2 Engineering model

The engineering model describes a distributed system as comprising a set of interconnected nodes. A node corresponds to an abstraction of information processing resources together with its basic software. Each runs a nucleus. The set of cooperating nuclei constitutes a distributed platform that supports the execution of applications constructed in conformance with the rules of the computational model.

Communication between application objects is supported by channels. Channels correspond to the realization, in engineering terms, of computational binding objects. A channel is composed of stubs, binders, and protocol objects. Stubs provide marshalling and unmarshalling functions to enable communication between heterogeneous nodes. Binders are responsible for maintaining the integrity of a channel in the face e.g. of communication failures or of application object migration. Protocol objects provide an abstraction of communication resources and protocols used to interconnect nodes.

To enable flexible QoS support, the engineering model posits policy-free access to abstractions of low-level operating system resources and services, including notions of threads and events with time parameters. For a description of the main elements of the model and its implementation using the Chorus microkernel see [13].

2.3 QoS contracts and admission control

QoS assertions associated with computational objects can be understood as a contract between that object and its environment. This contract is put in place at the time of object instantiation and remains in place throughout the lifetime of the object[1]. Since we consider open systems, which objects will be created and when is not known. Object instantiation may thus fail due to lack of adequate resources to meet the required QoS constraints at the time of instantiation. In other words, object factories, i.e. objects that can instantiate other objects, are required to act as *admission controllers*, creating an object only when it can be guaranteed to honor its QoS contract[2]. Obviously, other schemes are possible, however they imply more complex QoS contracts and guarantees[3].

Ideally, admission control should work for whole object configurations and not just individual objects. It should also cover the possibility for object configurations to evolve over time. In this paper we do not consider these cases, which involve more complex QoS declarations and probably constitute intractable problems in general. Instead, we focus on the instantiation of single objects.

Admission control at object instantiation time for deterministic QoS contracts involves verifying that appropriate resources are available to support the newly created object at the required level of QoS. This in turn implies an analysis of the object implementation. It is unclear at the moment to what extent this analysis can be automated. Schedulability analysis provides some elements of answer but the general case remains an open issue. In this paper, we show

[1] Unless explicitly modified. In this paper, we restrict ourselves to non-modifiable contracts. In other terms, an object, once created, always meets the same constraints.

[2] Of course, as part of more general QoS requirements, volume bounds may be associated with object factories, requiring for instance that a minimum number of objects be instantiable.

[3] One could think e.g. of load shedding or downgrading schemes in case of overloads. The corresponding QoS guarantees are conditional.

how admission control can be conducted for a specific but fundamental class of binding objects where we posit the knowledge of an implementation model.

3 Admission control for simple binding objects

As an example of admission control on objects, we consider in this section a simple binding object linking two computational objects. We consider a QoS assertion that consists of a single end-to-end delay constraint in the form of a deterministic bound on the transfer of signals through the binding object. Compared to delay-jitter and throughput constraints, an end-to-end delay constraint is more difficult to warranty, because this kind of property cannot be easily decomposed into obvious local constraints. This simple binding object should be considered as a first example illustrating our admission control approach. A multimedia application provides an example real-time application that may require the creation of such binding objects. For instance, a multiconference application can ask at any time the admission controller to admit a new binding object when a new participant comes in.

3.1 Implementation model

The computational binding object is mapped on an engineering channel. Its implementation, in engineering terms, can be modelled as a succession of three elementary tasks (see figure 1):

- a local processing task consisting of producing and copying periodically a message m on the network adaptator card. This task will be called the emission task.
- a transmission task which periodically takes the new message m and sends it in the form of packets. This task ends when the last interruption signalling the arrival of the last packet of a message is raised. We use an FDDI scheme for transmission[4].
- a local processing task at the other end of the channel which takes the newly arrived packets corresponding to one message and processes them. This task will be called the receiving task. It starts when the interruption corresponding to the last packet of a message has been raised.

3.2 Fixed Priority scheduling background

In this section, we briefly summarize the principles of fixed priority scheduling which form the basis of our admission control scheme. These principles were derived in particular by Liu and Layland [8][14], Joseph, Pandya, Leung, Lehoczky [7][9], Sha [9], Rajkumar [9]. The major source of inspiration for the work reported in this paper has been recent work done by Tindell [15][16].

[4] We use an FDDI network in our actual implementation. Our analysis extends to arbitrary networks with deterministic transfer delay bounds.

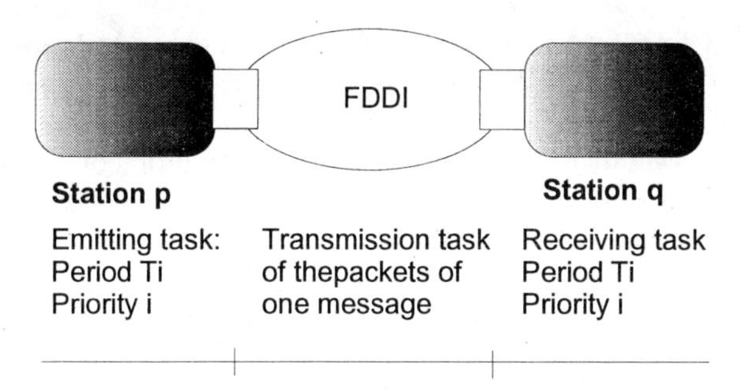

Fig. 1. Binding Model

Introduction and notations Various methods and algorithms exist to schedule tasks for real-time systems: statical approaches using calendars, dynamic approaches using either a fixed priority algorithm or an algorithm based on priority dynamically determined during scheduling (EDF is a typical example). Each of these approaches has advantages and drawbacks. The main advantage of fixed priority scheduling is the ability to analyse and compute rather easily the worst response time of a task and to extend this analysis framework to distributed sets of tasks such as our previous end-to-end channel. Classic scheduling techniques did not go beyond the local scheduling framework. There is a big lack of analysis of schedulability of distributed sets of tasks in scheduling literature. Taking into account the recent extended GRMS (Carnegie Mellon University) approach and thanks to the techniques developed by Ken Tindell from York University (UK), it is possible to bound end-to-end delays on a distributed system. We first briefly recall classic results of fixed priority scheduling from Liu and Layland simple local model to the quite general extensions of Joseph, Pandya, Lehoczky and Tindell. Since Tindell gathered all these results in a consistent way, we will further refer to the Tindell's analysis. We then explain how we apply this analysis to our specific problem.

The notations used thoughout this section are classic in the scheduling literature and especially in the Tilborg and Koob reference book (see [14]).

C_i :	execution time of task i
T_i :	period of task i
J_i :	release jitter of task i
d_i :	relative deadline of task i
B_i :	blocking time of task i caused by a priority inversion
$w_{i,q}$:	busy period of level i starting qT_i before the current task i release time
r_i :	response time of task i

From Liu and Layland periodical model to Lehoczky's analysis In Liu and Layland periodical model, tasks are periodic and deadlines are equal to release times of later tasks. Joseph and Pandya analysis is more general because it allows deadlines smaller than periods. It is the foundation of the global analysis which enables us to obtain a bound on the end-to-end delay of a distributed binding.It is summarized by the following recursive equation:

$$r_i = C_i + \sum_{j \in hp(i)} \left\lceil \frac{r_i}{T_j} \right\rceil C_j \tag{1}$$

where r_i is the worst response time of a task i, hp(i) denotes the set of tasks of higher priority than task i.These recursive equations can be solved by successive iterations starting from $r_i = 0$. Indeed, it is easy to show that r_i^n series is increasing. Consequently, it converges or exceeds T_i. In the latter case, task i is not schedulable. Joseph and Pandya analysis is still not sufficient since we need to cover the case when periods are smaller than deadlines. Lehoczky gives such an analysis. It is essentially based on the notion of busy period. By definition a busy period of level i is defined as the maximum interval of time during which a processor runs tasks of higher or equal priorities than task i priority. Lehoczky shows that to determine the worst response time of a task, it is possible to look successively at several busy periods, each one starting at a particular arrival of task i. If $w_{i,q}$ denotes the width of the busy period starting qT_i before the current activation of task i, the analysis can be done by the following equation:

$$w_{i,q} = (q+1)C_i + \sum_{j \in hp(i)} \left\lceil \frac{w_{i,q}}{T_j} \right\rceil C_j \tag{2}$$

These equations can also be solved by successive iterations and the worst response time of task i is given by :

$$r_i = max_q(w_{i,q} - qT_i) \tag{3}$$

The number of busy periods that need to be examined is bounded by the LCM of the tasks periods. In practice, it is bounded by :

$$\frac{\sum_{j \leq i} C_j}{T_i - C_i} \tag{4}$$

As such, this approach cannot be applied when tasks are no longer independent. In some cases, when tasks share a data structure, this data structure has to be protected by a semaphore. In such cases, it can happen that a low priority task blocks a higher priority task, it can even produce a deadlock. To avoid that problem, Rajkumar introduced the Priority Ceiling Protocol using priority inheritance. This protocol avoids deadlocks caused by priority inversions and bounds the priority inversion time. Combining the Priority Ceiling Protocol and the Rate-Monotonic analysis, we get the following sufficient condition of schedulability when $D_i = T_i$:

$$\frac{C_1}{T_1} + ... + \frac{C_n}{T_n} + \frac{B_i}{T_i} \leq i(2^{1/i} - 1) \, , \, 1 \leq i \leq n \tag{5}$$

where B_i denotes the longest blocking time of task i by a lower priority task.

Following Tindell's approach, this same enhancement can also be applied to the Lehoczky busy period analysis. It only means to add a term B_i in equation 2.

Finally, the last refinement we keep from Tindell's analysis is the notion of release jitter. In many cases, a task cannot be scheduled as soon as it arrives. In some cases, for instance, the exact release time of a task depends on the reception of a specific message which can take some bounded but unknown time. This jitter has an impact on Lehoczky's equation (see eq. 2). It must be replaced by the following one:

$$\begin{array}{rcl} r_i & = & max_q(w_{i,q} + J_i - qT_i) \\ \text{where } w_{i,q} & = & (q+1)C_i + B_i + \sum_{j \in hp(i)} \left\lceil \frac{w_{i,q}+J_j}{T_j} \right\rceil C_j \end{array}$$

3.3 Admission control for a simple binding object

In this section, we use the results from Tindell's analysis recalled in the previous section to derive a bound of the end-to-end delay offered by a computational binding object. We follow the engineering decomposition given in the binding model paragraph (see 3.1). The notion of release jitter will appear very useful to relate several periodic tasks which belong to the same chain of tasks. For instance, in our binding model, let us suppose the emitting task is periodic. Then, the release jitter of the next related task which is the transmission task happens to be equal to the response time of the emitting task. In this section, we implicitly suppose that the network adaptator memory card is big enough, which is not the case obviously for real traffics such as video traffics. This aspect is studied in paragraph 3.4 dealing with some refinements.

Technological infrastructure The technological infrastructure we use for our actual implementation is composed of an FDDI ring connecting several PC workstations running the Chorus operating system micro-kernel. The deterministic delay we consider implies that the binding objects use the synchronous bandwidth as opposed to the asynchronous FDDI bandwidth which does not provide any deterministic warranty.

Worst end-to-end delay analysis The first of the three elementary tasks of our binding model – the emission task E – is characterized by its execution time and its response time. We apply to it the Tindell's analysis. We suppose that the emission task E can be preempted by other periodic local tasks. For simplicity, we don't consider here other interfering tasks such as interruptions caused by reception of packets on the network card. The response time is given by examining the busy periods starting before its current release time.

The response time is given by:

$$r_i^E = max_q(w_{i,q} + J_i - qT_i)$$

where:

$$w_{i,q} = (q+1)C_i + B_i + \sum_{j \in hp(i)} \left\lceil \frac{w_{i,q}+J_j}{T_j} \right\rceil C_j$$

where hp(i) denotes the set of local tasks of higher priority than i on the station p that supports the emission task E.

For the moment, they are supposed to be periodic. Further in this section, we will add the interference due to interruptions caused by arriving messages.

The second of the three elementary tasks is the transmission task, T. From the workstation network adaptator point of view, the deterministic FDDI network can be seen as a processor running the system of tasks made of packets waiting for being sent and a more prioritary task P modelling the rotating token when it is away from the emitting station. This specific task interferes in the computation of the transmission response time. Let SA_p be the synchronous allocation of station p. During a period of time w, a property of the FDDI protocol is that the token will visit station p at least:

$$\left\lfloor \frac{w}{TTRT} \right\rfloor \text{ times} \tag{6}$$

where TTRT (target token rotation time) is the target rotation time determined during the FDDI loop initialization. For the packets transmission task, being preempted by P means that the workstation has finished its allocated synchronous time and thus that the token stayed at least SA_p on station p. Therefore, the time needed by the network to serve the other workstations is bounded by:

$$CI(P,w) = w - \left\lfloor \frac{w}{TTRT} \right\rfloor SA_p \tag{7}$$

CI denotes the computing interference of task P on the current transmission task.

During the same busy period w, the interference due to the other higher priority emitted messages is given by:

$$\left(\sum_{j \in hp(i) \cap out(p)} \left\lceil \frac{w_{i,q} + J_j^T}{T_j} \right\rceil m_j \right) \rho \qquad (8)$$

where ρ stands for the duration of a packet emission, m_j denotes the number of packets of message j, $out(p)$ denotes the set of emission tasks indices on workstation p which supports the emission task E. T can be modelled as a periodic task of period T_i inherited from the emitting task and the jitter corresponds to the response time of the emitting task. The power of the release jitter notion appears here very cleary since it enables us to chain tasks and compute their response times incrementally . Once the transmission of a packet has started, it cannot be preempted. Therefore, the priority inversion duration is bounded by ρ. From all this, the response time of the transmission task can be derived:

$$r_i^T = max_q(J_i^T + w_{i,q} - qT_i) + \psi + \tau$$

where:

$$w_{i,q} = (q+1)m_i\rho + \rho + \left(\sum_{j \in hp(i) \cap out(p)} \left\lceil \frac{w_{i,q} + J_j^T}{T_j} \right\rceil m_j \right) \rho + CI(P, w_{i,q})$$

$$J_i^T = r_i^E$$

where ψ denotes the optical and electrical propagation time on the network, τ denotes the worst time taken by the network adaptator at the other end to recognize the packet arrival and raise the corresponding interrupt.

The third of the three elementary tasks, the receiving task R, can be modelled as a periodic task inheriting the period from the emitting task. Its release jitter is equal to the difference between the biggest response time of the transmission task and the smallest one, ie:

$$J_i^R = r_i^T - \rho - \psi$$

Finally, a bound on the end-to-end delay is equal to the highest response time of the receiving task plus the smallest response time of the transmission task.

To find the response time of the receiving task, we can apply the same Tindell's analysis. We need to bound the computing interference due to the raise of the interruptions corresponding to the arriving packets. This can be done by modelling the arriving packets by a pseudo sporadically periodic task. According to Tindell, a sporadically periodic task is a periodic burst with an outer period corresponding to the burst period and an inner period corresponding to the packets period within a burst. In the case where the network adaptator card can access directly to a DMA without preempting the workstation processor, then this is no longer useful. If we consider the first case, the global response time is:

$$r_i^R = max_q(J_i^R + w_{i,q} - qT_i)$$

where:

$$w_{i,q} = (q+1)C_i^R + \sum_{j \in hp(i)} \lceil \frac{w_{i,q}+J_j}{T_j} \rceil C_j + \\ + CI(pt(i), w_{i,q})$$

where $hp(i)$ denotes the set of tasks on the receiving site which are of higher priority than the receiving task i. $CI(pt(i), w_{i,q})$ denotes the computing interference of a sporadically periodic task (this – complex – formula can be found in [16]).

Admission algorithms Based on the previous analysis, we can define admission control algorithms for our simple binding objects. Admission control algorithms are to be executed by the binding object factory, itself a distributed object (see [13] for details). In this paper we outline two different algorithms: a simple one, and a global one.

The simple admission algorithm uses a simple RPC between the site supporting the emission task E (hereafter site E) and the site supporting the receiving task R (hereafter site R)[5]. We assume that the algorithm is triggered by a request to the binding object factory to create a new binding between two remote interfaces located, respectively, on sites E and R, with a guaranteed end-to-end delay no worse than r_i. Let us assume that the emission task E has index i. The algorithm can be succinctly described as follows:

1. site E computes r_i^E and r_i^T, assuming that the priority of the new binding is lower than that of all already established ones supported by E (i.e. bindings for which E is either the emission site or a receiving site).
2. site E sends r, r_i^E and r_i^T to site R, together with the lowest priority of its existing bindings.
3. site R computes r_i^R, assuming that the priority of the new binding is lower than that of all already established ones supported by R.
4. If $(r_i^R + m_i\rho + \psi + \tau) \leq r_i$, then site R sends back to E an indication of success (the new binding can be established), together with the priority assigned to

[5] In terms of implementation, the algorithm can be run with only processing overhead since the information that needs to be passed between sites E and R can be piggy-backed on the requests of the binding establishment protocol. See [13], or [11] for a similar generic binding protocol.

the new binding, chosen to be lower than that of any of the existing bindings in E and R. If $(r_i^R + m_i\rho + \psi + \tau) > r_i$, R sends to E an indication of failure (the new binding cannot be established).

The above RPC must be executed as an atomic transaction to avoid conflicts between concurrent binding creations. The simple algorithm requires an assignment of priority to the new binding that is not known a priori. Optimality of local scheduling algorithms may thus be compromised.

The global admission control algorithm involves the following logical steps:

1. site E assigns a priority for the new binding (e.g. as per the rate-monotonic policy);
2. let G be the acyclic directed graph verifying the following properties:
   (a) the nodes of G are bindings;
   (b) the root of G is the new binding to establish; other nodes in the graph are already established bindings;
   (c) vertex $a \rightarrow b$ belongs to the graph, if and only if a is a node in G, a and b share a common site, and the priority of b is lower than that of a.

   The set S of bindings that are nodes in G is ordered by priority. Starting from the new binding, the algorithm computes for each binding j in S, in decreasing order of priority, the triplet (r_j^E, r_j^T, r_j^R) and verifies that its deadline is satisfied (as per the simple algorithm).
3. if all deadlines for the bindings in S are verified, the new binding is admitted.

The global algorithm can be implemented in a centralized or a distributed fashion. The centralized version involves a control site maintaining the set of exisiting bindings in the system. The distributed version involves computing the set S of existing bindings, e.g. using some graph traversal algorithm.

Compared to the simple algorithm, admitting a new binding with the global algorithm is more complex. However, it does not compromise the properties of local scheduling (e.g. optimality).

3.4 Refinements

We are currently implementing this admission control approach on several PC running an object-based platform using the Chorus operating system micro-kernel. On our way to implement admission control for simple binding objects we are confronted to several problems. The first one we kept silent in the previous paragraph is the question of the bounded network adaptator card memory. The second one deals with complexity aspects. The third one deals with the system interference we did not consider for simplicity in our model. We consider in this section the first two issues. System interference is in principle simple to deal with but requires an extensive analysis of the run-time behaviour of the Chorus micro-kernel. We do not consider this issue in this paper.

Bounded memory network adaptaptor To tackle this problem, we introduce a new task in our binding model: the network adaptating task which is in charge of periodically copying a sample of the messages already copied by the emitting tasks in the workstation RAM memory. The sample is determined as follows. Each time the adaptating task is activated, it examines a succession of queues, one for each binding object, and moves a specific quota to the network adaptator card. The quotas are chosen so that their sum does not exceed the adaptator card size. Only already filled queues (materialized with a flag on) are moved. When a queue is completely moved, its flag is put off by the adaptator task. The adaptator task is modelled by a periodic task, ending by a token request. A token request means that the adaptator card is no longer writable and that next time the token arrives, the adaptator card sends its frames. The adaptator task period needs to be chosen so that it is of course bigger than the total copy time of the card. Its release jitter is bounded by 2 TTRT since once the token is requested, it can come in less than 2 TTRT. Finally its priority must be chosen bigger than all other priorities.

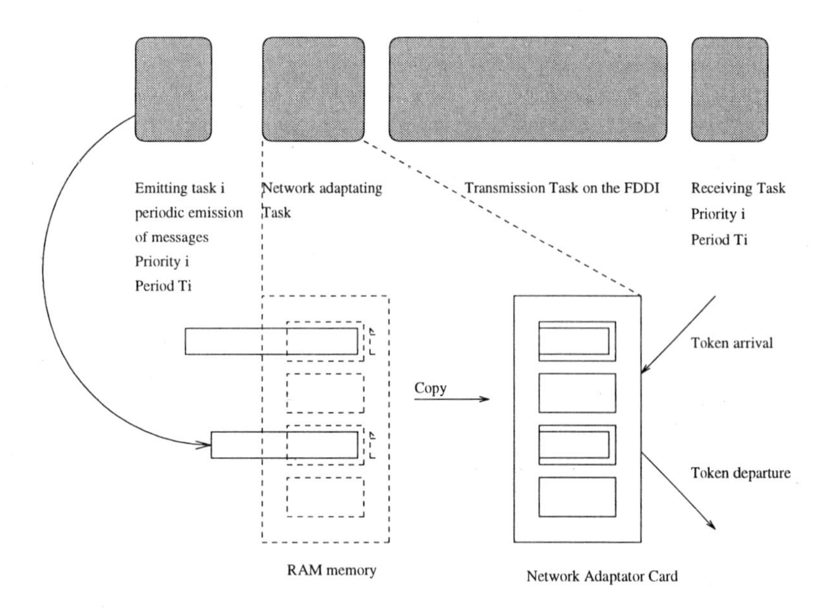

Fig. 2. Bounded Network Adaptator Card Model

Complexity aspects Our target applications are multimedia applications such as multiconference systems. With such applications, many simple binding objects need to be admitted on-line, each one during a certain period of time. Our main concern is to be sure that the admission control will be reasonably tractable. Figure 3 shows as an example several bindings already admitted and a new one (in dashed line) trying to be admitted during a time interval. The admission test based on fixed priority scheduling (see paragraph 3) has to be made in each phase of the concerned time interval. The phases are determined by the projection on the concerned time interval of the other time intervals. The admission controller needs also to retry the admission test on each of the already admitted bindings. This means that an admission test implies a maximum of 2N - 1 elementary admission tests.

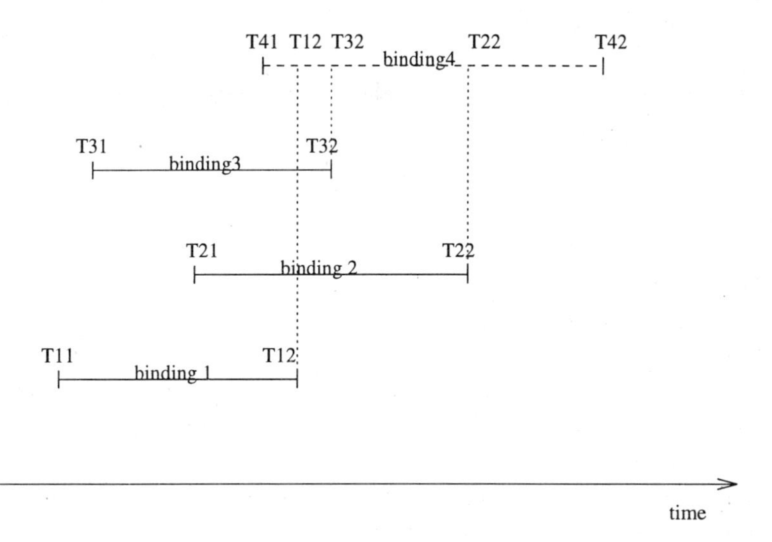

Fig. 3. Admission of a new binding object during a time interval

In order to have an idea on the complexity of the admission test, we can consider first the admission test for only local periodic tasks. The extension to distributed tasks will be considered in a near future. We suppose for simplicity that tasks are sorted in decreasing priority order. Figure 4 describes $w_{i,q} - (q+1)T_i$ as a function of q. We can understand, that in some cases, the determination of $r_i = Max_q(w_{i,q} - qT_i)$ can take a long time. In general, the number of busy periods to be examined is bounded by $\frac{\sum_{j\leq i} C_j}{T_i\left(1-\sum_{j\leq i}\frac{C_j}{T_j}\right)}$. To prove this bound, we consider the following inequation:

$$w \leq (q+1)C_i + \sum_j (\frac{w}{T_j}+1)C_j \tag{9}$$

We know that the maximum of $w_{i,q} - qT_i$ is positive. Therefore this maximum is reached before $w_{i,q}^{max} - qT_i$ becomes negative where $w_{i,q}^{max}$ is the root of:

$$w = (q+1)C_i + \sum_{j<i} (\frac{w}{T_j}+1)C_j \tag{10}$$

i.e.:

$$w_{i,q}^{max} = \frac{(q+1)C_i + \sum_{j<i} C_j}{1 - \sum_{j<i} \frac{C_j}{T}_j} \tag{11}$$

Thus, the maximum number of busy periods to be examined is bounded by:

$$q_{max} = \frac{C_i + \sum_{j<i}(1+\frac{J_j}{T_j})C_j}{1 - \sum_{j<i} \frac{C_j}{T}_j} \tag{12}$$

We can also prove that solving the Lehoczky's equation (2) by iterations can be done in less than:

$$\sum_{j<i} \lceil \frac{w_{i,q}^{max}}{T}_j \rceil \tag{13}$$

Therefore when $\sum_{j<i} \frac{C_j}{T_j}$ is rather small compared to 1, then, the complete admission test is bounded by :

$$K\ N\ Max_i \left(\log \frac{\sum_{j\leq i} C_j}{T_i \left(1 - \sum_{j\leq i} \frac{C_j}{T_j}\right)} \right) \tag{14}$$

where N is the number of bindings. Figure 4 shows the difference $w_{i,q} - (q+1)T_i$ versus q, the busy period number. It can be proved that as soon as this difference becomes negative once, then it is no longer useful to explore the future busy periods. On the figure two straight lines show the two approximations we get when $\lceil x \rceil$ is replaced by x or $x+1$. All the points are between these two straight lines. In this example, the values of Ci and Ti are:

$$C1 = 11, C2 = 4, C3 = 3, C4 = 11, T1 = 21, C2 = 25, C3 = 33, C4 = 49$$

The test was made for task 4. The figure 4 shows that the maximum response time corresponds to the 10th busy period.

4 Conclusion

We have presented an admission control scheme for elementary binding objects with bounded end-to-end transfer delay. The scheme is based on adaptations of recent results on fixed priority scheduling. We are currently in the process of implementing this scheme in an object-based engineering platform that conforms to the computational and engineering models mentionned in this paper. The platform relies on a modified version of the Chorus distributed operating system micro-kernel [13]. We hope to experiment it and evaluate its applicability in the context of several multimedia applications built on the object-based platform.

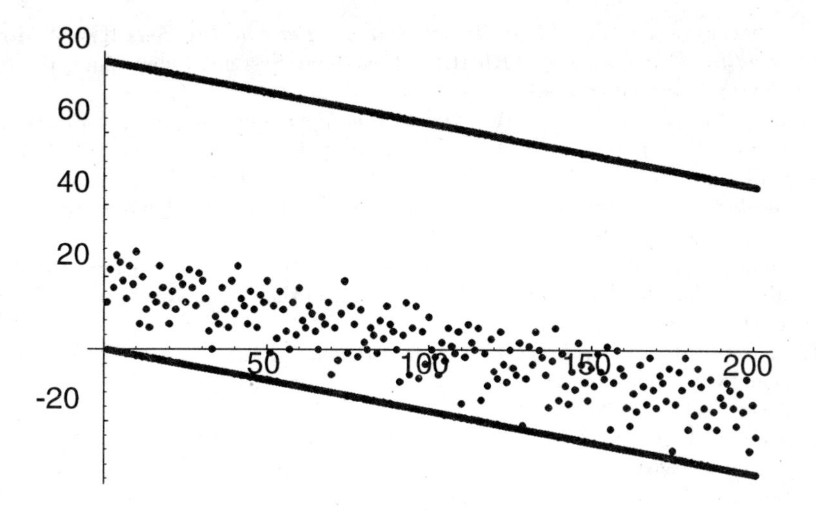

Fig. 4. Response time versus busy period number

The scheme presented in this paper is limited in a number of respects: it considers only deterministic constraints; it relies on fixed priority scheduling which can prove inappropriate for many real-time applications; it handles only a specific form of binding objects. Extending the scheme to overcome these limitations constitute as many research directions. For the time being, in cooperation with researchers at INRIA, we are currently investigating an alternate admission control scheme based on dynamic priority scheduling techniques. We also plan, in a near future, to investigate extending these schemes to multipoint binding objects.

References

1. UIT-T Draft Recommendation X.902 — ISO/IEC International Standard 10746-2, *"ODP Reference Model: Foundations"*, February 1995.
2. UIT-T Draft Recommendation X.903 — ISO/IEC International Standard 10746-3, *"ODP Reference Model: Architecture"*, February 1995.
3. Coulson, G., G. S. Blair, F. Horn, L. Hazard, and J. B. Stefani, *"Supporting the Real-Time Requirements of Continuous Media in Open Distributed Processing"*, Computer Networks and ISDN Systems, Vol. 27, No 8,July 1995.
4. Ferrari, D., *"Client requirements for real-time communication services"*, Research Report ICSI TR-90-007, Berkeley, California, USA, March 1990.
5. Hazard, L., Horn, F., Stefani, J. B., *"Toward the integration of Real Time and QoS handling in ANSA architecture"*, Research Report no NT/PAA/TSA/TLR/3498, CNET, Issy, France, June 1993.
6. A.J. Herbert, *"The Advanced Networked Systems Architecture"*, in S.J. Mullender, ed: "Distributed Systems", ACM Press, New-York, 1989.

7. Lehoczky, J. P., *"Fixed Priority Scheduling of Periodic Task Sets With Arbitrary Deadlines"*, Proceedings 11th IEEE Real-Time Systems Symposium, pp. 201-209, 5-7 December 1990.
8. Liu, C. L. and Layland, J. W., *"Scheduling Algorithms for Multiprogramming in a Hard-Real-Time Environment"*, Journal of the ACM 20(1), pp. 46-61, 1973.
9. Rajkumar, R., Sha, L., and Lehoczky, J. P., *"Real-Time Synchronization Protocols for Multiprocessors"*, Proceedings of the IEEE Real-Time Systems Symposium, pp. 259-269, 1988.
10. Sha, Rajkumar, Sathaye (Carnegie Mellon University), *"Generalized Rate-Monotonic Scheduling Theory"*, Proceedings of the IEEE, Vol. 82, n. 1, January 1994.
11. Shapiro, M., *"A binding protocol for distributed shared objects"*, 14th International Conference on Distributed Computer Systems (ICDCS), Poznan, Poland, June 1994.
12. Stefani, J. B., *"Computational aspects of QoS in an object-based, distributed systems architecture"*, 3rd International Workshop on Responsive Computer Systems, Lincoln, NH, USA, September 1993.
13. Stefani, J. B., Hazard, L., V. Perebaskine, Horn, F., Auzimour, P., Dang Tran F. *"A real-time distributed processing environment on top of Chorus"* , Research Report no NT/PAA/TSA/TLR/4179, CNET, Issy, France, January 1995.
14. A. Tilborg, G.M. Koob, *"Foundations of Real-Time Computing: Scheduling and Ressource Management"*, Kluwer Academic Publishers,1991.
15. K. Tindell, *"Holistic Schedulability Analysis for Distributed Hard Real-Time Systems"*, Euromicro Journal, Special Issue on Embedded Real-Time Systems, February 1995.
16. K. Tindell, *"Fixed Priority Scheduling for Hard Real-Time Systems"*, PhD thesis, University of York, United Kingdom, 1994.

An Enhanced Admission Control Scheme for Deterministic and Predictive Services

Wilko Reinhardt, Dirk Trossen

Technical University of Aachen
Dept. of Computer Science – Communication Systems (Informatik 4)
52056 Aachen – Germany
email: wilko@informatik.rwth-aachen.de

Abstract: Providing guaranteed Quality of Service in a packet switched network is still a research issue. In particular, real-time video and audio applications require deterministic end-to-end delay bounds and a limitation of the delay jitter. The scheduling disciplines used within the network nodes must be able to provide local delay bounds. Admission control algorithms verify whether the scheduler can accept an additional channel without violation of guarantees made for channels already established.

Within this paper we concentrate on admission control algorithms. First we introduce and compare two promising existing approaches: The Tenet scheme, introduced by the Tenet Group at the University of California in Berkeley, and the CSZ scheme that is a possible scheme for the use within the Internet. Our comparison shows that both approaches have disadvantages that we want to overcome with our own admission control algorithm.

We provide a deterministic service class, that allows data transmissions without packet losses and delay bound violations. This class is well suited for applications that know their traffic characterisation in advance. A second class are predictive services that just requires a rough peak rate allocation. The traffic behaviour is measured, so that for new admission requests not the peak load but the measured usage is used. We combine the deterministic service of the Tenet approach and the predictive service of the CSZ scheme. The admission control for the predictive service is extended to work together with a Rate Controlled Static Priority scheduler. Furthermore, we have improved the measurement scheme by using PI regulators which provide a better estimation of the current load behaviour than the mechanism of the CSZ scheme.

1 Introduction

Distributed multimedia applications have become more and more popular in recent years. This imposes several problems on the network environment, primarily related to the provision of support for realtime video and audio transmissions, which require both, high bandwidth and end-to-end control of delay and delay jitter. These Quality of Service (QoS) parameters need to be guaranteed by the network.

Due to variable queuing delays and congestions, it is not possible to guarantee constant delay values for an application within common packet switched networks. To solve these problems the network nodes must be aware of the data streams they have to handle and reserve adequate resources. These resources include bandwidth on the links, CPU time and buffer spaces.

Several proposals have been made about how a router architecture should be designed to support QoS in packet switched networks (e.g. [1],[2],[9]). All proposals lead to almost the same basic components that the routers should consist of:

- The *flow specification (FlowSpec)* describes the flow characteristics of the sender's data stream. In general, a FlowSpec should carry information on the characteristics like sensitiveness to delay, delay jitter, accepted types of loss and the service class.
- The *admission control algorithm* is responsible for verifying whether or not a new connection can be accepted without violating the guarantees made for existing connections. It is invoked at each intermediate router to make local decisions. Possible admission control algorithms used in the Tenet approach [6] or the CSZ Scheme [9] are briefly described and compared within this paper.
- It is the task of the *routing algorithm* to find a path through the network that allows the support of the QoS specified in the FlowSpec. Current routing decision are only based on one metric, for instance the number of hops to the destination or current load on the connected links. In particular, the characteristics of the data streams are not included. A possible solution to provide QoS based routing is introduced in [13].
- The *packet scheduler* is responsible for the forwarding of the data packets in accordance with the reservations made for the flow. This includes the classification of incoming packets into the related output queues as well as the fulfilment of the delay bound constraints made by the node for this data stream.
- It is the responsibility of the *reservation protocol* to negotiate the necessary resource reservations along the paths from the senders to the receivers. Possible reservation protocols are RCAP [1], RSVP [18] or ST-2 [4].

For the provision of guaranteed performance parameters different *type of services* were introduced. They mainly differ in the reliability of the offered guarantees. Besides hard deterministic bounds like in the deterministic service, the statistical and predictive services provide the possibility to accept packet losses or violations of delay bounds. The statistical service is even able to calculate an upper bound of the loss probability.

By accepting a *deterministic service* the network commits itself to guarantee performance parameters agreed at connection establishment time with no violations, as long as the stream's load does not exceed the requested traffic parameters. Each packet is delivered on time even if all channels passing through a node are sending with their peak rate.

While providing a *statistical service* the agreed bounds for delay and delay jitter are expressed in statistical terms. The network commits itself to guarantee the delay at least for a negotiated amount of packets. Similar to the deterministic service the user must specify average and peak packet rate, the request delay bound per packet and additionally the probability that a packet must fulfil the deadline.

In contrast to the deterministic and statistical services the *predictive service* requires only a rough description of the data stream's peak rate and the desired delay bounds. In accordance to this specification the network verifies whether these requirements can be met if already accepted connections keep their current load behaviour, which is permanently measured. As soon as the new connection is accepted its behaviour is measured and influences the accumulated load of all active connections. The application must not specify an exact description of the flow's behaviour. It can rely on the delay bounds as long as itself does not violate its peak rate specification, and the behaviour of all other active channels will does change significantly.

The *datagram service* does not provide bandwidth or delay constraints but a best effort support for all kinds of non real-time applications. A widely held opinion is to reserve a small amount of resources to allow the transmission of non real-time packets even if the scheduler is saturated by channels of other service classes.

The remainder of the paper is organised as follows: Within *chapter two* we briefly describe the existing approaches of the Tenet and the CSZ scheme. *Chapter three* compares and discusses the existing approaches. Within *chapter four* we introduce our own admission control mechanism that overcomes the major disadvantages of the discussed approaches. *Finally*, we provide simulations of the new algorithm for the bandwidth usage estimation of the predictive service.

2 Existing Approaches

A number of existing approaches have considered the problem of reserving resources for the transmission of real-time data streams. In the following we consider the most interesting approaches that are currently the subject of research. The Tenet group of the University of California in Berkeley introduced the *Tenet scheme*, a complete architecture for the establishment of real-time channels. The *CSZ scheme* is discussed as a possible admission control scheme for the Internet architecture providing integrated services [2].

2.1 The Tenet Scheme

The *Tenet scheme* provides a complete mechanism to establish end-to-end channels with real time constraints. Two service classes are available for real-time transmission, providing deterministic and statistical services as described in the introduction. The connection establishment is separated into two phases. The first phase establishes the channel by reserving the amount of resources necessary to forward the packet as quickly as possible through the network. At each router the admission control verifies whether enough resources are available in order to fulfil the requirements of the new stream, without violating guarantees already made for other connections. The admission control includes tests to check whether enough bandwidth is available, and determines the smallest delay that the scheduling algorithm can provide considering the bandwidth requirements of the new connection and the delay constraints made for already established connections. The amount of buffer space is calculated, depending on the delay calculations and the required bandwidth. A final test is performed to check, whether the currently available buffer space is sufficient to support an additional connection.

If the connection establishment procedure reaches its final destination the minimum guaranteed delay at each router and the propagation delays between the routers are known. Furthermore, the maximum possible delays for each individual router have been calculated. The destination verifies, whether the requested end-to-end delay can be guaranteed by the chain of routers. If the test fails, the connection establishment is rejected, and resources already reserved at the intermediate routers are released. If the test was successful any extra resources are released as far as they still satisfy the requirements. The second phase performs the relaxing of the resources and notifies the origin that the channel is established.

The offered source load is described by

- the minimum packet interarrival time *Xmin* on the channel,
- *Xave* the average interarrival time over an interval of duration *I*,
- and the maximum packet size *Smax*.

The source is allowed to send with a peak rate up to *Xmin* as long as it does not exceed the upper bound of *1/Xave* packets within the interval *I*. Besides the load parameters, performance parameters allow the specification of the end-to-end delay bound and the acceptable probability for a delay violation.

The admission control depends on the flow specification and the used schedulers within the routers. The connection establishment procedure and the flow specification of the Tenet scheme are specified to work in an heterogeneous scheduler environment. As long as the scheduler is able to guarantee a local maximum delay and delay jitter, and the required bandwidth, no special scheduling strategy is required. The Tenet group for example introduced adaptations of their concept to number of different scheduler disciplines like *Rate Controlled Static Priority* (RCSP) or *Earliest Due Deadline* (EDD).

Besides a number of publications, summaries of the Tenet protocol suite and the admission control algorithm can be found in [1] and [6].

2.2 CSZ Scheme

The *CSZ scheme* [9] provides a guaranteed service based on peak rate allocations as well as a predictive service. Within the CSZ scheme a special combination of scheduler disciplines provides a separation of deterministic, predictive and datagram traffic. For each individual deterministic channel a single queue is maintained within a *Weighted Fair Queuing (WFQ)* scheduler. A pseudo WFQ stream is reserved for the predictive and for the datagram service. This stream is the result of a priority scheduling of the *FIFO+* queues that serve the various predictive service classes. The lowest priority is assigned to the datagram service.

The admission control algorithm of the CSZ scheme is specially designed to co-operate with this specific scheduler design. In contrast to the Tenet approach, the CSZ scheme focuses on the admission control at a single node. A general concept of establishing channels throughout the network is not planned.

3 Discussion of the Tenet and CSZ Schemes

The major drawback of the Tenet approach is the need of an exact flow characterisation. Deterministic as well as statistical service require a precise description of the source's load behaviour a priori. Average and peak rates must be specified at connection establishment time. It is assumed that the user or the application knows the exact values, otherwise the reserved resources are not sufficient or not all resources are used. The load, resulting for example from compressed live video, makes an exact estimation of the required average and peak rates nearly impossible. The rate differ widely with the transmitted programme content as shown in [8] or [12]. From our point of view it will be quite difficult to provide an exact specification of the flow characteristics of live sources, and therefore a precise estimation of the required resources is nearly impossible.

With the principle of *graceful adaptation* an approach is made to overcome this major disadvantage [15]. If an unsuited channel is detected, its parameters can be adapted by establishing a parallel channel providing more realistic characteristics. After channel establishment the data transmission is switched to the new channel. The problem with this approach is the availability of resources for the new channel. If the node is already saturated it will be impossible to install a new channel with less resources to replace an existing one. However, this should be possible by reducing or increasing the allocated resources of the channel without the installing a separate connection.

With an admission control based on *maximum-average allocation*, the Tenet group introduced a possibility to avoid the peak rate allocation for the deterministic service. This algorithm assumes that the load behaviour is known exactly. In this case, a traffic constrained function can be calculated that describes the behaviour of the source. The problem is the precise determination of the function and its transmission over the network. [12] gives an good overview of this research issue. In general, this method is only suitable if the complete data stream is available in advance, e.g. in the case of movies. However, for live sources the peak rate allocation is required.

Another major drawback of the Tenet service model is the *complexity of the statistical test*. The test verifies whether the acceptance of an additional channel violates the probability guarantees for the delay bounds of existing statistical channels. Therefore the following computations are performed: Assume that p_k is the probability that the channel k is active at a certain moment within the interval I_k. Let H be a combination of active channels that exceed the node speed. Further let H_n be the set that includes all possible combinations of H. The probability P that the delay bounds are violated is then given by

$$P = \sum_{H \in H_n} \left(\prod_{i \in H} p_i \prod_{j \notin H} (1 - p_j) \right).$$

It is assumed, that if the node speed is exceeded, the delay bounds of each channel are violated. It has been proved by the Tenet group that the probability of delay bound violations can be determined by this method.

The statistical test is a mathematical method to provide the overbooking of nodes with guarantees of a maximum delay bound violation. The main drawback is the high effort of computing the set H_n. We will illustrate this by an example. Let c_n be the number of accepted channels. We assume that all channels have the same flow characteristic and that c_s is the number of channels that saturate the node. In this case

$$\binom{c_n}{m-c_s}$$

possibilities exist to select m ($c_s \le m \le c_n$) channels that saturate the node. The cardinality of the set H_n is determined by

$$|H_n| = \sum_{i=c_s}^{c_n} \binom{c_n}{i}$$

The order of H_n increases rapidly with the number of accepted channels and the degree of overbooking. Assuming that a node is saturated with 50 channels. To calculate the probability of delay bound violations for 56 channels, the set H_n includes over four million elements. H_n has to be calculated at each intermediate node. If we further consider that the amount of 50 active channels is rather small for a backbone router, our opinion is that the effort of calculating H_n is far to high to provide a fast connection establishment. Furthermore, the statistical service assumes, like the deterministic service, a precise a priori description of the data flow.

The *predictive service* introduced in the CSZ scheme overcomes Tenet's major drawback of an exact a priori specification of the traffic flow. To process the admission control for a new channel, a rough peak rate value is sufficient. Based on this peak rate value and behaviour measurements of existing channels, it is possible to estimate whether the new channel can be accepted or not.

One disadvantage of the CSZ scheme is the absence of a general concept about how a channel can be established from end to end by passing several intermediate nodes. The approach concentrates only on fulfilling local, node specific traffic guarantees, without considering that the node is just an element within a chain of routers. Furthermore, the complete admission control is suited only to the introduced scheduler discipline. Besides a number of sorted queues, a WFQ-scheduler is used for the separation of the predictive and the deterministic channels. [14] proved that if all channels use their full peak rate capacity, WFQ can only provide a local delay guarantee of one second. Considering a chain of routers the cumulative end-to-end guaranteed delay will be in the range of a few seconds, which is not acceptable for real-time transmissions.

The delay estimation for the predictive service depends on the quality of the measured behaviour of already accepted channels. Within the original CSZ scheme, the actual bandwidth usage is estimated by the ratio of past and actual measurements. The ratio is expressed by a single parameter. If the past is overestimated, the resulting approximation is not able to react quickly enough on changes of the accumulated load. If the past is underestimated, the resulting approximated usage is much too bursty, since each temporarily increasing rate influences the measured bandwidth. For the measurement of the current delay past values are not considered. The reason for this design decision is not obvious.

To summarise the discussion, both approaches have disadvantages. While the Tenet scheme provides precise calculations of the delay bounds along the path from the sender to the receiver, the major drawback is the required traffic specification. The quality of the delay calculations depends on the exact determination of the traffic characteristics. We doubt whether this will be generally possible for all kinds of applications.

The predictive service provides a method to overcome this disadvantage. However, the CSZ scheme is not sufficient to establish a channel with end-to-end delay constraints. Furthermore, the used scheduler has rather bad possibilities to guarantee delay bounds and is too complex for an efficient implementation.

In the following chapter we introduce an approach that combines the guaranteed service of the Tenet approach with the maximum-average allocation of resources and the predictive service. The measurements of the used bandwidth and the delay are improved by an adaptive PI regulator based approach. Furthermore, the predictive service is adapted using a rate controlled static priority scheduler.

4 A new Approach to Provide Deterministic and Predictive Services

This chapter introduces our proposal of an admission control algorithm. It provides a combination of Tenet's deterministic service, based on the maximum average allocation, and a modified predictive service. The latter is based on an enhanced load measurement scheme, and is integrated into the end-to-end negotiation scheme of the Tenet approach. Furthermore, the predictive service is integrated into the RCSP scheduling discipline.

4.1 Service Model

Our service model includes *deterministic*, *predictive* and *datagram services*. Since the deterministic service is based on maximum average allocation, it is mainly intended for any kind of non-live sources which will be transmitted as real-time streams. However, also live sources can be transmitted, if the average value are set equal to the peak rate value. This could be used for live transmissions which do not accept any violations of delay bounds.

The predictive service is intended for applications that are not able to specify exactly their communication requirements exactly, and where just a rough idea of the peak rate is available. Furthermore, these applications agree that not all packets must be delivered in time. To request a predictive service the application must specify a priority class for the data stream. For instance classes will be available for high or low priority video and audio.

Besides the two real-time services the datagram service provides the classical best effort service.

The *FlowSpec* for both deterministic and predictive service is based on the (*Xmin*, *Xave*, *Smax*, *I*)-model as introduced by the Tenet group. Table 1 explains the parameters.

$Xmin_i$	minimum inter-packet arrival time
$Xave_i$	average inter-packet arrival time
I_i	averaging interval used in the definition of Xave
$Smax_i$	maximum packet size
$Dmax_i$	end-to-end-delay bound
k	service class of predictive service

Table 1: Traffic specification for a channel i

There are no guarantees of delay violation probabilities or for packet loss probabilities, as it is provided by the statistical service. This avoids costly tests of the statistical service. Instead of this the enhanced predictive service provides the possibility of overbooking the nodes, to increase the bandwidth usage and to avoid an over estimation of the requested bandwidth for each single channel.

4.2 Scheduler

Our admission control scheme is independent of the underlying scheduler, as long as the used scheduler is able to guarantee local delay bounds and provides rate control. However, we will explain our approach with help of an example scheduling algorithm.

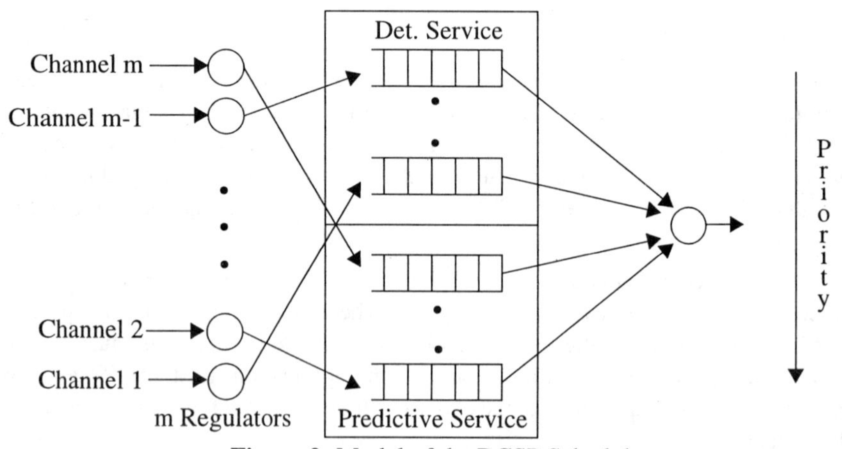

Figure 2: Model of the RCSP Scheduler

Some rate controlled servers are introduced in [5],[10],[16],[19] and [20]. A rate controlled server consists of two components: The *regulator* shapes the input traffic from each connection into the desired traffic pattern by assigning an eligibility time to each packet. The *scheduler* then orders the transmission of the eligible packets from all connections. Since the traffic pattern is restored at each single node, the traffic characteristics at the entrance of the network can be used for calculating local delay bounds. In our simulations we have applied a *Rate Controlled Static Priority (RSCP)* scheduler (Figure 2) as introduced in [19]. The n priorities are subdivided in m_1 priorities for the deterministic and m_2 priorities for the predictive service. The priorities of the predictive service are lower than the deterministic service priorities. Thus, the guarantees of the deterministic service can be met, even if channels of the predictive service violate their estimated bandwidth allocation. By definition 1 is the lowest priority in each service class.

The classification of priorities is variable for the guaranteed service. A flow is assigned to the priority which can hold the local delay at the time of establishment. For the calculation of the local delay bound we have used the estimations presented in [12]. For the predictive service the desired class is fixed and requested within the FlowSpec. The flow is assigned to the requested priority, if the priority class is able to provide the requested delay bound. If the node is overbooked, packet losses, resulting from buffer overflows, or delay bound violations are shifted from the higher to the lower priorities.

4.3 Admission Control

In this section we discuss the admission control tests. The channel establishment algorithm of the TENET approach described in section 2.1 is used to realize the tests accross the network. The tests are divided into the server and destination tests. The *server tests* are performed at each intermediate node forwarding the connection establishment request in the direction of the destination. The *destination tests* verifies, whether the sum of local delay bounds can satisfy the requested end-to-end delay. In the case that reserved resources over-satisfy the requirements, the reservations are relaxed on the way back to the sender. Table 3 summarizes the variables used in the tests at intermediate and destination nodes.

4.3.1 Server-Tests

The establishment request passes four tests at each intermediate server. If a single tests fails, the request is rejected and the resources at already processed nodes are released.

Usage Test

The *usage test* is performed for deterministic and predictive services. The utilization of the node including the new channel is estimated by

$$U_{ges} = \sum_{j=1}^{c_{m1}} \frac{Smax_j}{Xave_j} + \frac{Smax_i}{Xave_i} + \sum_{j=1}^{m_2} \hat{v}_j$$

The average utilization is used for the deterministic service while the measured bandwidth is used for predictive service. The test succeeds, if

$$U_{ges} \le l$$

that means, that the usage of the link including the new channel have to be less or equal than the node capacity.

m_1	number of classes for the guaranteed service (1=lowest priority)
m_2	number of classes for the predictive service (1=lowest priority)
c_{m1}	number of connections for the guaranteed service
c_{m2}	number of connections for the predictive service
C_q^1	set of connections for guaranteed service class q
C_q^2	set of connections for predictive service class q
$\hat{d}_j$	delay measurements of predictive service class j
$\hat{v}_j$	bandwidth measurements for the predictive service class j
v_G	total used bandwidth of all guaranteed services
B_n	buffer space at the local node
T_n	factor for Buffer Space Test
$b_{n,i}$	buffer space of connection i
$d_{n,i}$	local delay bound of connection i
$d_{prop,n}$	propagation delay of the link from server n-1 to server n
l	node capacity

Table 3: Parameter of node n

Computation of the minimum delay for guaranteed service

For a new deterministic channel the node calculates the minimum local delay bound that can be achieved. The estimation of the local delay bound depends on the used scheduling algorithm at the server. [12] provides the delay calculation for maximum average allocation and using the RCSP scheduler. The condition is that no guarantees of existing deterministic or predictive channels are violated when establishing the new one. The new delay bounds of predictive services are based on actual delay measurements. The calculated minimum local delay bound $dmin_{n,i}$ and the propagation delay $dprop_n$ between the actual and the previous node n-1 is forwarded to the destination.

Computation of the minimum delay for predictive service

If a new channel applies for a predictive service, the new delay bound for the requested class must be determined. For a given class of the predictive service, the minimum delay $dmin_{n,i}$ is the sum of the measured delay $\hat{d}_k$ of class k and the maximum local delay of the new channel in a worst case situation. The bandwidth, which is assigned, is the link speed minus the assigned bandwidth for classes with higher priority. Therefore $dmin_{n,i}$ can be calculated by

$$dmin_{n,i} = \hat{d}_k + \frac{\frac{Smax_i}{Xmin_i}}{l - v_G - \sum_{s=k+1}^{m_2} \hat{v}_s}$$

The new delay bounds for classes with lower priority are calculated according to the equation for j=1,...,k-1. If the condition holds, that no guarantees of classes with lower priority are violated, the minimum delay $dmin_{n,i}$ is forwarded to the destination.

Buffer Space Test

The last test determines the maximum delay caused by the buffering of packets in the server. Define

$$bmax_n = B_n - \sum_k b_{n,k} \quad \text{with } k \in \bigcup_{q=1}^{m_1} C_q^1 \cup \bigcup_{q=1}^{m_2} C_q^2$$

$$dmax_{n,i} = \min\left\{d\colon d = \frac{b_{n,i} \cdot Xmin_i}{Smax_i} \geq dmin_{n,i} \text{ and } \frac{bmax_n}{T_n} < b_{n,i} \leq bmax_n\right\}$$

The size of the buffer space for the new connection is determined to store packets for a delay minimum which is greater than $dmin_{n,i}$. The starting buffer size is defined as $b_{n,i} > bmax_i / T_n$.

While the minimum is computed the used buffer must be less than $bmax_n$, the remaining free buffer space. Otherwise, existing guarantees are violated. If the test is successful, the maximum delay $dmax_i$, that can be achieved by buffering the packets is forwarded to the destination node.

4.3.2 Destination Tests

After the arrival of an establishment request, destination tests are performed on basis of the server test results. The outcomes of the destination tests are backwarded to the sender. The revisited nodes adapt the reservations in accordance to the calculations of the destination tests.

Delay-Test

The *Delay Test* checks if the sum of the minimum local delays plus the propagation delays is less than the requested end-to-end delay bound. Define

$$Dmin = \sum_{j=1}^{N} dmin_{j,i}$$

$$Dprop = \sum_{j=1}^{N+1} d_{prop,j}$$

$$Dmax = \sum_{j=1}^{N} dmax_{j,i}$$

If the condition

$$Dmin + Dprop \le Dmax$$

holds, the end-to-end delay requirement of the application can be fulfilled.

Computation of local delay

The local delay bounds are calculated. There are two possibilities: Assuming the proposed delays or not. If the sum of the maximum local delay bounds plus the propagation delay is less than the required end-to-end delay, the calculated bounds are chosen by

$$d_{n,i} = dmax_{n,i} \qquad \forall n = 1, \ldots, N$$

Otherwise the jitter is distributed in a well-balanced fashion to the local delays of the servers, according to

$$d_{n,i} = dmin_{n,i} + \frac{Dmax_i - Dprop - Dmin}{N} \qquad \forall n = 1, \ldots, N$$

The result $d_{n,i}$ is send back to each intermediate node.

Computation of local buffer space

The local buffer space is adopted to store packets that arrives during the local delay time $d_{n,i}$. The buffer space is calculated according to

$$b_{n,i} = \left\lceil \frac{d_{n,i}}{Xmin_i} \right\rceil \cdot Smax_i$$

The buffer space $b_{n,i}$ is sent back to the intermediate nodes.

4.4 Load and Delay Measurements

The reliability of the predictive service depends on measurements of the network load. The estimation of the load can be more progressive or more conservative.Two factors influence the measurement quality, namely

- the adaptation of the load estimation when a new connection is accepted and
- the measurements of the network load itself.

Table 4 shows the bandwidth and delay adaptations when a new connection is established. The adaptation of the guaranteed service is done on basis of the maximum average allocation instead of the peak rate behaviour.

Deterministic Service	New utilization: $v_G = v_G + \frac{Smax_i}{Xave_i}$ New max. delay: $\hat{d}_j = \hat{d}_j + \frac{\frac{Smax_i}{Xave_i}}{l - \sum_k \frac{Smax_k}{Xave_k}}$ $\forall j = 1, \ldots, m_2$ $k \in \bigcup_{q = j+1}^{m_1} C_q^1$ j = priority of connection i in the RCSP-Scheduler
Predictive Service of class k	New utilization: $\hat{v}_k = \hat{v}_k + \frac{Smax_i}{Xmin_i}$ New max. delay: $\hat{d}_j = \hat{d}_j + \frac{\frac{Smax_i}{Xmin_i}}{l - v_G - \sum_{s = k+1}^{m_2} \hat{v}_s}$ $\forall j = 1, \ldots, k$

Table 4: Adaptation of estimated load and delay after the acceptance of a new channel

The estimation of the network load changes. In the CSZ scheme this value is computed from a combination of the actual and past load. Two parameters are necessary to adjust the estimation, the measurement interval length T and the parameter a, which provides a ratio between actual and past measurements [9]. Before we introduce our approach to estimate the network load and the local delay, we define the following requirements for the estimation:

a) Short bursts should not influence the estimation of the bandwidth.

b) If bursts are greater than a defined interval length, the estimation should reflect the utilization change.

c) The height of a burst have to take into account with an adjustable factor. The estimation should follow the bursts with variable speed dependent of the speed of the load change.

The fulfilment of the introduced requirements allows a precise adjustment of the estimation and therefore increases the reliability of the provided service. The measurement mechanism of the CSZ do not meet the requirements. In our approach we introduce a method to estimate the bandwidth and the delay more correctly.

4.4.1 PI-Control

The demands for the estimation of the network load corresponds to the control of a parameter. The estimation of the load is adjusted by the utilization of the network (real value) in a measurement interval. A useful controlling mechanism is the *proportional-plus-integral control* (PI control). The regulator consists of a proportional factor (*P*) and an integrating part (*I*) of the current load. The proportional factor models our third requirement, whereas the first and second requirement are modelled by the integrating part, especially the adjustable time of following the burst. The estimation of the network load $u(t)$ is generally defined by

$$u(t) = K_R \cdot \left(e(t) + \frac{1}{T_I} \cdot \int_0^t e(\tau)\, d\tau \right) \quad [3]$$

with

$u(t)$: estimation of the network load

$e(t)$: difference of estimation and current load

K_R: proportional factor

T_I: integrating part.

Since our model works with discrete measurement intervals, the integration part can be replaced by a sum. The discrete difference of estimation and current load is defined as $e(k)$. The regulator is based on the previously computed values of $e(k)$ and $e(k\text{-}1)$. The discrete version $u(kT)$ of the estimation is given by

$$u(kT) = u((k-1)T) + K_R \cdot \left(e(kT) - \left(1 - \frac{T}{T_I}\right) e((k-1)T) \right)$$

where T is the length of the measurement interval. Different approaches exits to determine the *P*- and *I*-Part of the regulator. One method is the impulse response to sharp rise of the real load. A change of the load is given to the input of the regulator and its reaction is observed. Figure 5 shows a typical response of a regulator. Two sizes can be defined as characteristic parameters of the regulator's reaction: The time T_U is defined as the point of intersection of time axis and the straight line through the point of inflection of the curve. $T_U + T_G$ is defined as the point of intersection of impulse height and the straight line through the point of inflection. According to these parameters there are different possibilities of tuning the regulator parameters K_R and T/T_I.

4.4.2 Adaptive PI-Control

Controlling the speed of the adjustment is a desirable feature. The integrating part T/T_I is responsible for adjusting the speed of the adaptation. If the parameter is fixed, higher bandwidth changes are adjusted in the same way like small changes. This can be done by a continuous adjustment of the integrating part, depending on the relative error of input and output bandwidth. Such a regulator is called *adaptive PI regulator.* This regulator type provides an accelerated bandwidth adaptation. To estimate the bandwidth usage of the predictive service classes, we have integrated an adaptive PI regulator in our admission control scheme.

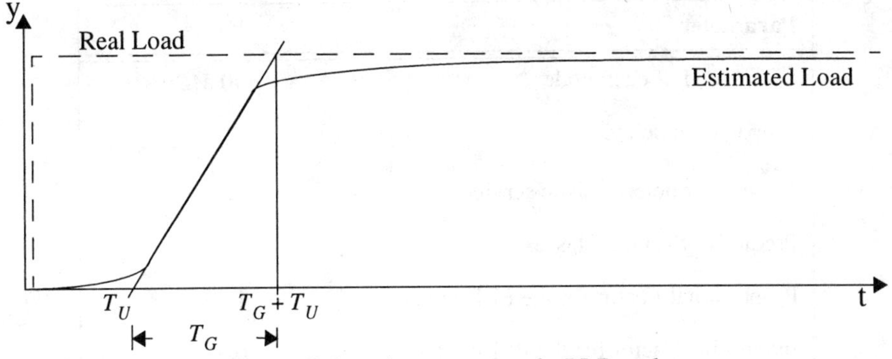

Figure 5: Impulse response of a PI-Regulator

5 Simulations

For the validation of our admission control scheme we have performed several simulations. In the following the used load generator and the simulation scenarios are briefly described. For the simulations we have used a video and audio load generator designed and implemented by Karabek [11]. The audio part models a bidirectional connection between two partners. The peak bandwidth is about 128 *kBit/s* with a burstiness factor of 3. The video load is based on the results of Heyman [7]. The peak rate is about 1.6 *MBit/s* with a burstiness factor of 3.85.

The choice of suitable scenarios is important for the validation of the model. Because of the already proved analytical bounds for the deterministic service, we refrain from its simulations. Thus, the scenarios are concentrating on the predictive service. The reliability of the predictive service depends on the measurements of the network load. This can be tested in one node only. Having a more realistic scenario, we used a network with a number of connections.Table 6 shows the network parameters. Each node has the same performance specification for the link speed, the propagation delay and the scheduler.

5.1 Scenario 1: Worst case considerations

For a first experiment consider the following scenario: The load estimation of the predictive service is observed in a particular node. When the load decreases the estimation decreases, too. At this time a peak load request is initiated. The bandwidth allocated to the peak load is the remainder of the bandwidth not allocated by other connections.

Packet losses caused by overloading a node are of special interest for the validation of the predictive service. The peak load was established at a lower priority than the active connections. Thus, if packet losses occur then only packets of the peak load are lost. The simulations have been performed for different proportional factors of the regulator, while the PI regulator of the adaption have been driven with constant parameters. The results are compared with the outcomes of the original CSZ scheme.

Parameter	**Value**
Link speed of each node	40 *MBit/s*
Propagation delay	5 μs
Queues for deterministic services	20
Predictive Service Classes	5
Proportional Factor for the PI-Regulator	1
Integrating Factor for the PI-Regulator	0,9
Measurement Interval	0,1 *s*
Duration of the Simulation	20 *s*

Table 6: Simulation parameters

Table 7 shows the existing load at the time the peak load is established. The existing video connections of the predictive service have been established every 0.15 seconds starting from 0.2 seconds. Thus, the current load before establishing the peak load is about 27.5 MBit/s. After 9 seconds the peak load was established when a minimum usage of the predictive service was observed. The delay bound of the peak load is 1.15 seconds so the connection is certainly established.

Quantity	Type	Class	$\frac{Smax}{Xmin}\left[\frac{kBit}{s}\right]$	$\frac{Smax}{Xave}\left[\frac{kBit}{s}\right]$
10	Video guaranteed	-	1600	416
12	Video Predictive	5	1600	1600
3	Audio Predictive	4	128,64	128,64

Table 7: Basic bandwidth usage at the observed node

Figure 8 shows the resulting packet losses and bandwidth allocation of the worst case scenario when using the adaptive PI regulator for the bandwidth estimation. 15% over-booking is possible without any considerable packet loss (<5%), in the case that the regulator is adjusted more conservative. An increase of the allocated bandwidth results in an increase of the packet loss rate.

Figure 9 shows the results of the same experiment using the CSZ measurement model. Both curves are similar to the results of the adaptive PI-Regulator. The results of this experiment are used to derive optimal parameter adjustments for the next scenario.

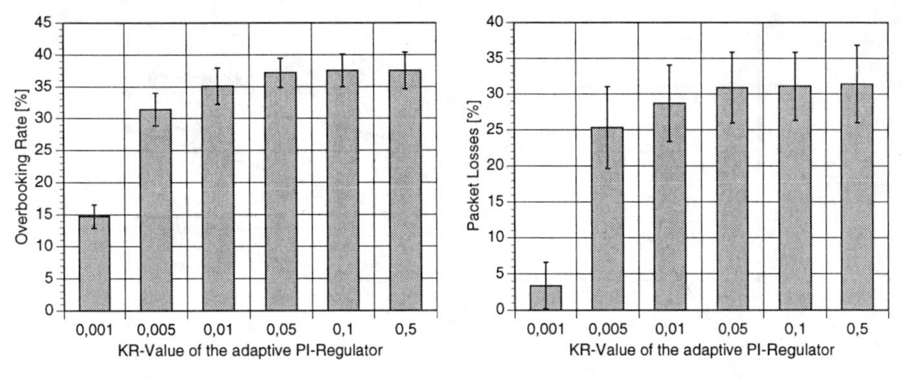

Figure 8: Overbooking and packet losses when using the PI regulator

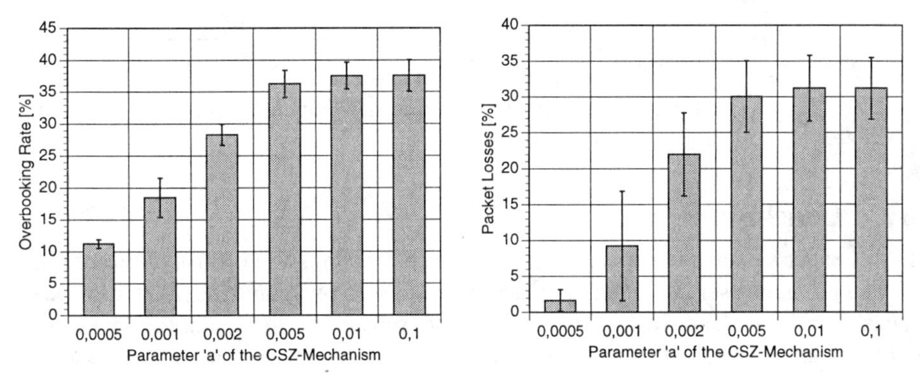

Figure 9: Overbooking and packet losses when using the CSZ scheme

5.2 Scenario 2: Acceptance Speed of new Connections

A second scenario considers the speed at which new connections can be accepted. Starting with an idle node, 100 video connections requests for acceptance for the predictive service. The inter arrival times between two single connection requests are varied from 10 up to 600 milliseconds. The parameter adjustment for both measurement algorithms was derived from the previous scenario, since they provide the best ratio between overbooking rate and packet losses. The K_R value 0,005 (0,001) of the PI regulator and the parameter a=0,001 (0,0005) of the CSZ scheme lead to comparable results, which are depicted in Figure 10.

In general, the adaptive PI-regulator accepts new connections more quickly than the CSZ scheme, since the adjustment of the load estimation follows the real load more precise. Packet losses due to buffer overflows were under 1% for both mechanisms, so that we refrain from depicting these results.

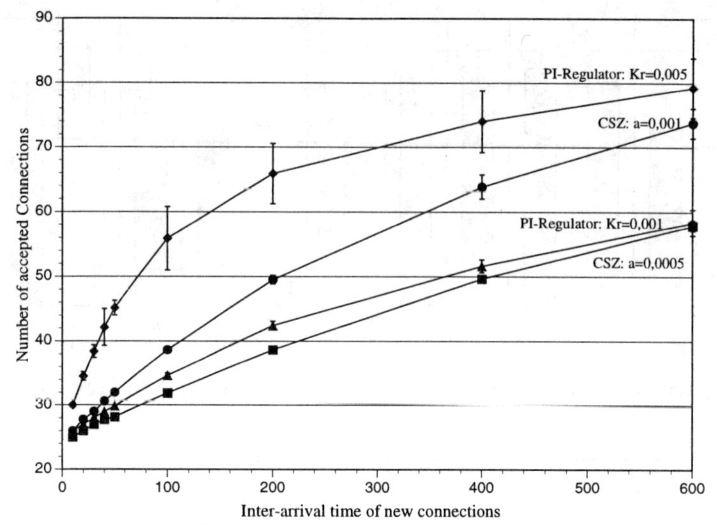

Figure 10: Acceptance speed of new connections

6 Conclusions

In this paper we have introduced a new approach for admission control of deterministic and predictive services. Our approach is intended to overcome the major disadvantages of the Tenet and CSZ-schemes. The Tenet scheme assumes, that the application provides an exact specification of a source's load behaviour. Based on this specification it is possible to provide a deterministic guarantee without any packet losses and delay bound violations. By using a maximum average allocation to estimate the necessary bandwidth, the allocated bandwidth is lower as for a peak rate allocation, while still providing absolute guarantees. With the statistical service it is further possible to calculate exact loss probabilities in case of a channel overbooking. The major disadvantage of the statistical service is the high effort to calculate the packet loss probabilities.

The predictive service, which was introduced in the CSZ scheme for the first time, is able to provide a channel overbooking without the need of an exact source specification. However, the CSZ scheme does not provide mechanisms for the end-to-end establishment of real-time channels. Furthermore, the applied scheduling discipline is not sufficient to provide optimal local delay bounds at each router.

Our approach provides a combination of the maximum average allocation of the Tenet scheme and the predictive service. The predictive service was integrated into an end-to-end negotiation scheme and was adapted to the RCSP-scheduler. Moreover, we have introduced a new measurement mechanism to estimate the bandwidth usage of active channels.

With this approach we are able to provide deterministic guarantees, if an exact traffic specification is available. If no precise traffic value specification is possible (e.g. in the case of live sources), the predictive service provides guarantees as long as the behaviour of the active channels is constant. Furthermore, our simulations have shown that 15% overbooking of a node is possible without any considerable packet losses.

The new bandwidth usage measurement scheme leads to similar results like the CSZ scheme, but the time between two single connection request can be shorter, due to a more precise estimation of the actual network load.

References

1. A. Banerja, D. Ferrari, B.A. Mah, M. Moran, D. Verma, H. Zhang: The Tenet Real-Time Suite: Design, Implementation and Experiences; The Tenet Group, University of California at Berkeley, Technical Report TR-94-059, November 1994.
2. R. Braden, D. Clark, S. Shenker: Integrated Services in the Internet Architecture: An Overview; RFC 1633, 1994.
3. W. Büttner: Digital Control Systems; (In German), Vieweg Publisher, ISBN 3-528-13041-5, 1992.
4. L. Delgrossi (ed.): Internet STream Protocol Version 2 (ST2), Protocol Specification, Version ST2+; Internet-Draft ST Working Group, November 1994.
5. D. Ferrari: Distributed Delay Jitter Control in Packet Switching Networks; Technical Report TR-91-056, International Computer Science Institute, Berkeley, 1991.
6. D. Ferrari: A New Admission Control Method for Real-time Communication in an Internetwork; in S. Son (ed.), Advances in Real-Time Systems, Ch. 5, pp. 105-116, Prentice-Hall, Englewood Cliffs, NJ, 1994.
7. D. Heyman, A. Tabatabai, T.V. Lakshman: Statistical Analysis and Simulation Study of Video Teleconference Traffic in ATM Networks; in IEEE Global Telecommunications Conference, GLOBECOM`91 Phoenix, Arizona, December 1991.
8. M. Hamdi, P. Rolin, Y. Duboc, M. Ferry: Resource Requirements for VBR MPEG Traffic in Interactive Applications; in D. Hutchison, A. Danthine, H. Leopold, G. Coulsen (Editors): Multimedia Transport and Teleservices, Springer Publishers, Lecture Notes in Computer Sciences No. 88, 1994.
9. S. Jamin, S. Shenker, L. Zhang, D. Clark: An Admission Control Algorithm for Predictive Real-Time Service (Extended Abstract); in P. V. Rangan (ed.) Network and Operating System Support for Digital Audio and Video, Lecture Notes in Computer Science No. 172, Springer-Verlag, 1992.
10. Kalmanek, Morgan, Restick: A high performance queueing engine for ATM networks; in Proc. of 14th International Switching Symposium, Oct. 1992.
11. R. Karabek: Audio-/Video-Real-Time Communication in Packet Switched Networks; Diploma Thesis (in German), Technical University of Aachen, Lehrstuhl für Informatik IV, July 1993.

12. E. W. Knightly, H. Zhang: Traffic Characterization and Switch Utilization using a Deterministic Bounding Interval Dependent Traffic Model; in Proc. of IEEE INFOCOM '95, April 2-6, 1995 Boston, USA, 1995.
13. S. Neuhauser: QoS-driven Routing in Multimedia Communication Network; in Fachgruppe Informatik, Technical University of Aachen, (editor): Graduiertenkolleg Informatik und Technik, Aachener Informatik Berichte (1993), Nr. 7, pp. 189-215, Aachen, Germany, 1993.
14. A. Parekh, R. Gallager: A Generalized Processor Sharing Approach to Flow Control - The Single Node Case; Lab. for Information and Decision System, MIT Technical Report LIDS-TR-2040, 1991.
15. C. J. Parris, G. Ventre, H. Zhang: Graceful Adaption of Guaranteed Performance Service Connections; in Proc. of IEEE GlobeCom'93, Houston, Texas, USA, November 1993.
16. D.C. Verma, H. Zhang, D. Ferrari: Delay Jitter Control for Real-Time Communication in a Packet Switching Network; Proceedings TriComm'91.
17. H. Zhang: Service disciplines for integrated services packet-switching networks; PhD Dissertation, University of California at Berkeley, November 1993.
18. L. Zhang, S. Deering, D. Estin, S. Shenker, D. Zappalla: RSVP: A New Resource Reservation Protocol; in IEEE Networks, No. 5, 1993.
19. H. Zhang, D. Ferrari: Rate-Controlled Static Priority Queueing; in Proceedings of IEEE InfoCom `93 San Francisco, March 1993.
20. H. Zhang, D. Ferrari: Rate-Controlled Service Disciplines; in Journal of High-Speed Network, Feb. 1994.

Equitable Conditional Access and Copyright Protection for Image Based on Trusted Third Parties

J.-M. Boucqueau[1], S. Lacroix, B. Macq[1], J.-J. Quisquater

Laboratoire de Télécommunications, Université catholique de Louvain,
Place du Levant 2, B–1348 Louvain-la-Neuve (Belgium)
E-mail: {Boucqueau, Macq}@tele.ucl.ac.be

Abstract. Image communication needs new cryptological tools for conditional access, copyright protection and image authentication. The aim of this paper is to overview the corresponding systems' features. The design of efficient copyright protection by watermarking images and image authentication by signatures is also briefly discussed. It is shown that equitable systems need the use of a Trusted Third Party, and a possible solution is presented. The implementation of this new function in the multimedia world strongly differs from that in the Electronic Data Interchange environment where it is already well known. This difference will also be detailed.

1 General Introduction

Pay-TV services exist since a lot of years. Today, the Information Highways and the digitalization advent broad considerably the multimedia horizons with new types of services, but in the same time sweeps new types of problems along with. The digitalization and the images/services quality and variety requirements imply the data flow compression. The broadcasting of Pay-services implies the scrambling of the data-flow . These issues will briefly be addressed in section 2.

The large amount of participants and the variety of technical implementations joint to the fact that most of the trade will be held remotely arise non-negligible security and interoperability issues (in the scope of this paper, interoperability means either compatibility with existing systems or compatibility between different techniques). Security encompasses the protection of information as well as the protection of the participants. Consequently a strong Access Control system is necessary for ensuring aspects like privacy, protection against non-desired services enforcing, billing or non-repudiation. Moreover, most of the contracts between the participants have to be acknowledged remotely. Unlike usual trades, it is impossible to compare signatures or to see the interlocutor and judge if he is the right one. Efficient and dynamic management of contracts will require an additional trusted entity: the Trusted Third Party (TTP). Section 3 addresses this issue.

Copyright protection and images authentication are also part of the issues emphased by the informations highways development. The success of a service is depending on its content. The *Copyright Owner* provides a part of this content. The delivery system has to provide him reliable guarantees of his property's use. We will briefly discuss this topic in sections 4 and 5, presenting the techniques and a way to manage these. Section 6 describe a way to manage the previously described Conditional Access and Copyright Protection systems. A whole section, 7, will be devoted to the main introduced tool, the Trusted Third Party.

2 Scrambling digital images

Digital images are very correlated processes. Compression of digital TV is essentially based on decorrelative coding. A direct scrambling of the images before compression would suffer from two major drawbacks:

- The output of an encryption algorithm seems to be random, making any further compression impossible.
- The correlation of the plaintext can be used for efficient attacks: even if no messages of the plaintext are known, the encryption algorithm could be inverted by maximizing the correlation of the estimated plaintext.

These two factors require that scrambling be performed after compression and before channel coding.

Applying a classical encryption method after a compression coding of the image seems attractive since the output of the coding is more or less random and already encoded at the required bit rate. This method is however not satisfactory for at least two reasons:

- A user could intend to protect his images independently from the nature of the transmission channel, i.e. independently from the compression algorithm in use in this channel. More precisely, if a scrambled MPEG-2 multiplex has to be *transcoded* (e.g. satellite contribution to the head-end of a CATV network at 10 Mbit/s and then, transcoding to 3 Mbit/s for distribution), the knowledge of the scrambling key has to be given at the transcoding points in order to decrypt, decompress, and re-compress at lower bit rate and re-encrypt. The classical approach is, therefore, not very secure for *transcodability* (the transcoder must be able to decrypt).
- In many applications, the encryption has to be somewhat transparent (we will call *transparency* the property, such that $E_K(I)$, the encrypted image, "looks like" I):
  - A broadcaster of pay-TV does not always intend to prevent unauthorized receivers from receiving his program, but rather intends to promote a contract with non-paying watchers.
  - The access to the icons of a secret image bank could also remain unprotected.

  Classical encryption of MPEG-2 bit streams gives no transparency.

Two direct approaches may solve these problems (and also the possible attacks through synchronized and end-of-block words). The first approach is to envisage multi-resolution encryption before compression. The aim is to produce an encrypted image which is still compressible (the scrambling produces a ciphered correlated image). In [5], it was proposed to do this by using a lossless multi-resolution transform and by permuting details of multi-resolution transform coefficients. However, such permutations should be performed in order to obtain a cipher-image resistant enough against correlation attacks.

A second way of encryption could be obtained by permutations of Huffman words, letting low-frequency coefficients be unchanged for transparency. In this case, Huffman decoding and changing in the quantification steps for transcoding is possible. This approach is straightforward and the more or less decorrelative properties of the DCT make a correlation attack not very easy.

These remarks show the interest in developing new approaches, in which the source coding is developed in combination not only with channel coding but also with cryptographic coding. *Combined source-cryptographic coding*, including features as transcodability and transparency which are efficient and resistant against attacks, remains an open issue.

In any case, the problem of transcodability exists in any real-life network. A solution is to use secure smart cards (see [2] for a complete description of smart cards. This paper also gives interesting technical information about the use of smart cards for analog access control systems for pay-TV) to store and use the secret keys at the transcoding nodes together with a unique IC (or a secure box) to perform the transcoding.

3 Conditional access for pay-TV

The solutions proposed to date as the basis for the design of pay-TV systems by standardization organizations and by all proprietary schemes (although the terminology differs sometimes) consider this problem as an *access control* problem. Let us briefly recall here that, in its general or abstract form, the access control problem implies the presence of:

- *service providers*: this concept has quite a general meaning and can refer both to the proprietor of a theater or to the person in charge of the protected area in a nuclear plant;
- *users*: people wishing, at a given time, to be granted access to some protected place, goods or service.

In the pay-TV systems discussed here, one usually observes the presence of a third type of actor. In addition to service providers (SPv) and users (U), the system also comprises an *issuer* (I). The issuer is seen as the general organizer of the services. An essential part of his role will be to introduce new operators or new services to the system and to establish the initial links between operators and users. We shall discuss later the role of the issuer in the system security. Basically, the role of the issuer should be limited to a bare minimum and, in the philosophy of the system designers, most of the operations will directly take place between operator and user[1]. The introduction of an issuer, i.e. of a supreme system authority, is certainly a fundamental option having specific consequences both for the security and for the cost of the system.

The usual solution that appears in access control problems is to interpose, between the user and its target, an *entitlement*, also called *access title*, that is nothing other than a customized authorization, as for example, the ticket giving access to some numbered seat in a concert hall. This manner of proceeding, furthermore, offers the service provider additional freedom of time and space. The entitlement clearly implements a link between the service provider and the user. A basic aspect of the system security will consist in checking this two sided link: "the entitlement indeed originates from the operator" and "the user fulfills the required conditions".

At this preliminary stage of our discussion, we can already make a second important observation about entitlements: *the validity of an entitlement should always be limited in time.* In this case, if the user stops fulfilling the access conditions, he simply will not receive his new entitlement. This *absence* of action is clearly more robust, from the security point of view, than the *active* attitude that consists in invalidating the entitlement. Such an action can indeed always be "intercepted" or "filtered out".

A third characteristic of the entitlement is specific to the pay-TV environment. The simplicity of the system management implies the possibility of granting, renewing or modifying the entitlements of each specific user by directly addressing over the transmission channel itself (over the air addressing).

A system offering all the above characteristics would be called *conditional access system* or *access control system* (ACS).

The definition of a conditional access system generally contains no assumption of its implementation. It is easy, however, to perceive that the system designers were quite aware of the available technologies and, more precisely, that they had in mind the implementation of the security processor by a smart card.

As a matter of fact, the discussion of ACS has led us to consider the use of a "plug-in" security processor[2] delivered to the users. Besides the arguments already mentioned to support this design choice, let us also mention the *multi-faceted* nature of the application:

- many competing system managers may issue security processors of the same type;
- distinct, competing service providers can use the same security processors.

The ACUs are shared between several entities, different service providers and a user. These entities have to trust its design conformity. The ACUs are build following strict rules. There must then be checked from the hardware as the software points of view. This checking is followed by a certificate and a unique

[1] The issuer role could, for example, be played by the proprietor of the signal transmission channel (cable, satellite channel, ...).

[2] Or, more generally, of a removable unit including the security processor.

identification number insertion. This operation is achieved by a *Certification Authority* (CA). The checking of the ACU certification and validity are achieved by the *Trusted Third Party* and will be discussed later.

4 Copyright protection by digital image stamping

The issue of copyright protection of digital broadcasted sources is being studied more and more. As an example, the European project CITED of the CEC ESPRIT project line has developed a complete model for the management of the user rights for interactive multimedia programs, where software programs are checking and marking the users actions. Here we will restrict ourselves to the problem of marking a digital image with the copyrights. The goal is to give the author (or the owner) of a digital image the possibility to technically *attest* to the origin of the image, i.e. to attest that he is the owner of the image or that some copied image is a fake. Such a similar approach, injecting radioactive particles into the canvas, is said to be used to protect great painters' works. The canvas is then still accessible but permanently stamped.

It is clear that an *electronic stamp* must be *holographically inlaid* over all the picture, being

- indelible by a hacker,
- perceptually invisible, i.e. it is a *watermark*,
- statistically invisible, i.e., it cannot be deleted by statistical analysis (like Kalman filtering) and the stamped picture is similarly compressible as the original one,
- fully resistant to any additional noise (compression, transmission, ...).

This problem is studied in Japan (e.g. see [6] and [4]) where some preliminary solutions are proposed.

This topic is also dealt with by the authors.

One of our approaches to the stamping issue may be formulated as the following description:

Let I be the original image and ϵ be the stamp. We define the stamp procedure as $\mathcal{S}$, such that

1. Q is a procedure which extracts essential characteristics of I.
2. $\mathcal{S}(Q(I)) \mapsto \epsilon$.
3. $I \oplus \epsilon$ is the stamped image.
4. $C_{\mathcal{S}}(Q(I \oplus \epsilon)) \mapsto$ stamped by $\mathcal{S}$ = "YES".

The $\mathcal{S}$ procedure is secret and only known by the owner of the image. The property number 4. allows an owner to prove that he is the image owner. The two further properties are

5. $I \oplus \epsilon$ is perceptually identical to I.

6. For small enough η, (typically, compression distortions), $C_{\mathcal{S}}(\mathcal{Q}(I \oplus \epsilon + \eta)) \mapsto$ stamped by $\mathcal{S}$ = "YES". In this case, the stamp is resistant to attacks or to noise.

The $\mathcal{Q}$ procedure has been derived from psychophysics. It is known that slight variations of the pixels around contours are not noticeable and this effect is denoted in the literature as *masking* [8]. The LSBs of the pixels around a contour can therefore be considered as a data transmission channel available for the transmission of a stamp. $\mathcal{Q}$ is therefore a contour detector. Taking into account the fact that the picture will be compressed and decompressed perhaps many times, this associated channel is very noisy and has to be strongly protected against transmission errors. Furthermore ϵ must be hidden. This has led us to the use of a pseudo-random generator for the $\mathcal{S}$ procedure. Procedure $C_{\mathcal{S}}$ is simply a correlation procedure, verifying whether the image has been stamped with the pseudo-random sequence generated by $\mathcal{S}$ and hidden around contours detected by $\mathcal{Q}$.

5 Authentication of pictures

As mentioned in the introduction, digitalization of images makes a lot of post-processing possible [7]. Image manipulations can be used for many illegal purposes. This has led Friedman [1] to propose a scheme for image authentication by a certified signature.

A stamp is not a signature since its rules are totally different. The stamp aims to protect the author while the signature aims to protect the receiver. For example, a receiver would not have any advantage in deleting the signature. Actually, the properties of a stamp and a signature are exactly symmetrical. The signature is the header, function of the image, that cannot be produced by anyone other than the sender. At the opposite end, the stamp is the inlaid information that cannot be deleted by anyone else. Of course, a cryptographic system can use both by inlaying the signature, for example.

As opposed to a handwritten signature, the form of a digital signature must depend upon the data to be signed and the signer.

6 Common Functional Model

The way we follow to present the Conditional Access problematic and the Copyright Protection solution paths lead naturally to notion of *Open system*. This means the ACU is shared between different, equally important, service providers and used by different equal users. Here follows a brief description of a *Common Functional Model* supporting this notion.

6.1 The Basic Multimedia Chain

The following list describes the basic functions that are necessary to establish the flow of information from the creator to the consumer.

- The function of copyright owner (CO) is played by creative people involved in the construction of services. It is the most upstream job in the multimedia chain and consists in its artistic contribution.
  Examples: a photograph, an actor, a programmer, a singer, a journalist.
- The function of *service producer* (SPd) is played by an assembler of media objects. Most of the time he is the first maker of multimedia services able to charm non professional users.
  Examples: a TV station, a radio station, a database owner, a film producer.
- The function of *service provider* (SPv) consists in the multiplexing of several services into what we call a *bouquet.* In its simplest version, a bouquet can be a single service. Bouquets can be very attractive and can act as an added value to the services. So, the SPv is an integrator of services.
  Examples: a TV station, a cable operator, a satellite operator, a national PTT.
- The function of *carrier* (CR) deals with the network management and consists in fulfilling the infrastructure capacity with bouquets and to assure the feeding in services to the subscribers. We assume that a CR does not look inside a service or a bouquet and does not change anything inside the bitstream he has to transmit.
  Examples: a TV station, a cable operator, a satellite operator, a national PTT.
- The function of *user* (U) is played by common people interested in watching TV or accessing databases or any kind of services. He receives these services by a connection to the cable, the switched network or through a satellite channel. He access the services through his terminal and the entitlements contained in his ACU.

The five functions described here form what we call the *basic multimedia chain.* Neither access control nor copyright protection features are included; this requires the definition of additional functions.

6.2 Conditional Access Necessities

As this is an *Open* system, the service provider is interested in managing the entitlements for his service either himself or by some representative. This leads to the following definition:

- The function of *access manager* (AMa) consists in the marketing and the management of the entitlements for some bouquets. So, he takes the billing and the initialization of the ACUs in charge.
  Examples: a TV station, a cable operator, a satellite operator, a national PTT, a service provider.

The requirement of managing the entitlements as well as the introduction of new services over the air requires a secure communication between the user and the service provider. From this we deduce another function:

- The function of *trusted third party* (TTP) is played by a fair authority whose role consists, among others, in updating a list of certified and valid ACU's.
  Examples: the public authority, the FCC in the USA.
  Remark: It is not realistic to have one world-wide (or even Europe-wide) TTP. There will probably be at least one TTP for each country, and even in that case problems may occur for time-critical actions, e.g. when there is not enough time to ask a validation from the TTP. In order to reduce the amount of requests to a TTP, it might be useful to have "mirror sites" where an exact (and verified) copy of the list of ACUs is located. The mirror must be updated each time a new ACU is introduced or an ACU is blacklisted.

Because the ACU will likely be a *multi-service ACU*, the service providers have to use ACUs that were not necessarily chosen by themselves. However, a service provider needs a confirmation that a user's ACU is valid. This requires that

1. each new ACU is certified by some trusted authority, and
2. each certified ACU needs a unique identification number.

That is why we deduce one more functions:

- The function of the *Certification Authority* (CA), previously introduced, consists in the certification and numbering of newly issued ACUs. The certification is the same for each ACU, the numbering makes each ACU unique.

6.3 Copyright Protection Necessities

From the copyright protection requirements, we deduce five more functions:

- The *Agent* (Ag) manages the copyright owner's business. He has a contract with one (or several) copyright owners which allows him to act in his name. Thus, he is the central function in the copyright protection part.
- The *Watermarker* (Wm) encrusts in each image the copyright information, according to the watermarking algorithm that has been chosen by the copyright owner or his agent.
  The watermarker has only contacts with those agents he works for.
- The *Registrar* (Rg) manages a database containing information about image marks, owners and beneficiaries. He is responsible for the registration of all sold images.
- The *Controller* (Ct) keeps the tools which are used to detect the watermarks. He is able to retrieve the signature from each image and has control over the broadcast images at any time in order to detect possible misuse. We could

split this function into two parts by separating an additional function called the *retriever*.
The controller and the registrar form the central security instance in the copyright protection CFM. In fact, we could replace these two functions by the TTP. However, in order to underline the different tasks, we have chosen to keep them separate (nevertheless, in real life the controller and the registrar can be the same person since different functions is not necessarily identical to different persons).

- The *Pirate* (Pir) makes illegal copies of images he receives and distributes them (possibly after editing the image in order to hide its origin). The role of the pirate can be played by any other function of the multimedia chain, even by the copyright owner himself, whose identity might be faked.
The pirate acts in an invisible way until the day he is discovered. From that moment, some particular actions should be taken by the other actors of the scenery.

6.4 Miscellaneous

The functions defined in the previous sections allow us to describe all actions that take part in access control and copyright protection. However, there are some actions that are performed during a system session which belong to neither of these categories. Although most of these actions (e.g. audience metering) are not further discussed in this CFM, we would like to complete the list of functions appropriately.

- The function of *terminal provider* (TP) deals with the management of a stock of terminals. The terminal provider can sell and/or rent terminals to subscribers. He can also provide the terminal with a service.
Examples: an infrastructure provider, a national PTT, a terminal constructor, a service provider, an investment company.
- The function of *audience meterer* (AM) consists in the evaluation of the audience of bouquets or services. This kind of business exists today and will continue to exist. With the arrival of the digital technology, some facilities will raise. This can be a tool for evaluating the copyright fee and redistribute them.
To rule the game we need a kind of legal authority or, more precisely:
- The function of *regulatory authority* (RA) is out of the business scenery, but groups together a lot of *trusted* organizations implicated in the application of the law in several fields. We attempt to list exhaustively their jobs:
  - to update and certify a list of intellectual property rights related to their owners;
  - to collect and distribute royalties;
  - to play the role of judge in case of intellectual property violation;
  - to collect the license fee for TV or radio;
  - to authorize (or prohibit) a service (censure);
  - to certify terminals or other related devices (VCR for instance).

Examples: the public authority, the Sabam in Belgium, the Cisac in Europe, the FCC in the USA.

6.5 The interactions between the functions

From the preceding description we can attempt a scheme of the connections between the functions. For reasons of clarity, we provide a separate diagram for the copyright protection part (figs. 1, 2).

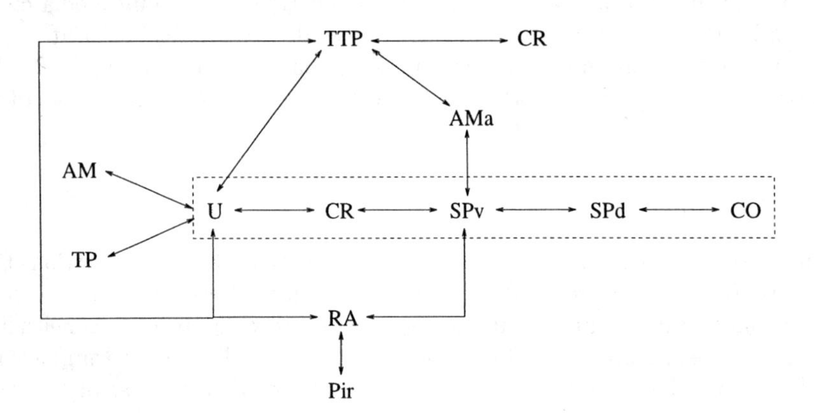

Fig. 1. Functions concerned with access control

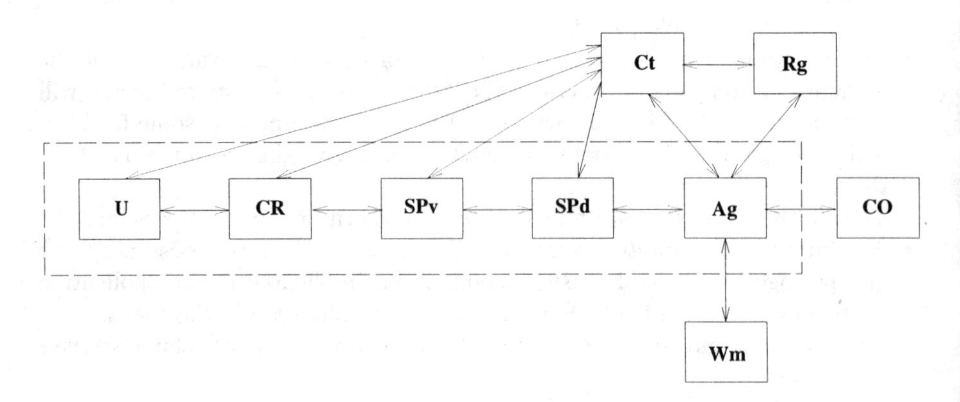

Fig. 2. Functions concerned with copyright protection

7 Conditional Access implementation through Trusted Third Parties

7.1 Access Control Unit

The ACU is the security module aimed to manage and store the entitlements, the contract between SPv and U. It can be embedded in the decoder, but in most cases, it is detachable (e.g. smart-cards technology, see below). The ACU is characterized by a unique serial number, an identity linked to this serial number and a couple of ciphering/deciphering public cryptography key (E_N-D_N). It can also contain some information about the U profile, including his preferences, ability to pay and in some case credit card information.
In some case, like parental blocking, the use can be restricted by mean of a PIN code. This code should be asked and verified, if necessary, by the host verifier. This restriction is associated with the ACU, since it is portable and usable in any host, protected or not by a blocking. It concerns either entitlement already inserted in the card and entitlements which could be bought later if they are related, e.g., to the level of a film. The ACU is bought by the end-U.
In order to have the system as open as possible, it is necessary to minimize the number of Access Control Units (one per module is the goal). For interoperability purpose, these ACUs are implemented by a multiple usage component: the smart card. Some advantages of those cards are: low price, portability, easy entitlements' management, easy IPR management and multiple functions, like billing (in further extensions they could also act in the billing process). A main advantage is the high level of security they provide. They allow to store and run the cryptographic algorithms safely. It is necessary to incorporate those cards in the multimedia system and it should be possible to share a same card between several SPvs. To do this in a secure manner, an additional entity, the Trusted Third Party (TTP), is required in order to manage such a registration in the system. Those entities are necessary in order to insure the security for anyone who manages information in the card.

7.2 The TTPs

What: A person or a society which would like to obtain the TTP status would have to fulfill at least the following conditions before being allowed to play an active role in the system.
First, it must have a secure device. This device will have to be able to resist to breakdowns as to attacks. For the breakdowns resistance, it must be a reliable device having, among others, an autonomous emergency supply and the possibility to be easily replaced by an other TTP. The attacks can be achieved inside as outside the system. In the first case, filters and other elements like this must be used to check the received data before treating them. In the second case, the device(s) will have to be placed inside a securised building.
The TTP must be connected to a huge network containing, among others, service providers, users and the other TTPs as we will see later.

The TTP must have a huge computation capacity to compute simultaneously a lot of cryptological functions. It must be completed by a huge addressing capacity and an ability to manage large databases.
Finally, if all these conditions are fulfilled, the potential TTP must be introduced in the system and must obtain its own identity. This means, in a first time, to be accepted by the entity it wants to be part of. It must accept the area own rules and obtain the area legal TTPs status. Then, its last step will be to obtain clients...

Who: To decide which one's are allowed to play the TTP function is not part of the kernel design. We are just able to establish a non exhaustive list. The final decision is the area leaders responsibility. The Government could be the TTP. It could also appoint an important organization more or less linked to the Government like *France Telecom* or the *CCETT* in France. It could also be any person or society responding to the previously defined conditions.
Without being necessarily, competition between TTP would improve the system. Its minimal set of functionalities will be defined. This lets a lot of other possibilities which could be used to complete or to improve the client service, so that some TTP becomes more attractive then others for potential clients.

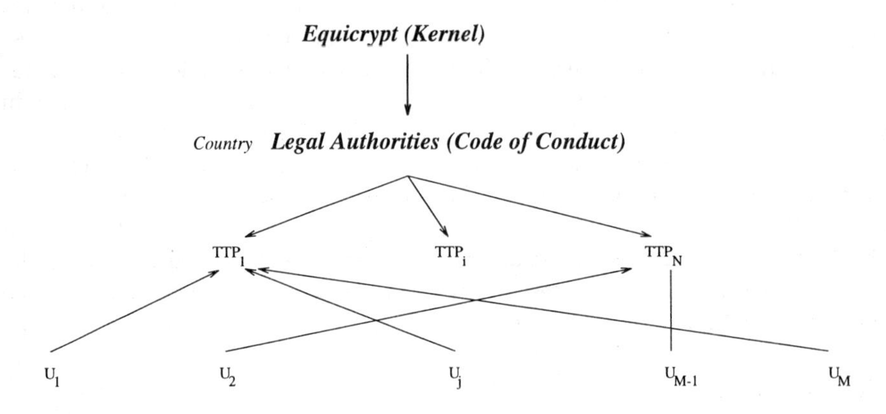

Fig. 3. TTP Competition

TTP functionalities: The TTP functionalities could be divided into four parts. The first one concerns the certification and its control. the second one deals with the entitlements distribution problematic. The third one is optional. The TTP could be responsible for the authors rights management. Finally, the TTP could play a role in the accompanying measures like the audience metering or the billing. In the scope of this paper, we will limit us to the two first aspects:

- Their actions are limited to the Entitlements Management Messages and they act in the introduction of an external component in the system, a main task resides in the certification. There need some entities able to blacklist the non-valid components and to manage their removal from the system. This task can not be done by an isolated SPv. The issue is very similar to the problem of the credit card, but in the audio-visual world, transactions must be held remotely. It is consequently necessary to replace the physical link which exists in usual trade (e.g. the vendor can examine your signature on your Visa card to be sure of your identity).

- The TTPs approve any request (opening of a service, ...) between a U and a SPv which are faraway from one another and ensure a secure transaction for both parties. TTPs are aimed to work in network, in the same way as the bank networks. They can also be seen as the Kerberos entities. Any safe transaction is impossible to imagine without this kind of entity, it is the same in the audio-visual world.
- They play an important role in the INTEROPERABILITY event, allowing any customer from a country to request a service to any SPv in another country. More precisely, if a trusted link exists between the TTPs, this kind of organization already exists (see CCITT X.509 standard for network), all transactions will be safe.
- In addition, consequently to the different existing laws in different countries, some cryptographic algorithms can be allowed in some places and forbidden in other sites. The TTPs could be the intermediaries entity, necessary to process the transformations on the encrypted data in order to remain in accord with the law and to insure the security all over the transition.
- They can be the responsible security administrator to perform the suitable reaction in case of security alarming. (e.g. repeated wrong authentications, attempts to get access or to perform actions without authorization,...).
- Since the TTPs act in network, they can be in competition and if a SPv does not make confidence any more in a TTP, it can simply refuse all certification issued from this TTP. This case is to be envisaged, but should be extremely rare since the TTPs just complete a protocol between two entities in order to close the security loop between them. Consequently, the total amount of secret shares by two distinct functions is minimized and the access control is more efficient. By this way, anybody, including the TTP, is never able to get enough information to simulate an other one. If a really unfair TTP is detected, it shall always be possible to retrieve it from the network, i.e. there should always exist different possible ways between any couple of participants.

  Each participant will be able to choose how many TTP it wants or accepts. This number will, among other, depend on the U and SPv quantity in this area, so that the function stays financially viable. The TTP revenues will depend on the number of achieved functions.

  All the TTPs will form a cellular network[3]. It will be a plane structure introducing no systematic hierarchy between TTP. The network must be so that it is always possible to establish a secure communication between any

[3] Like the GSM in mobile communications.

TTP couple. This will be a "way" using a certain TTP number. This way must not be unique. So, the system will not be in trouble if a TTP is unfair or unusable.

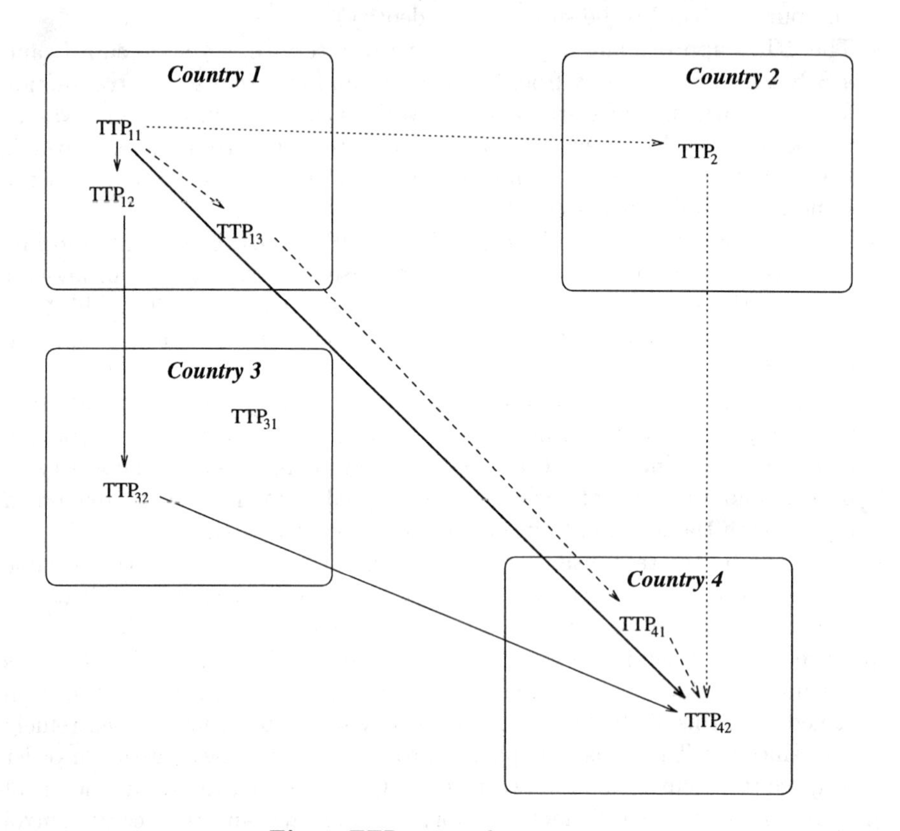

Fig. 4. TTPs network organization

- More than providing a link between SPv and U, the TTP's:
  - avoid a direct physical contact between U's ACU and SPv when a new service key is required in the ACU. This operation would need a secure local, or at least a secure channel network between each U and SPv.
  - allow compatibility between a TTP's secure system and a proprietary system.
  - allow the sharing of a unique ACU by several SPv's.
  - protect the ACU from "heretic" writing by pirates.
  - protect the service keys from piracy.
  - allow the use of pseudonyms in order to protect the U's privacy.

If those requirements are fulfilled, authenticity and conformity of the ACU are assured and any SPv can be sure of an ACU [1].

8 Conclusions

The Conditional Access as the Copyright Protection has to be provided in an Open environment. This paper gives one way to design such an open system which, furthermore, should be Equitable and Interoperable. This implies the acceptance of different concepts like the *Access Control Unit*. The most unusual concept is that of the *Trusted Third Party*. The integration of this function in the future Multimedia chain allows us to develop an Open, Equitable and Interoperable system, compatible with the proprietary ones.

References

1. G. L. Friedman: The Trustworthy Digital Camera: Restoring Credibility to the Photographic Image, IEEE Trans. on Consumer Electronics, vol. 39, no.4, 1993, pp. 905-910.
2. L. C. Guillou, M. Ugon and J.-J. Quisquater, *The Smart Card: A Standardized Security Device Dedicated to Public Cryptology*, *in* Gus J. Simmons (Editor) Contemporary cryptology. The science of information integrity, IEEE Press, 1992, pp. 561–613.
3. M. E. Hellman: An extension of the Shannon Theory Approach to Cryptography, IEEE Trans. on Information Theory, vol 23, no 3, May 1977, pp. 289-294.
4. N. Koike, T. Matsumoto and H. Imai: A Scheme for Copyright Protection of Digital Images, SCIS93 (The 1993 Symp. on Cryptography and Information Security, Shuzenji, Japan, January 1993, paper 13C (in Japanese).
5. B. Macq and J.-J. Quisquater: Digital Images Multiresolution Encryption, The Journal of the Interactive Multimedia Association Intellectual Property Project, vol. 1, no 1, pp. 179-186, January 1994.
6. K. Matsui and K. Tanaka: How to Secretly Embed a Signature in a Picture, The Journal of the Interactive Multimedia Association Intellectual Property Project, vol. 1, no 1, pp. 187-206, January 1994.
7. W., J. Mitchell: When Is Seeing Believing ?, Scientific American, February 1994, pp. 44-49.
8. A.N. Netravali and B. Prasada, Adaptive quantization of picture signals using spatial masking, Proc. of IEEE, vol. 65, 1977, pp.536-548 1990, pp. 387–394.
9. C. E. Shannon: Communication Theory of Secret Systems, Bell System Technical Journal, vol 28, no 4, 1949, pp. 656-715.

Designing Point-to-Point Interactive Video Applications

Maher Hamdi Hossam Afifi Pierre Rolin Luis M. Rojas

{hamdi,afifi,rolin,rojas}@rennes.enst-bretagne.fr
Télécom Bretagne
BP 78, 35512 Cesson Sévigné, France

***Abstract*:** *High speed networks provide sufficient bandwidth for VBR video applications. Interactive point to point video such as tele-teaching needs a reduced gap between the coding and decoding procedures in software based tools. Efficient algorithms for video coding do not satisfy this requirement.*

This paper presents a video compression algorithm based on multi dimensional Discrete Cosine Transform and particularly suited to interactive video. This algorithm offers stable and equivalent computation delay for both coding and decoding procedures along with a satisfactory network bandwidth consumption.

First we give the design elements from both the network and the software point of view. We describe then the idea behind the extension of bi-dimensional to three-dimensional transform. Then we show the gain achieved in network bandwidth by the algorithm compared to conventional video compression techniques. We give finally the computational complexity and image quality for different classes of video clips.

1 Introduction

High speed networks availability combined with powerful computer systems allow new interactive distributed applications. Interactive video is one of these services, commonly found in LAN environment. Some trials are even done on the Internet Mbone[21]. A large effort attempts now to port them on national ATM test-beds for public use [9],[15],[20]. Despite powerful hardware video interfaces products we observe that software implementations remain extensively used for video applications. Main reasons stand in the capability to select more easily the appropriate configuration parameters.

The efficiency of an interactive video system (one considers only Variable Bit Rate Codecs) primarily depends upon two factors: network ability to handle the generated traffic and coding/decoding algorithm.

ATM proposes traffic shapers (e.g. Generic Cell Rate Algorithm) [13] to control the user flow. In case of excessive burst size, user's data is deliberately dropped by the network. As a consequence, traffic conformance is a key issue to end user image quality. On the other hand it is well known that, in the Internet, bursts dramatically perturb other connections traffic flow. The video system should handle delays and shape the traffic to take into account those network considerations. In particular within ATM networks, generated traffic has to conform to negotiated traffic contract.

Cells would otherwise be discarded. Needed network resources mainly depend on generated traffic characteristics such as traffic burstiness [6], [9] which is a key parameter to networks resource management. Burstiness is not only caused by the video contents [9] but also by the coding algorithm [17]. It can hence lead to a classification of various algorithms with respect to their ability to generate smoothed traffic.

A second classification can be made according to the compression technique used in the coding algorithms. The Discrete Cosine Transform (DCT) [1] is quite popular and is used in the MPEG [5], [7], [9], [12] and H.261 [8] recommendations. MPEG standard is based on Intra and Interframe coding. The Intraframe coding consists in a 2 dimensional DCT. Motion estimation and compensation are used for Interframe coding. MPEG-1 is suited for mass video storage and retrieval systems and is adapted to rates up to 1.5 Mbits/sec. More recently MPEG-2 has been standardized as a broadcast TV quality recommendation. Generally, MPEG is designed to be used in point to multipoint applications (Video-On -Demand, TV Distribution...). In practice, hardware implementation are used for MPEG coders.

This paper we proposes a video coding algorithm and evaluate its performance from a network point of view. It is known that motion estimation and compensation techniques (in software environment) are responsible for a non negligible gap between coding and decoding delays and hence are not suited to interactive software based video applications.

The proposed algorithm uses the DCT technique for both Intra and Inter frame coding. It is especially suitable for interactive video on high speed network. The study fits in the current exploration phase for the MPEG IV standard for new low bit rate coding techniques. Other coding algorithms are studied in this group where the idea is globally to use different coding algorithms per zone according to a "best fit" evaluation. 3 D DCT can hence be used in video-conferencing situations.

This paper is organized as follows, section 2. presents the terms and technical background necessary to analyse the numerical results. Section 3. presents the algorithm and codec architecture. Section 4. presents the performance of the 3D DCT video system and gives numerical results.

2 Interactive video and network considerations

Video coding for communications cannot be studied apart from its impact on network performance. Clearly, there is a trade-off between network efficiency and image quality.

Packet loss effect on end to end image quality is not yet well understood. For example, early MPEG reference models suggested cell loss rates lower than 10^{-9} but rates near 10^{-4} are currently being considered as acceptable. Cell loss effect is not only dependent on the average cell loss rate but also on the cell loss distribution over time. Periods of high cell loss due to network congestion can have a serious detrimental impact on image quality.

Delay requirements clearly vary depending on the application. For interactive video communication applications a maximum multiplexing delay of some 50 ms is appropriate while a much longer delay would be tolerable for a user simply watching a video sequence. Another delay to be considered is the time spent by packets in a playback buffer necessary to smooth out random network delays and resynchronize the received signal: the first packet arriving to the buffer is delayed a certain time T, and packets are afterward extracted in arrival order periodically. To fix the initial delay and to dimension the buffer it is necessary to be able to characterize the range of network delays due to waiting in multiplex buffers. For these reasons number of video applications (especially interactive video) require a guaranteed service that ensures the cell loss rate and multiplexers delay to be kept within some negotiated values.

On the other hand, studies on the performance of ATM multiplexers handling variable bit rate traffic show that there are broadly two types of congestion leading to cell delays (e.g., [24], [25]). The same considerations extend to packet switched networks but we prefer to use ATM teminology. If the combined rate of all multiplexed sources is less than multiplexer output rate, delays occur due to the coincident arrival of cells from different sources i.e, caused by the residual ATM jitter. These delays have a short duration, generally less than the time to transmit a few tens of cells (less than 1 ms). This type of congestion is referred to as *cell scale congestion*. An example of a multiplexing scheme allowing for cell scale congestion only is presented in section 2.1. The second type of congestion occurs when the combined rate exceeds the output rate for a period greater than the duration of a video frame. The delays here are typically much longer than those occurring in cell scale congestion. This type of congestion is known as *burst scale congestion*. Multiplexers operating at the burst scale can be controlled as suggested in 2.2.

While both delay and loss are key parameters for video QoS, we consider the following extreme cases:
i) the application is very sensitive to delay (requires a CBR like channel) but is tolerant to data loss (for example through loss recovery mechanisms).
ii) the application is very sensitive to loss (requires very low loss probability) but can suffer long delays. This is the case of non interactive broadcasting application where there is an economical advantage to keeping decoders simple and able to play out the received signal with a minimum of processing.

2.1 Bufferless multiplexing

In this section we consider video applications that cannot suffer network multiplexing delays (i.e, the only tolerated delays are those introduced by the coder/decoder buffers). In this case, multiplexer performance can be controlled by ensuring that the rate overload probability leading to burst scale congestion is negligible. The only delay to be taken into account is then the small delay due to residual ATM jitter (i.e, cell scale congestion delays). If the buffer is long enough to absorb cell scale congestion only, periods of burst scale congestion lead to cell loss. The cell loss ratio is given by the following fluid approximation.

$$\text{(2.1)}\quad \mathrm{CLR} = \frac{E\{(\Lambda_t - c)^+\}}{E\{\Lambda_t\}}$$

where $E\{\ \}$ is the expectation function, Λ_t is the source bit rate at time t (we assume the notion of bit rate is clear from the source type, e.g., a video source may have a constant bit rate throughout a given frame) and c is the network channel bit rate .

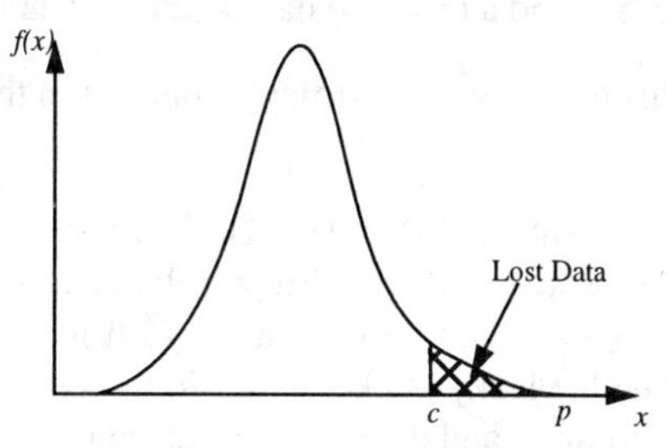

Fig. 1 Data Loss in Bufferless Multiplexer.

If the source bit rate distribution (denoted by $f(.)$) is known, the cell loss ratio can be easily derived as a function of the network bandwidth c (obtained from 2.1):

$$\text{(2.2)}\qquad \mathrm{CLR} = \frac{\int_c^p x f(x)\,dx}{E\{\Lambda_t\}}$$

where p is the source peak rate (see figure 1).

For a given channel capacity, the tighter the distribution function, the lower is the data loss ratio. Since the coding algorithm is responsible for the distribution function shape, it can be used to compare coding algorithms suitability to delay sensitive video applications. Experimental results are presented in section 4.

2.2 Lossless Multiplexing

For economical reasons, it may be preferable to keep the decoder design very simple (in large scale deployment for instance). In this case, the decoding algorithm cannot suffer data loss but the application can still negotiate a multiplexing delay.

Delay and loss are much less easily controlled when the multiplexer has a large buffer designed to absorb burst scale congestion. It has been shown that the delay distribution and buffer saturation probability depend significantly upon detailed source traffic characteristics, including autocorrelation, which are generally unknown at the start of a communication initiation and uncheckable by the network in real time [27]. In particular, the long term dependence observed in certain types of video sequences [26], [23] can lead to extreme congestion which can hardly be avoided by buffer dimensioning [28].

Burst scale congestion can be controlled by imposing constraints on the input traffic. In particular, if the amount of data $N(s,t)$ offered to a multiplexer by a video source in any interval (s,t) satisfies the burstiness constraint:

(2.3) $N(s,t) \leq c(t-s) + b$

and the multiplexer output is greater than c, then the queue length is bounded by b. If we use a buffer of length b, and a reserved bandwidth c, delay is bounded by $\frac{b}{c}$ and the system is lossless. This property is of particular interest in the ATM traffic control context.

Apart from the source Peak Cell Rate (PCR), the only parameters so far agreed on are the Sustainable Cell Rate (SCR) and Burst Tolerance (BT) which can be controlled by a leaky bucket (specifically, the Generic Cell Rate Algorithm [13]). For present purposes we define the leaky bucket LB(c, b) to be a counter incremented at rate c bits/s up to the maximum b and decremented as data is admitted to the network by the corresponding number of bits. Data is not accepted when the counter would otherwise decrease below zero. Traffic passing through LB(c, b) satisfies the burstiness constraint (2.3). In this case, loss is avoided by allocating a SCR equal to c and a BT equal to b.

More generally, for a given video traffic, $b(c)$ is called *burstiness function* [6] if the traffic conforms to LB(c, $b(c)$) for c lying between the average and the peak rate. If Λ_t is known, the burstiness function $b(c)$ can be computed using the following:

$$(2.4) \qquad b(c) = \max_{s,t} \int_s^t (\Lambda_t - c)\, dt$$

The advantage of knowing $b(c)$ is the ability of choosing the right values of SCR and BT to meet any delay requirement with zero loss rate. If d is the maximum multiplexing delay required by the application, SCR is set to $D^{-1}(d)$, where $D(c) = b(c)/c$, and BT is set to b(SCR).

The burstiness function of a video traffic gives a global idea about the application greedyness in terms of network resources i.e, for a given value of c, the lower values of $b(c)$, the smaller is the multiplexing delay. Again, the coder is responsible for traffic burstiness and the $b(c)$ function can be used for comparing the suitability of cod ing algorithms to loss sensitive video applications. It will be used in this paper to evaluate the efficiency of the 3D DCT algorithm from a network point of view.

2.3 Algorithmic Issues

Coding and decoding algorithm design depends on the target video application architecture. For instance MPEG-1 is aimed to be used in video storage/retrieval and VOD applications where movies are compressed off-line and distributed to a large number of user equipped with MPEG-1 decoders. In this case, the decoding algorithm needs to be much more simpler than the coding one. This principle was followed in the design of MPEG-1 standard. MPEG-2 can be used for real time TV

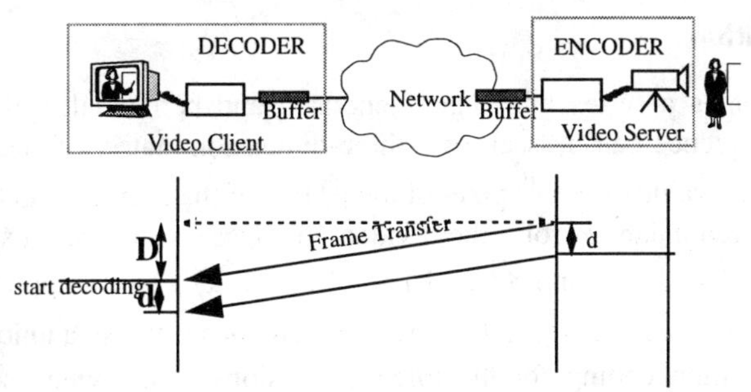

Fig. 2 A point to point interactive video application.

broadcasting where the signal is coded by a powerful hardware device and distributed (via satellite or cable) to a large number of clients. Again the number of decoders in this application is orders of magnitude greater than the number of coders. These typical applications architecture (point-to-multipoint) gives good reasons for the unbalanced algorithms complexity/cost between coders and decoders.

In point to point video application, like video-conferences, tele-teaching and remote-surveillance, a coder and a decoder are operating simultaneously and there is not necessarily an advantage to be gained if the coder is more complex than the decoder. Moreover, coding and decoding delays limit the temporal resolution of the video application. Figure 2 shows that the temporal resolution (in frames per second) is limited by the maximum value of the coding and decoding time per frame (noted *d*). An efficient interactive video application has hence to reduce this delay. Ideally, frame coding and decoding time should be identical. That was a principal motivation while designing the algorithm presented hereafter.

3 The 3D DCT algorithm

3.1 Introduction

The three dimensional DCT has been proposed for image coding [2], [16], [19] as well as for HDTV [18]. While the general idea around a three dimensional transform is the same, each method did adopt a different choice in what concerns the third dimension. Compared to other Inter frame techniques 3D DCT was not adopted.

The principle arguments that were raised against it were i) its computational cost and ii) its massive memory requirements. Although this may be true in a public TV broadcast environment (economical reasons) it is no longer the case in the computer networks environment. Moreover, compared to other interframe coding techniques, the 3D DCT needs much less computing than motion compensation and estimation based methods (as shown later). Secondly, memory is no longer a constraining limit in workstations and several images can easily be stored in.

3.2 The Algorithm

A video session is a series of images generated and periodically displayed on a screen. This sequence can be seen as a three-dimension-matrix of size Rx*Ry*Rz. Let S(i,j,k) be the value of the ith pixel of the jth line of the kth image in the sequence. The spatial redundancy of the video images can be expressed as: $S(i, j, k) \cong S(i+1, j, k)$ and $S(i, j, k) \cong S(i, j+1, k)$. The temporal redundancy can be written: $S(i, j, k) \cong S(i, j, k+1)$. The main idea is to use a unique technique of lossy information coding for the three dimensions of the video sequence. We choose the DCT transform and a quantization step which are performed on small cubes of a*b*c pixels. This involves a three dimensional DCT which can be expressed as follows:

$$F(u, v, w) = \frac{8c(u)c(v)c(w)}{abc} \sum_{i=0}^{a-1} \sum_{j=0}^{b-1} \sum_{k=0}^{c-1} S(i, j, k) * \cos\left(\frac{(2i+1)u\pi}{2a}\right) \cos\left(\frac{(2j+1)v\pi}{2b}\right) \cos\left(\frac{(2k+1)w\pi}{2c}\right)$$

with $c(\alpha) = \frac{1}{\sqrt{2}}$, for $\alpha = 0$ and $c(\alpha) = 1$, for $\alpha \neq 0$.

Obviously, this is performed on each pixel component (e.g. YUV). The main advantage of this coding scheme is that we significantly reduce the temporal redundancy with a simple method that has shown satisfactory results in both computing time and in compression ratio when associated with the adequate quantization stage. We thus adopt a homogenous technique for intra-frame and inter-frame coding. Compared to MPEG we avoid the use of complex and time consuming methods such as motion estimation and compensation. We note however that 3D DCT algorithm introduces a delay of c (typically 8) images in the coded stream. Such delay is also found in MPEG and corresponds to the number of images between I and P frames. It is a variable parameter set by users and depends on each sequence. Efficient DCT algorithms [18] have been developed and widely used both for software and hardware codecs.

3.3 Codec Architecture and Implementation

The computation of the DCT on a axbxc cube can be performed by a set of 1 dimension DCT (using the Chen algorithm [4] for example). Let S(*,j,k) the x-axis parallel arrays of the cube (of size a), there are bxc arrays of this kind. In the same manner we define S(i,*,k) the axc y-axis parallel arrays of the cube (size b) and S(i,j,*) the axb z-axis parallel arrays (size c). The whole DCT can be obtained if we perform a 1 dimensional DCT on each S(*,j,k), then on each S(i,*,k) and finally on each S(i,j,*). The complexity of the whole procedure is evaluated later.

The coder general architecture is shown in figure 3. The decoder follows the inverse path of the coding procedure. Obviously, the coder and the decoder should have the same processing delay.

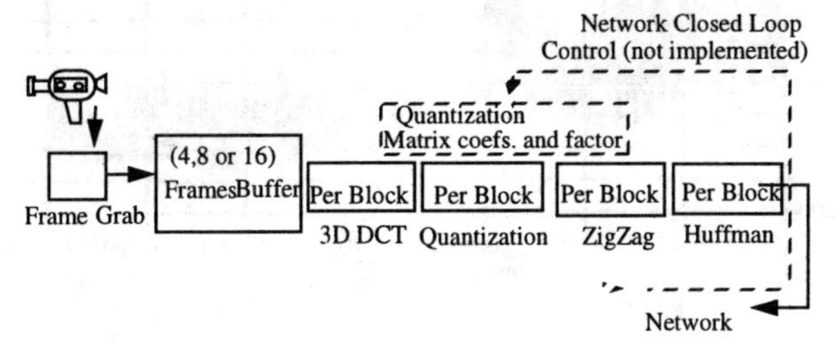

Fig. 3 Stream Generation Pipeline.

Cube size is an important issue at this stage. It is the main trade-off parameter between compression ratio and computation time. The choices are in multiple of 4 since resulting matrices are more easily evaluated:

- 8x8x8 cubes consist in performing a 2D DCT stage 8 times as in MPEG standards followed by 1D DCT on third dimensional vectors (temporal axis). This configuration is used when the coding/decoding delay is more important than the compression ratio e.g. the application does not tolerate more than 8 images delay in the coder.
- 16x16x8 gives better compression ratio. Since the interframe coding algorithm in the 3D-DCT case is also based on DCT, it has to be applied on large areas to track the block motion across successive frames. We may be limited however, as mentioned in [11] by blocking effects that affect the restituted quality. In fact, the 16x16 two dimensional DCT has been shown to be the best compromise [14], taking into account the statistical distribution of images, quantizers and noisy channels.
- 16x16x16 is a good solution for moderately interactive application (tolerating a few seconds delay). This gives a good image quality and a smoothed bit rate (since it is averaged over sixteen subsequent images). We don't give however any measurements for this DCT size.

After the DCT has been performed on the cube, the coefficients are arranged in a linear way to concentrate the zeros at the end of the stream for a better Huffman compression. This is done using a 3 dimensional zigzag so that the most important of them has the lowest ranks in the array, this is useful for the Huffman coding step. The zigzagging in the cube is based on the same principle than the one used in a 2 dimensions 8x8 bloc. The idea is to wander through the cube, visiting the points according to their distance from the origin which must not decrease at each step.

We can see that implementing the 3D DCT in both software and hardware is relatively simple. In hardware a pipeline processing is possible since each of the consecutive frames has to be transformed with a 2D DCT and correlated afterwards with the third dimensional one.

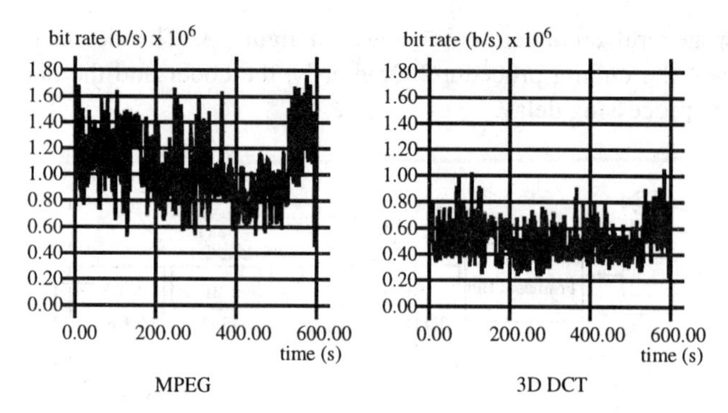

Fig. 4 Instantaneous Traffic for the "Spitting Images" sequence for MPEG and 3D DCT.

4 Results

In this section we show results for the needed network bandwidth, compression ratio, coding delay and image quality. We also give the same measurements using MPEG-1 coding. The purpose of comparing the two algorithms is not to know whether 3D-DCT is better or worst than MPEG because 3D-DCT algorithm is aimed to be used in point-to-point software applications only. Comparison is presented for illustration purposes as 3D-DCT performances can be more easily appreciated when compared to a well known coding algorithm. Point-to-multipoint video in an open-loop network environment is equivalent to the point-to-point situation.

4.1 Network performance

The first series of measurements concerns the bandwidth required for the 3D DCT algorithm. We believe that measurements should be done on stationary video sequences. Small numbers of frames as used in [2] are hence insufficient in the 3D-DCT case. The first sequence we have chosen is a 10,000 frames long clip corresponding to the well known British "Spitting Images" TV program. The second video sequence has the same length and is a touristic documentary clip on french Islands "Mayotte". Due to disk space constraints we restricted the frame size to (384x288) format and used a Parallax interface on SunOS 4.1.3 system to grab the sequences at a rate of 15 images per second.

The first curves correspond to the instantaneous traffic generated by 3 D DCT and MPEG for both clips. The two algorithms were run in a VBR mode without any rate control. Figure 4 shows the curve for the "Spitting Images". In this bursty 10 minutes long scene the average traffic in MPEG coding was 1.05 Mb/s and the standard deviation was 0.243 Mb/s. In 3D DCT the average was 0.488 Mb/s and the standard deviation was 0.136 Mb/s. It should be noted that MPEG traffic has been smoothed by averaging each 8 successive frames since 3D DCT traffic is naturally averaged over 8 frames (in the case of a 16x16x8 cube).

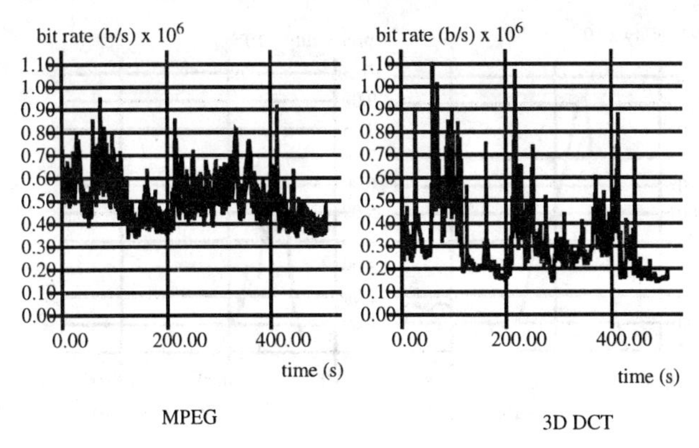

Fig. 5 Instantaneous Traffic for the "Mayotte" Sequence for MPEG and 3D DCT.

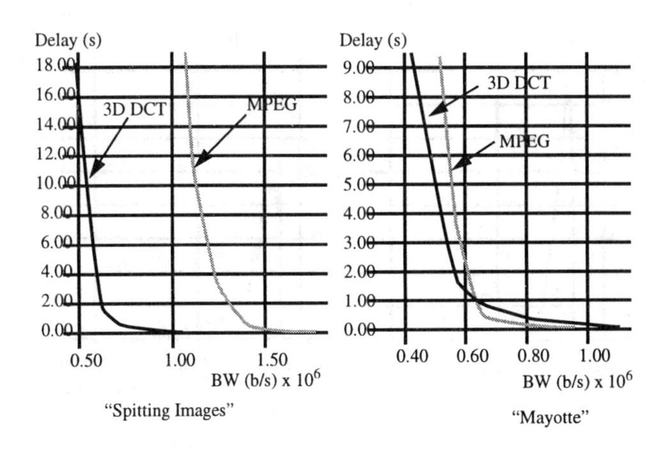

Fig. 6 Burstiness curve.

The second clip is shown in figure 5. It confirms that the average traffic for 3D DCT (0.320 Mb/s) is inferior than MPEG (0.517 Mb/s).The variance is slightly better in MPEG than in 3D DCT. There are however more variance in "Mayotte" than in "Spitting" scene for the 3D DCT. The reason for these sparse peaks in "Mayotte" comes from the Huffman coding that is not well suited to this class and to the selected block size that should be increased for the 3D DCT to cover more motion in successive frames. These two factors are currently being studied.

The second series of curves analyses the Burstiness as defined previously. Figure 6 traces the buffer delay experienced by a video stream when accommodated in a certain bandwidth. This is also an indication of the buffer size that should be used at both sides of the connection to absorb traffic burstiness (in a loss free way). The results show that 3D DCT suffers less delay than MPEG. We also see from figure 6 that 3D DCT can be accommodated in a very narrow bandwidth and still work.

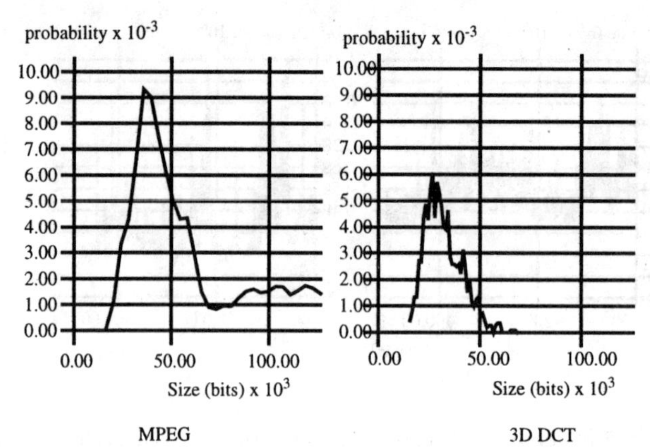

Fig. 7 Image size density for 3D DCT and MPEG in the "Spitting" Sequence

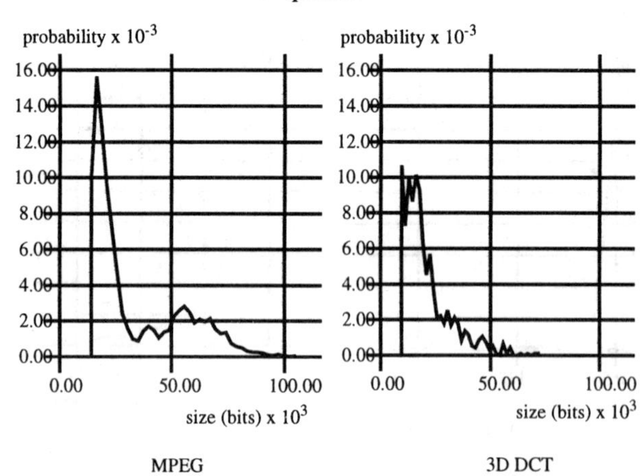

Fig. 8 Frame size density for 3D DCT and MPEG in the "Mayotte" Sequence.

The third group of curves shows the frame sizes density. In figure 7 we show the densities for the "Spitting" sequence. The vertical axis shows the probability per image size and the horizontal axis shows the image size. figure 7 shows the densities for "Mayotte" sequence. We see for both figures and especially figure 7 that MPEG generates larger frames than 3D DCT. These large frames correspond to I pictures.

From the different groups of curves we can conclude that the 3D DCT offers a satisfactory compression ratio. The transmission delay and hence the needed buffers for both emitter and receiver are of the same order of magnitude than with MPEG recommendations. We still have to study the compromise between DCT on larger blocks, and the corresponding time complexity and image quality.

4.2 Computing time performance

Time performance in the 3D DCT[1] algorithm is stable. Coding and decoding consume approximately the same amount of time. One slight variation happens during the huffman tables lookup.

MPEG		
Configuration	Additions	Multiplications
IBBPBB	4448	4288
IBBPBBPB-BPBB	3706	3573
3D-DCT		
Configuration	Additions	Multiplications
8x8x8 based	624	384
16x16x8 based	1184	704

Table 1 : Per Block Computational Cost

In order to show that 3D-DCT algorithm is faster than MPEG's one when implemented in software, we have computed the number of operations per block of 8x8 for both algorithms. A summary is given in table 1 and shows that in all cases the 3D DCT is less time consuming than MPEG coding algorithm.

In figure 9 we show the measured time for the different clips. The vertical axis shows the time on a log scale and the horizontal axis corresponds to the number of frames.

The MPEG coding time was smoothed in order not to show the variance due to I, P and B coding differences. The measurements were made on two different machines (SS5 and SS10 Sun workstations). The figure shows the differences in ceilings for coding and decoding for both methods.The results are encouraging. We confirm the large gap between MPEG coding and decoding times. Moreover we see that 3D DCT behaviour is very close to MPEG decoding which is a good result. We confirm finally that coding and decoding in 3 D DCT are equivalent. It should be noted as explained previously that the small variation found in DCT videosystem is not due to the DCT computation but to the huffman coding procedure. Our conclusion for this step is that 3D DCT algorithm is a good solution for implementing interactive video application in software.

1.Sources are available by simple mail to the author.

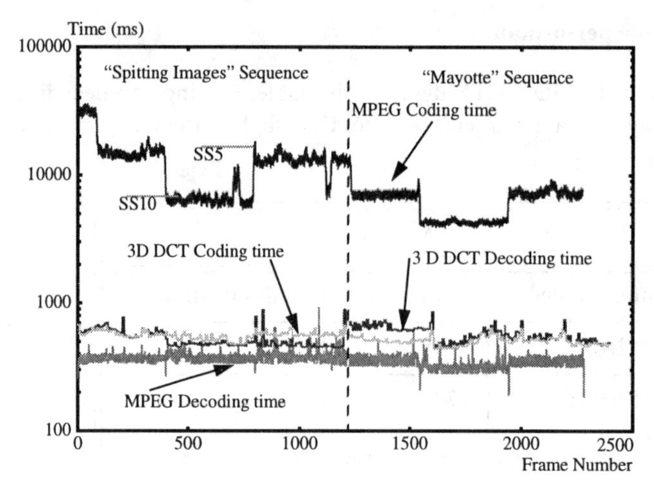

Fig. 9 Coding/Decoding Time for the two video sessions in Log scale.

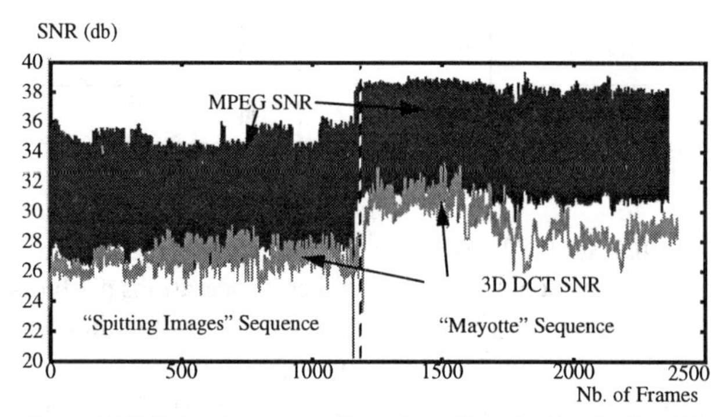

Fig. 10 SNR for image coding/decoding for both algorithms.

4.3 Image Quality

As mentioned in [11] various quantitative criteria are not always good evaluators of the performance and restitution quality of a coding algorithm. Visual images inspection is the most significant judgment. 3D DCT offers satisfactory decoded images visual restitution. The selected quantization factor in our measurements was the value of 4. We also used the same quantization matrix used in MPEG and extended to 16x16 points to compare the results on equivalent parameters basis. In figure 10 we show the results for the same sequences previously used. The vertical axis indicates the (Signal to Noise Ratio) SNR [18] ratio in decibels for both DCT and MPEG. MPEG demonstrates a better SNR. We should note however that the SNR in MPEG is very variable. This is due to I,B and P frame types.

Since our main concerns during development of the videosystem were to reduce time complexity and increase network efficiency we can conclude that even with these SNR results, the 3D DCT is globally a satisfactory way for video coding. We emphasize on studying the factors that contribute to an improved SNR, i.e. the quantization matrix.

5 Conclusion

Designing compression algorithms for networking clearly depends on the target application architecture. In this paper we analysed different design elements which are determinant for interactive point-to-point video applications. In particular, frame size distribution and burstiness function are key parameters in performance evaluation of compression algorithms. If the coder and the decoder are implemented in software, it is necessary to have a symmetric algorithm complexity to be able to build interactive video.

We have described a coding algorithm for software based interactive video applications. The algorithm is based on the three dimensional discrete cosine transform. The results obtained in software are satisfactory and give insight of a good adaptability of the method to video networking applications like tele-teaching and remote-surveillance. This study was focused on the adaptation of the algorithms to network constraints. We have shown that the 3D DCT algorithm is a good solution for software based interactive video application from both the network and the system point of view. Our next step will be to improve the performance of the coder and reduce the signal to noise ratio through adaptive quantization and better compression techniques than Huffman coding.

References

[1] N.Ahmed, T. Natarajan, and K. Rao. Discrete Cosine Transform. IEEE Computer Transactions C-23:90-93 (1974).

[2] A. Bascurt, R. Goutte. 3 Dimensional Image Compression by Discrete Cosine Transform. Proceedings of Eurasip 88, Elsevier Publisher (Norh-Holland)

[3] O.E. Besset, W. Shaming. A two dimensional discrete cosime transform video bandwidth compression system. "NAECON". Dayton OH. May 1980.

[4] W. Chen, C. Smith, S. Fralick. A Fast Computational Algorithm for the Discrete Cosine Transform. No. 25 IEEE Transactions on Communications (1004-1009) 1977.

[5] CCITT SG XV Working Party XV/1. Experts group on ATM Video Coding. Coded Representation of Picture and Audio Informatio. MPEG 92/N0245.

[6] R.Cruz. A Calculus for Network Delay, Part I: Networks Elements in Isolation. IEEE Transactions on Information Theory, Vol 37, No. 1. January 1991.

[7] D. Le Gall. MPEG: A Video Compression Standard for Multimedia Applications. Communications of the ACM, 4(34): 305-313, April 1991.

[8] Recommendation H.261.Video Coder for audiovisual services at p* 64 Kbits/sec. CCITT White Book 1990.

[9] M. Hamdi, D. Curet, J. Roberts, G. Madec, Statistical Multiplexing of VBR MPEG2 streams. Contribuion to ISO/IEC/JTC1/SC29/WG11, Singapore Nov. 1994.

[10] M. Hamdi, P. Rolin, Resources Requirements for VBR Mpeg Traffic in Interactive Applications, Cost237 Confer ence On Multimedia Transport and Teleservices, Vienna. Springer-Verlag LNCS 882. Nov. 1994 .

[11] H-C. Huang, Huang J-H, J-L Wu. Real-Time Software-Based Video Coder for Multimedia Communication Systems. Proceedings of the First ACM Multimedia Conference. 1993.

[12] ISO-IEC/JTC1/SC29/WGII, MPEG Test Model II. July 1992.

[13] ITU-T. *traffic Control and Resource Management in B-ISDN.* I-371 Recommendation. 1992.

[14] J.W. Modestino, D.G. Daut, A. H. Vickus. Combined Source Channel coding of images using the Block cosine Transform" IEEE Transactions on Communications. Vol. Com29 pp 1261-1274, Sept 1981.

[15] Le Moign. Architecture des services multimédias sur réseau ATM. ATM Developments'94. Rennes France. March 1994.

[16] R. Natarajan. On Interframe Transform Coding. IEEE Transactions on Communications Vol. Com.25 No. 11 Nov.77.

[17] P. Pancha and M. El Zarki. Bandwidth Requirements of Variable Bit Rate MPEG Sources in ATM Networks. In INFOCOM'93, pages 902–909. IEEE Computer Society Press Los Alamitos, California, March 1993.

[18] K. R. Rao P. Yip. Discrete Cosine Transform. Algorithms, Advantages and Applications. Academic Press, Inc. Harcouet Brace Jovanovich, Publishers. 1990.

[19] J. Roese, W. Pratt. G. Robinson. Interframe Cosine Transform Image Coding. IEEE Transactions on Communications Vol. Com.25 No. 11 Nov.77.

[20] S. Travert. Le Transport de flux de données MPEG2 sur réseau ATM. ATM Developments'94. Rennes France. March 1994.

[21] T. Turletti . H.261 Software Codec for Videoconferencing over the Internet. INRIA Report 1834. Inria Sophia Antipolis. January 1993.

[22] G. Wallace The Jpeg Still Pictures Compression Standard. Communications of the ACM, Vol. 34. No. 4. pp 33-40, April 91.

[23] M.W. Garrett and W. Willinger. Analysis, Modeling and Generation of Self-Similar VBR Video Traffic. In *Proceedings of SigComm.* ACM, September 1994.

[24] I.Norros, J.Roberts, A.Simonian, and J.Virtamo. The Superposition of Variable Bit Rate Sources in an ATM Multiplexer. *To appear in IEEE JSAC special issue*, April 1991.

[25] H. Kroner. Statistical multiplexing of sporadic sources - exact and approximate performance analysis. In *ITC-13, Copenhagen.* Elsevier Science Publisher B.V. (North-Holland), 1991.

[26] J. Beran, R. Sherman, M. S. Taqqu, and W. Willinger. Variable-Bit-Rate Video traffic and Long-Range Dependence. *Accepted for publication in IEEE Transactions on Communications*, 1992.

[27] B.Bensaou, B.Guibert, J.Roberts and A.Simonian. Performance of an ATM Multiplexer in the Fluid Approximation using the Benes Approach. *Annals of Operation Research*, 49:137-160, 1994.

[28] I.Norros. A Storage System with Self-similar Input. *Queueing Systems* 1994.

RIVUS: A Stream Template Language for Capturing Multimedia Requirements

Donna Lindsey and Peter F. Linington
(*email: {D.L.Lindsey, P.F.Linington}@ukc.ac.uk*)

Computing Laboratory
University of Kent
Canterbury, Kent, CT2 7NF
United Kingdom

Abstract. Object based architectures, such as the standardized reference model for Open Distributed Processing (ODP), must not only provide for the simple client-server styles of interaction found in data processing, but must also support the rich variety of different communication configurations and dependencies that are found in multimedia applications. To do this, ODP defines both simple two-party communication and an explicit binding model which can describe a wide range of more complex situations. It encapsulates the complexity within a visible binding object.

This paper explains the ODP view of binding, introducing the idea of a binding template which represents a class of communication activity, and then introduces a language called RIVUS for the specification of these templates. It demonstrates, by the use of a number of examples, that such a language can capture much of the behaviour required in multimedia application designs.

1 Introduction

ODP provides a framework for describing distributed systems and for specifying their properties in an open way. It has been defined to support standardization activities, to express the relationships between different kinds of specification and to describe open distributed systems in a uniform vocabulary. The main parts of the ODP reference model are now being published as international standards [1, 2, 3]. Part 2 provides a general object model, associated definitions and a basis for conformance testing, while Part 3 defines the architectural constraints which make a system open, giving the basis for interoperability and portability of components.

The ODP architecture is organized around the idea of a set of viewpoints. These allow different properties of a system to be grouped together, so that a design can be thought of as a set of specifications covering different areas of concern, rather than as a single monolithic whole which would generally be too large to comprehend and maintain.

The five viewpoints are:

a) the enterprise viewpoint, which is concerned with the objectives, business environment, and policies which affect the system.
b) the information viewpoint, which is concerned with a common information model to be shared by the specifications of the various system components, allowing a consistent interpretation of the material communicated.
c) the computational viewpoint, which is concerned with the functional structure of the system, describing it in terms of interacting objects. These objects interact at interfaces, which provide the boundaries at which distribution can take place. Thus the definition of stable interfaces forms an essential part of the specification of an open system.
d) the engineering viewpoint, which is concerned with the way in which object interaction is to be supported. A computational interface may be supported by a variety of engineering mechanisms, depending on the quality and reliability of communication needed.
e) the technology viewpoint, which is concerned with the constraints implied by the implementation environment, and with the extra detail needed to test the system.

The ODP architecture is expressed by defining an abstract language for each of these viewpoints, with each language representing the architectural constraints imposed. The standard does not lay down a specific form for these languages; however, specific notations can be defined corresponding to the viewpoint languages for use in any particular development environment.

2 Object Interaction

This paper is primarily concerned with the way objects interact. As such, it focuses on the computational and engineering languages.

2.1 The Computational Language

The simplest form of object interaction is a direct interaction between two objects, bound at a single interface, in which the objects involved have clear client and server roles. This corresponds to familiar language ideas like method invocation, and to the support provided by many current middleware platforms. A prerequisite of such an operation is that the client should have some form of reference to the server, which can be used to establish a binding between them.

In a particular implementation language, the binding process may be implicit, in that the allocation of the necessary client resources (the client interface) and the establishment of a client-server binding are carried out within the supporting infrastructure. This happens in response to a client request for invocation of a named operation at the server referenced. The management of the binding process is hidden from the application programmer, with the infrastructure taking responsibility for the properties and lifetime of the bindings it creates.

However, this approach may be somewhat limiting in cases where the application wishes to engage in more complex forms of communication, or to take more direct control of the resources used. Thus, for example, a client may wish to establish a binding to be used for a whole series of interactions, specifying initially the resources to be associated with it. Or, the client may wish to control multiparty or fault tolerant communication. To support such direct control during the lifetime of the binding, the ODP reference model introduces the idea of explicit, compound bindings.

A compound binding is created by an explicit "bind" action, which creates a binding object. The introduction of such binding objects into the architecture greatly increases the expressive power of the model, since it allows the direct representation of a set of communication resources, and encapsulates them, so that they can be manipulated as a whole.

Once a binding object has been created by the bind action, it is just like any other object; it can interact with other objects, and has defined interfaces and behaviour. The result of the bind action is a control interface which is returned to the binding object, and through which the properties of the binding can be modified, and via which the binding can be instructed to delete itself. In this way, the control of communication resources can be included within the computational model of the application at a minimal cost in terms of new modelling concepts.

Depending on the defined behaviour of the binding object, new interfaces can be added to the binding, or existing ones removed, so that a suitable binding object can describe, for example, a dynamic multicast group or a multi-way conference call. The process by which a new interface is added to the binding is itself, of course, just binding. There is a potential recursion here, which is terminated, at some stage, by appeal to the primitive two-party form of binding described above. However, by requiring that this primitive binding be carried out within a single physical location, the compound binding can be made to capture all the timing and quality of service information relating to the communication.

The information needed to create a binding with a particular class of behaviour is called a binding template. This template identifies the types of interfaces which can participate in the binding, the roles they play and the way behaviour at the various interfaces is linked. Thus a binding template might define that information made available at one interface is communicated directly to one or more others, or that the passage of certain kinds or quantities of information results in the generation of operations reporting the fact from within the binding.

2.2 Engineering Language

In the engineering language, the behaviour of computational binding objects is mapped onto some engineering channel structure. This channel must guarantee at least the behaviour specified in the binding template, but it may contain additional mechanisms to enhance the communication capabilities. These mechanisms can take two forms.

Firstly, the channel may provide some mechanism specifically to overcome a particular kind of communication problem, enhancing the quality of communication by providing a distribution transparency. For example, there may be specific mechanisms to locate interfaces using logical names or to support the mobility of interfaces.

Secondly, there may be a need for some form of conversion service within a channel to support communication across domain boundaries. The objects that do this are called interceptors; they are introduced as a result of negotiation when the channel is created and they may be required to convert between encodings resulting from the use of different technologies, or to ensure actions or checks needed at administrative boundaries.

Not all applications will place the same requirements on their communication support. There can, therefore, be many possible channel structures to support a single form of computational behaviour, with the various solutions being chosen to meet the requirements of a specific application. One of the aims in emphasising the split between the computational and engineering viewpoints is to support the automation of the process which takes a binding template and its parameters and applies known recipes to construct a channel with the desired properties. Access to suitable software tools to support this process would significantly reduce the cost of application development.

3 Streams

A characteristic of multimedia applications is that certain kinds of information, for example video, audio or sensor data, are generated at a continuous and steady rate. This information flows between a source, such as a camera, microphone or video-store, and a sink, such as a display screen, or loudspeaker. In ODP, a stream is used to represent this continuous flow of information.

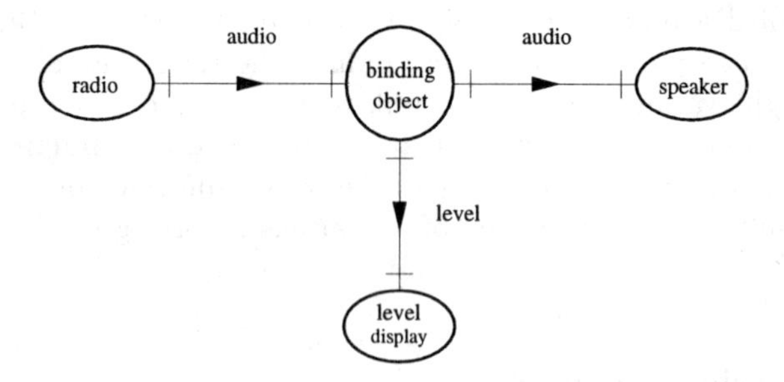

Fig. 1. An audio system stream

A stream can represent a simple flow of information, for example the flow of video from a camera to a screen. It can also represent more complex behaviour such as the flows of different related forms of information in an audio system.

Fig.1, shows a stream in which audio flows from radio to speaker, and information on the loudness of the signal is derived continuously to update a level display.

3.1 Computational View of Streams

At the computational level a stream, as shown in Fig.2, is described as a binding between a number of stream interfaces. Like any other binding, it is represented by a stream binding object which encapsulates the information flows between stream interfaces. The presence of the binding object allows stream configuration and quality of service details to be made explicit. The binding action which links the stream binding object and the computational object supporting the stream interface is considered to be primitive.

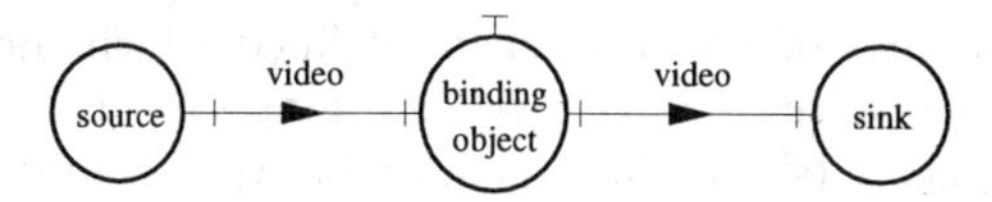

Fig. 2. A stream binding

A stream interface is the point of access to an object where information is generated or consumed and is composed of a finite collection of flows. A flow is an abstraction of a sequence of interactions between a source and a sink object. Each flow represents a basic continuous media type, for example audio, video or sensor data. A flow is represented by an action template which contains the name, information type and direction of the flow. The flow direction indicates if information passes into or out of the stream interface.

Stream interfaces have a type which describes the information properties of the interface and the direction of the information flows; for example, an interface with a pair of source and sink audio flows may have type *telephone*. The type definition may be specialised to a single application or group of applications and reflects their use of the interface. For example, an interface containing a single audio source flow may have type *cd_player* in one application and *tannoy* in another, depending on the quality. The type reflects the different properties of the audio stream interface in each application.

A stream interface can also be described in terms of already defined interface types to allow more complex flows to be modelled. For example, a cooperative work interface might be composed of a defined interface of type *tv* and a single video flow. The *tv* interface, which supports the human interaction, allows audio and video flows to be grouped and managed together whilst the video flow presents a view of the workspace.

A binding object is instantiated from a stream binding object template by a create action. The template contains a number of stream interface signatures which define the flows and interfaces that comprise a particular stream interface. This information is used by the supporting system to allocate resources on instantiation of the binding object. If the demand for resources is too great, the supporting system can refuse the create request or offer a subset of the requested resources. A binding object may support a number of stream interfaces which are not in general interchangeable. Where necessary correct pairings of interfaces in the template and interfaces to be involved in the binding are specified by associating roles with each element in the template. Each role has a type that reflects the content of the stream signature. A role allows the binding of a stream interface to be considered at a single point and the rest of the binding object ignored.

In some cases a binding object is instantiated with a set of control interfaces. Operations are defined at these interfaces which are used to change the membership of the binding. For example, an add operation introduces an interface to a binding in a named role and can succeed or fail depending on the requirements of the role. A subtract operation is used, at a later stage, to remove an interface from a binding. Further operations can be defined which monitor and control the binding to ensure stated service levels are maintained.

The binding object acts as an intermediary between bound objects; it receives information from the party producing a particular flow and presents it to one or more parties with a consumer role. A binding object will therefore only allow a binding to exist between interfaces if a meaningful communication is possible.

If an object with interface X is offered to fulfill a role Y and the flows in X have their type matched and their direction reversed by flows in Y, such that the overall flow direction is the same in any binding produced, then X and Y are said to be complementary interfaces, (Fig.3), and a binding is possible.

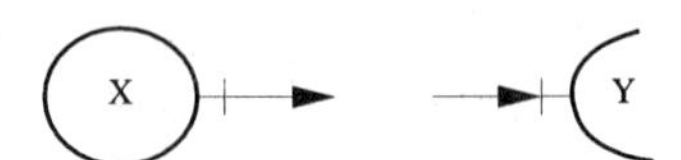

Fig. 3. Complementary interfaces

It is not necessary for X and Y to have the same type. If a suitable type relationship exists between X and Y then X can be used in the binding. However the components in X and Y which are required in the type relationship must be complementary for the binding to occur.

Subtypes. The ODP model states that if a subtype relationship exists between interfaces then a binding is possible. An informal subtype rule for streams is defined in the ODP Reference Model [2], with a complete definition included in Annex A. The informal rule states that an interface X is an interface subtype of a role Y if, for all flows which have the same names:

- the flow type in X is a subtype of the flow type in Y, if X is producing the flow.

– the flow type in Y is a subtype of the flow type in X, if X is consuming the flow.

A further assumption is made that X is a subtype of Y, if X can be substituted [4, 10] for Y without loss of functionality; that is X offers either the same or more capabilities than Y. It may be necessary for the binding object to filter out any extra capabilities offered by X.

Partial Match. An interface which has some but not all of the capabilities of a role Y may also be accepted by the binding object. The interface can not usefully interpret all the stream interactions that might occur but it may have sufficient functionality for the binding object to create a meaningful binding. For example, a *telephone* interface may be acceptable in a *video conference* role.

A further type relationship, called partial match, can be added to the subtype relationship and can be used by the binding object to make acceptance decisions. An interface X partially matches a role Y if:

– at least one flow in X corresponds to a flow in Y.
– where a correspondence exists, a subtype relationship exists.

The binding object needs to know that Y can be substituted for X and that they are both complementary to the interface role. It can maintain vast amounts of type information or, as suggested by W.Brookes et al. [5, 6], rely on a type repository. A type repository contains descriptions of all types currently known to a distributed system domain and statements of the relationships that exist between these types. If a number of distributed domains maintain their own type repositories then it should be possible for them to federate. It is then necessary for the type repository to translate between different domain representations of the same type to allow type matching across distributed platforms.

If an offered interface meets the type and service requirements of a named role the binding object instantiates the interface template associated with the nominated role and uses a primitive bind action to bind the interface to it.

3.2 Engineering View of Streams

At the engineering level, a stream is considered as a series of information channels between engineering objects. An engineering object will carry out some of the functionality defined in the binding object template. It joins individual flows in the accepted interfaces and handles any unexpected behaviour, for example, in Fig.4 the stereo provided by the source object is not required by the sink and that channel is therefore discarded. The engineering object also has the task of monitoring service levels.

At this level the representation of the physical data also becomes important. A video flow, for example, is considered as an *MPEG* or *JPEG* channel. It is possible that an engineering object acting as producer may have a different data representation to one or more of the consumer objects. At this level interceptor objects are provided to convert, if necessary, the data to the appropriate format for the consumer objects.

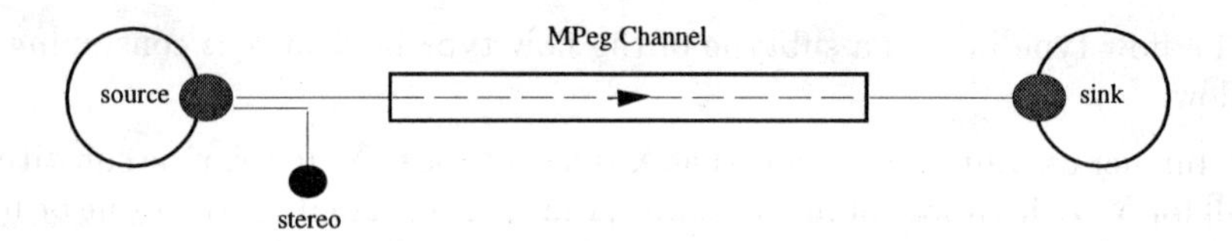

Fig. 4. Engineering stream

4 RIVUS - A Stream Template Language

Current research has highlighted the need to support continuous media in distributed systems.

Haltern et.al [8, 9] have proposed additions to the interface language of TINA-ODL to support continuous media. The additions allow stream interface templates to be specified with quality of service parameters. Quality of service values are given in the form of environment contracts which set out the service levels required by the stream for incoming flows and those offered to outgoing flows.

Additions to the ODP model have also been proposed in [7]. It is suggested that a new stream interface type be added which supports quality of service annotations. *Reactive objects*, specified in the Esterel synchronous language, are also introduced to support the synchronization of multimedia objects. Such objects respond to external events by producing new events and are used to constrain real-time communications between objects.

Much of the current work has addressed the problems of defining and supporting streams with quality of service guarantees. However little has been said about how the streams are configured and how, at binding time, compatibility between interfaces with both type and quality of service attributes is ensured. RIVUS attempts to address some of these issues.

RIVUS is a template language for specifying multimedia streams. An application programmer can define the behaviour of stream binding objects and allow the underlying infrastructure to create the stream. The language can be used to state:

- the resources that the binding object requires;
- the functionality required for named roles;
- rules for binding complementary interfaces; and
- any necessary type conversions.

The language assumes that the streams defined will be created using the explicit binding model.

A formal syntax [11] of the template language exists but its discussion is outside the scope of this paper.

4.1 Binding Templates

A binding template is a computational specification for a stream binding object. It is a formal definition of the complex behaviour of the binding object. The template of a video conference binding may be defined as follows:

$$video\_conf = [$$

ROLETYPES

...

BINDROLES

...

BINDREQ

...

BINDBEHAVE

...

BINDCARD

...

$$]$$

Each of the template components are described in detail in the following sections. It is from this template that a binding object will be instantiated.

4.2 Binding Types

The template includes type definitions of the roles that the binding object supports in the binding. The definition is given as a list of uniquely named components and their types. This level of detail enables the binding object to discover type relationships between interface and role types. If a binding object can query a type repository about type relationships then analysis of the definition only becomes necessary if an interface type can not be found.

A role type is defined in terms of the flows of information that the interface handles. Each flow description contains the data type and the direction of the flow of information.

$$videoconf\_prod = < a1 : (audio,\ SOURCE),\ a2 : (video,\ SOURCE) >$$
$$videoconf\_cons = < a1 : (audio,\ SINK),\ a2 : (video,\ SINK) >$$

By defining a role type in terms of already predefined stream types, flows are grouped and managed together.

$$videoconf = < b1 : videoconf\_prod,\ b2 : videoconf\_cons >$$

The type *videoconf* consists of two bundles of flows which map to the sender and receiver streams in the conference. If *videoconf_prod* and *videoconf_cons* are not defined within the template then the grouping of flows can be explicitly created.

$$videoconf = < b1 : < a1 : (...), a2 : (...)> , b2: < a1 : (...), a2 : (...) > >$$

It is also possible to define a role type in terms of an already defined type but remove the grouping such that each component in the defined type is managed individually.

$$tv = < COMPONENTS\ OF\ videoconf\_prod >$$

In this case the interface type *videoconf_prod* will be expanded such that the role type *tv* consists of a labelled audio and video flow.

4.3 Binding Roles

A programmer can define binding object roles in the template section *BINDROLES*.

$$\begin{array}{l} if_{\alpha} : videoconf\_prod; \\ if_{\beta} : videoconf\_cons; \end{array}$$

Roles define the type and behaviour of the stream. For example, the type of if_{α} and if_{β} signifies that the stream is a video conference where communication is in a single direction from producer to consumer.

The roles exist, on object instantiation, as implicit binding sites to which stream interfaces can be bound. An interface is offered with a named role, for example if_{α}. It is the task of the binding object to check that the interface meets the functionality of that role.

Several interfaces can have the same role in a binding. For example, there may be several interfaces in the role of conference producer, if_{α}. The binding object instantiates a binding site from a role for each interface offered whilst resources allow.

4.4 Binding Requirements

The role type defines the optimum functionality expected from an interface. If the subtype and partial match rules are used to determine the acceptability of an interface in a binding then it is possible that the created stream does not meet stated quality of service requirements. For example, a *video* interface can be accepted, using the partial match rule, into a binding representing a video conference. Certain participants in the conference may require both audio and video and have therefore had their requirements undermined by the binding object. If the service requirements are not strictly defined and a subtype or compatible interface are acceptable it would still make little sense to accept a video interface into the conference binding.

The functionality required for a role needs to be made explicit to ensure the binding object successfully meets the service requirements of the stream. The template language defines two constraints which allow the programmer to make these requirements explicit.

If no constraint is applied to a role then the assumption is that the offered interface must match the role functionality exactly.

ISSUBTYPE $\boldsymbol{if_\alpha}$

The subtype constraint states that only subtype interfaces are acceptable in the named role. This ensures that all the functionality of the role is provided. Any extra capabilities that the interface supplies are handled by the binding object at a later stage in the binding process.
The assumption is that the constraint also applies at the stream component level. The role, if_α, with a component of type *audio_mono* will accept a subtype interface with a component of type *audio_stereo.*

ISMATCH $\boldsymbol{if_\beta}$

The partial match rule states that an interface is acceptable if it matches at least one component in a named role. This constraint, for example, would allow a *radio* interface to bind to a role if_β of type *videoconf_conf.* It is the binding object's responsibility to handle the interface and role components for which correspondences do not exist.
It is possible that an interface that partially matches a role is only acceptable if it has a certain functionality. In most cases it makes little sense to participate in a video conference with only video capabilities. An optional argument to the constraint allows the programmer to list the components that are necessary for a meaningful binding.

$$\text{ISMATCH } if_\beta \ \{\text{a1}\}$$

In this case an interface will only be accepted in the role if_β if it provides at least the component *a1.*

ISSUBTYPE $\boldsymbol{if_\alpha.a_1}$

It is assumed that the constraint applied to a role also applies to its components. If if_α has type *videoconf_prod* and its audio component, *a1*, has stereo properties then the constraint

$$ISMATCH\ if_\alpha$$

allows an interface with mono audio properties into the binding. To maintain service requirements at the flow level, a constraint is placed on a named flow.

$$ISSUBTYPE\ if_\alpha.a1$$

The above constraint states that any interface that binds to if_α must provide at least stereo capabilities if it matches the component *a1*.

4.5 Binding Behaviour

Once an interface has been accepted, the binding object instantiates an interface from the named role and binds the offered interface to it.

The binding object creates connections between instantiations of complementary producer and consumer roles. The binding object needs to have its binding behaviour defined. It may support several groups of consumer and producer roles and therefore needs to know which roles complement each other. It may also support several instances of each supported role and these can be mapped together in various ways. The programmer has three options for defining the binding object's mapping behaviour. These behaviour definitions ensure that a variety of communications are catered for.

SINGLECAST $\boldsymbol{if_\alpha}$ $\boldsymbol{if_\beta}$

The binding object creates a one to one mapping between a single instance of role if_α and a single instance of role if_β.

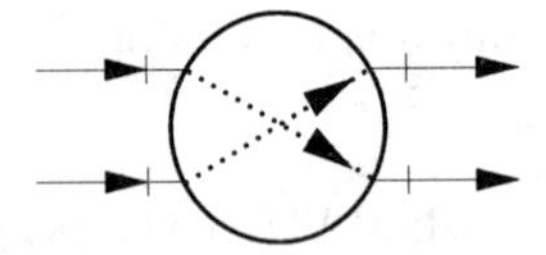

Such a binding might model a telephone call. The binding object may support several one to one bindings for example, to model a telephone exchange.

BROADCAST $\boldsymbol{if_\alpha}$ $\boldsymbol{if_\beta}$

The binding object creates a one to many mapping between a single instance of if_α and all instances of if_β.
This object behaviour can be used to model a video lecture.

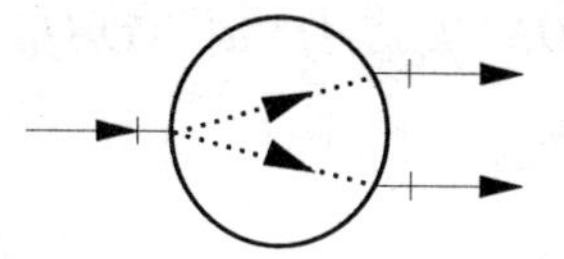

MULTICAST $\boldsymbol{if_\alpha}$ $\boldsymbol{if_\beta}$

The binding object creates a mapping between all instances of if_α and if_β. This behaviour is needed to support streams that model multi-party conferences.

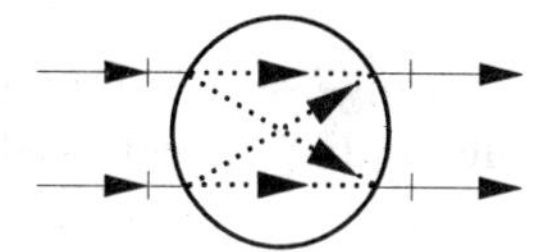

4.6 Type Conversion

Producer and consumer objects can be combined where the type of the information produced is different from the type of the information consumed. The binding object needs to convert, transparently, the information supplied by the producer to a form that the consumer understands. The template language is used to state the type conversions that the binding object supports.

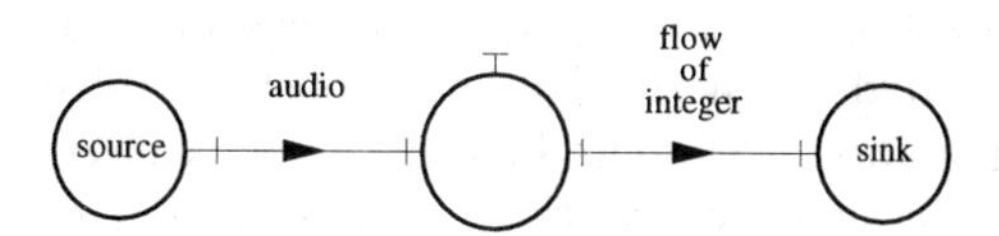

Fig. 5. Computational type conversion

In Fig. 5. a stream binding models the flow of information between a microphone, producing audio, and a graphic equalizer display which expects a series of integers relating to the level of bass, treble etc in the audio flow. In this case the roles, if_α and if_β have the types

$$microphone = < a1 : (audio, SOURCE) >$$
$$equalizer = < b1 : (sensor, SINK) >$$

The programmer states that type conversion is necessary between certain flows in the complementary roles if_α and if_β.

$$CONVERT\ if_{\alpha}.a1\ \ TO\ if_{\beta}.b1$$

4.7 Binding Cardinality

Binding cardinality is a statement of the number of a particular interface the binding object wishes to support. If no explicit statement is made then the cardinality defaults to one.

$$REQUIRE\ min\ ..\ max\ OF\ if_{\alpha}$$

The cardinality is given as a range. The minimum value states the number of a particular interface that the binding object needs for the stream binding to make sense. If the underlying system is unwilling to support this number then the binding object can not be instantiated. The maximum states the largest number of interfaces that the binding object is willing to support. If the supporting system has insufficient resources then the maximum can be reduced dynamically.

4.8 From RIVUS to the Engineering Level

At the engineering level, channels are created through which the continuous data passes. Each channel maps to a single flow binding and is uniquely identified to enable the binding object to effectively monitor and maintain quality of service levels associated with the channel. Groups of channels are further identified to enable the binding object to manage stream interfaces, for example to provide fine-grain synchronization, such as lip-synch, between audio and video flows.

Channels are created when the binding object has identified sufficient instances of complementary roles. The binding object constructs a mapping set which it uses to create logical connections. The behaviour rules in the template determine which roles are used to construct such a set. Each mapping set contains the names of corresponding components in the role instances.

A null connection is created for all components that exist in the binding but are not listed in the mapping set. If there are no correspondences, for example a *video* interface takes the producer role in a video conference binding and an *audio* interface takes the consumer role, then the result is an empty mapping set and a stream binding which results in a meaningless communication. Such a binding may be legal within the definition of the behaviour of the binding object and therefore can not easily be prevented.

If type conversion has been defined, the binding object locates an interceptor which can convert from component type to component type. If an appropriate interceptor can not be found then the binding will fail.

The binding object also handles conversions between flows with different representations of the same type, for example, as illustrated in Fig. 6, compressed

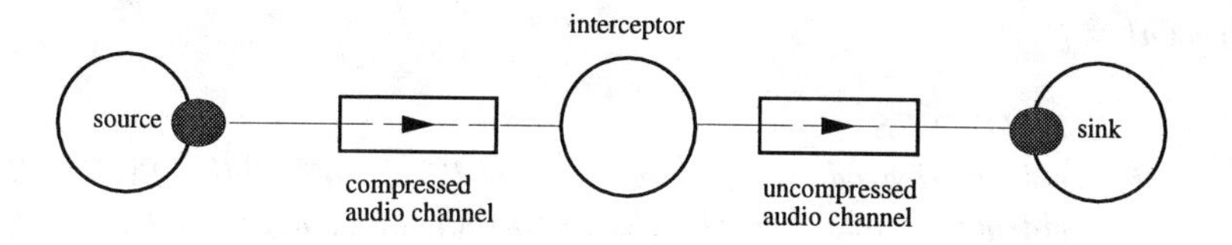

Fig. 6. Engineering type conversion

and uncompressed audio flows. The binding object carries out this conversion transparently to the programmer.

The supporting infrastructure is then used to build channels from these logical connections. Physical paths are created between the objects in the bindings or, where necessary, between them and selected interceptors.

5 A Video Conference Case Study

The case study describes the modelling and building of the simple multi-party video conference in Fig. 7.

The template defines a video conference with two complementary roles, *talk* and *listen*. It is expected that any participant in the conference will assume the roles of talker and listener. As stated earlier, binding is only possible between complementary interfaces; therefore a participant will need to bind to the talker role with a listen interface and to the listener role with a talk interface.

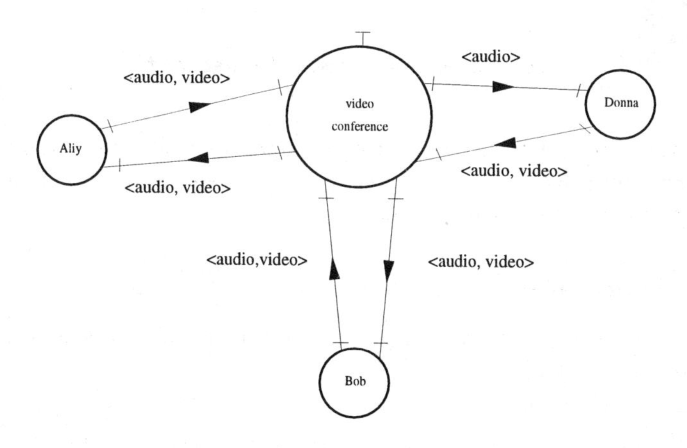

Fig. 7. A multi-party video conference

It is possible, with the current template definition, for a participant to assume a single role. This situation can be avoided by grouping the talk and listen roles together and applying the necessary constraints.

```
videoconf = [

    ROLETYPES
      videoconf_prod = <a1:(audio, SOURCE), a2:(video, SOURCE)>;
      videoconf_cons = <a1:(audio, SINK), a2:(video, SINK)>;

    BINDROLES
      talk : videoconf_prod;
      listen : videoconf_cons;

    BINDREQ
      ISSUBTYPE listen;
      ISMATCH talk {a1};

    BINDBEHAVE
      MULTICAST talk listen;

    BINDCARD
      REQUIRE 2 .. 6 OF talk;
      REQUIRE 2 .. 6 OF listen;
]
```

The binding requirements state that a participant needs to provide all the functionality of the talker but only a subset of the functionality of the listener.

A create action instantiates a binding object from the binding template.

```
ref = createstreamobj(videoconf)
```

The action can fail if the supporting system has insufficient resources. The reference `ref`, returned on successful creation, provides a management interface at which membership changes are made.

5.1 Donna Scenario

The application object *Donna* supports two stream interfaces *donna_talk* and *donna_listen* with type

```
donna_talktype   =< a1 : (audio, SOURCE), a2: (video, SOURCE) >
donna_listentype =< a1 : (audio_mono, SINK) >
```

It offers the interfaces into the conference binding with two separate calls made to the add operation supplied at the binding management interface.

```
ref$addtostream("talk", donna_listen)
ref$addtostream("listen", donna_talk)
```

The operations state that the participant identified as *donna* wishes to bind to the roles *talk* and *listen* with the stream interfaces *donna_listen* and *donna_talk* respectively. *donna_listen* and *donna_talk* are references to predicates which define the type and service capabilities of the stream interface. The operation returns the status success or failure.

Both operations succeed. The interface *donna_listen* does not provide all the functionality of the role *talk* but it meets the role's requirements; it provides at least a complementary audio flow. The binding object instantiates a stream interface from each of the roles and binds the offered interface to it.

5.2 Aliy Scenario

Object *Aliy* supports three stream interfaces with the following types:

aliy_video = < *a1 : (video, SOURCE)* >
aliy_tv = < *a1 : (audio_stereo, SOURCE), a2: (video, SOURCE)* >
aliy_videoconf = < *a1 : (audio, SINK), a2: (video, SINK)* >

Aliy first offers the interface of type *aliy_video* into the binding and is refused. *aliy_video* is not a subtype of the type of *listen*. At a later stage, *Aliy* offers two further interfaces of type *aliy_tv* and *aliy_videoconf* to the binding object.

The binding object now has a single instance of each complementary role. It finds all the component correspondences in the instances and creates logical connections for them. In this case it creates a connection between the audio flows provided by *donna_listen* and *aliy_tv*. The binding object filters out the unused video component of *aliy_tv*.

Data conversion is also necessary because the audio flow components have different properties. The binding object allocates an interceptor which carries out the conversion from stereo to mono transparently to the end objects.

The supporting infrastructure uses the logical connections to create physical paths across which data can be transferred. Physical paths exist between *Donna*, *Aliy* and the allocated interceptor.

As part of its management role, the video conference object signals to *Donna* and *Aliy* that they can now talk to each other.

5.3 Bill and Bob Scenario

During the conference, a further object *Bill* successfully adds two interfaces of type *videoconf_prod* and *videoconf_cons* to the *listen* and *talk* roles.

The binding object creates connections between *Bob*'s interfaces interface and all instances of the *talk* and *listen* roles. Again the video component of *Bob*'s *videoconf_prod* is filtered out in the connection made to *Donna*. With these connections made each participant is now able to talk and listen to every other participant. The binding object signals *Bob*'s arrival to *Aliy* and *Donna* and notifies *Bob* that he can participate.

Later *Bill* tries to add itself to the binding and is refused. The binding object has a full complement of *talk* and *listen* instances, as defined by the binding cardinality, and is unwilling to support any more.

6 Conclusion

The explicit, compound binding model defined in the ODP reference model can support the wide range of complex behaviour that occur in today's distributed applications. The representation of the quality and nature of communication by an explicit binding object derived from a pre-defined template, provides a powerful descriptive tool which allows such behaviour to be described and modelled.

Multimedia applications have increased the number of different kinds of interactions that can occur in distributed systems. The definition of interactions involving streams, based on continuous flow of information, allows such multimedia applications to be described within the ODP modelling framework.

By capturing much of the behaviour in multimedia applications, the language RIVUS allows the specification of binding templates. At present a language syntax and syntax checker exist for RIVUS. The language is at present being extended to support type inheritance.

Current work is aimed at the construction of an infrastructure in which streams are instantiated directly from a template description in RIVUS. Future developments will introduce type repositories to support this process and will provide common tools for the management of explicit binding objects.

References

1. ISO/IEC: ITU Recommendation X.902 — ISO/IEC 10746-2, Open Distributed Processing - Reference Model - Part 2: Foundations, Technical Report ISO/IEC 10746-2: 1995, International Standards Organization, Central Secretariat , Geneva, Switzerland, 1995.
2. ISO/IEC: ITU Recommendation X.903 — ISO/IEC 10746-3, Open Distributed Processing - Reference Model - Part 3: Architecture, Technical Report ISO/IEC 10746-3: 1995, International Standards Organization, Central Secretariat , Geneva, Switzerland, 1995.
3. P. F. Linington: RM-ODP: The Architecture, Proc. 3rd IFIP TC 6/WG 6.1 International Conference on Open Distributed Processing, eds. K. Raymond and L. Armstrong, Brisbane, Australia, February 1995, pp. 3-14.
4. A. Berry and K. Raymond: The A1 Architecture Model, Proc. 3rd IFIP TC 6/WG 6.1 International Conference on Open Distributed Processing, eds. K. Raymond and L. Armstrong, Brisbane, Australia, February 1995, pp. 55-66.

5. W. Brookes, J. Indulska, A. Bond and Z. Yang: Interoperability of Distributed Platforms: a Compatibility Perspective, Proc. 3rd IFIP TC 6/WG 6.1 International Conference on Open Distributed Processing, eds. K. Raymond and L. Armstrong, Brisbane, Australia, February 1995, pp. 67-78.
6. W. Brookes, A. Berry, A. Bond, J. Indulska and K. Raymond: A Type Model Supporting Interoperability in Open Distributed Systems, Proc. First International Conference on Telecommunications Information Networking Architecture, Melbourne, Australia, February 1995, pp. 275-289.
7. G. Coulson, G. S. Blair, J. B. Stefani, F. Horn and L. Hazard: Supporting the Real-Time Requirements of Continuous Media in Open Distributed Processing, Computer Networks and ISDN systems, 27(8), July 1995, pp. 1231-1246.
8. A. van. Halteren, P. Leydekkers and H. Korte: Specification and Realisation of Stream Interfaces for the TINA-DPE, Proc. First International Conference on Telecommunications Information Networking Architecture, Melbourne, Australia, February 1995, pp. 299-313.
9. P. Leydekkers, V. Gay and L. Franken: A Computational and Engineering View on Open Distributed Real-time Multimedia Exchange, Proc. NOSSDAV'95, Boston, USA, April 1995.
10. B. Liskov and J. M. Wing: A Behavioural Notion of Subtyping, ACM Transactions on Programming Languages and Systems, 16(6), June 1994, pp. 1811-1841.
11. D. Lindsey: RIVUS: A Template Language for Modelling Multimedia Streams, Technical Report 10-95, University of Kent, 1995.

Lecture Notes in Computer Science

For information about Vols. 1–975

please contact your bookseller or Springer-Verlag

Vol. 976: U. Montanari, F. Rossi (Eds.), Principles and Practice of Constraint Programming — CP '95. Proceedings, 1995. XIII, 651 pages. 1995.

Vol. 977: H. Beilner, F. Bause (Eds.), Quantitative Evaluation of Computing and Communication Systems. Proceedings, 1995. X, 415 pages. 1995.

Vol. 978: N. Revell, A M. Tjoa (Eds.), Database and Expert Systems Applications. Proceedings, 1995. XV, 654 pages. 1995.

Vol. 979: P. Spirakis (Ed.), Algorithms — ESA '95. Proceedings, 1995. XII, 598 pages. 1995.

Vol. 980: A. Ferreira, J. Rolim (Eds.), Parallel Algorithms for Irregularly Structured Problems. Proceedings, 1995. IX, 409 pages. 1995.

Vol. 981: I. Wachsmuth, C.-R. Rollinger, W. Brauer (Eds.), KI-95: Advances in Artificial Intelligence. Proceedings, 1995. XII, 269 pages. (Subseries LNAI).

Vol. 982: S. Doaitse Swierstra, M. Hermenegildo (Eds.), Programming Languages: Implementations, Logics and Programs. Proceedings, 1995. XI, 467 pages. 1995.

Vol. 983: A. Mycroft (Ed.), Static Analysis. Proceedings, 1995. VIII, 423 pages. 1995.

Vol. 984: J.-M. Haton, M. Keane, M. Manago (Eds.), Advances in Case-Based Reasoning. Proceedings, 1994. VIII, 307 pages. 1995.

Vol. 985: T. Sellis (Ed.), Rules in Database Systems. Proceedings, 1995. VIII, 373 pages. 1995.

Vol. 986: Henry G. Baker (Ed.), Memory Management. Proceedings, 1995. XII, 417 pages. 1995.

Vol. 987: P.E. Camurati, H. Eveking (Eds.), Correct Hardware Design and Verification Methods. Proceedings, 1995. VIII, 342 pages. 1995.

Vol. 988: A.U. Frank, W. Kuhn (Eds.), Spatial Information Theory. Proceedings, 1995. XIII, 571 pages. 1995.

Vol. 989: W. Schäfer, P. Botella (Eds.), Software Engineering — ESEC '95. Proceedings, 1995. XII, 519 pages. 1995.

Vol. 990: C. Pinto-Ferreira, N.J. Mamede (Eds.), Progress in Artificial Intelligence. Proceedings, 1995. XIV, 487 pages. 1995. (Subseries LNAI).

Vol. 991: J. Wainer, A. Carvalho (Eds.), Advances in Artificial Intelligence. Proceedings, 1995. XII, 342 pages. 1995. (Subseries LNAI).

Vol. 992: M. Gori, G. Soda (Eds.), Topics in Artificial Intelligence. Proceedings, 1995. XII, 451 pages. 1995. (Subseries LNAI).

Vol. 993: T.C. Fogarty (Ed.), Evolutionary Computing. Proceedings, 1995. VIII, 264 pages. 1995.

Vol. 994: M. Hebert, J. Ponce, T. Boult, A. Gross (Eds.), Object Representation in Computer Vision. Proceedings, 1994. VIII, 359 pages. 1995.

Vol. 995: S.M. Müller, W.J. Paul, The Complexity of Simple Computer Architectures. XII, 270 pages. 1995.

Vol. 996: P. Dybjer, B. Nordström, J. Smith (Eds.), Types for Proofs and Programs. Proceedings, 1994. X, 202 pages. 1995.

Vol. 997: K.P. Jantke, T. Shinohara, T. Zeugmann (Eds.), Algorithmic Learning Theory. Proceedings, 1995. XV, 319 pages. 1995.

Vol. 998: A. Clarke, M. Campolargo, N. Karatzas (Eds.), Bringing Telecommunication Services to the People – IS&N '95. Proceedings, 1995. XII, 510 pages. 1995.

Vol. 999: P. Antsaklis, W. Kohn, A. Nerode, S. Sastry (Eds.), Hybrid Systems II. VIII, 569 pages. 1995.

Vol. 1000: J. van Leeuwen (Ed.), Computer Science Today. XIV, 643 pages. 1995.

Vol. 1001: M. Sudan, Efficient Checking of Polynomials and Proofs and the Hardness of Approximation Problems. XIV, 87 pages. 1995.

Vol. 1002: J.J. Kistler, Disconnected Operation in a Distributed File System. XIX, 249 pages. 1995.

VOL. 1003: P. Pandurang Nayak, Automated Modeling of Physical Systems. XXI, 232 pages. 1995. (Subseries LNAI).

Vol. 1004: J. Staples, P. Eades, N. Katoh, A. Moffat (Eds.), Algorithms and Computation. Proceedings, 1995. XV, 440 pages. 1995.

Vol. 1005: J. Estublier (Ed.), Software Configuration Management. Proceedings, 1995. IX, 311 pages. 1995.

Vol. 1006: S. Bhalla (Ed.), Information Systems and Data Management. Proceedings, 1995. IX, 321 pages. 1995.

Vol. 1007: A. Bosselaers, B. Preneel (Eds.), Integrity Primitives for Secure Information Systems. VII, 239 pages. 1995.

Vol. 1008: B. Preneel (Ed.), Fast Software Encryption. Proceedings, 1994. VIII, 367 pages. 1995.

Vol. 1009: M. Broy, S. Jähnichen (Eds.), KORSO: Methods, Languages, and Tools for the Construction of Correct Software. X, 449 pages. 1995. Vol.

Vol. 1010: M. Veloso, A. Aamodt (Eds.), Case-Based Reasoning Research and Development. Proceedings, 1995. X, 576 pages. 1995. (Subseries LNAI).

Vol. 1011: T. Furuhashi (Ed.), Advances in Fuzzy Logic, Neural Networks and Genetic Algorithms. Proceedings, 1994. (Subseries LNAI).

Vol. 1012: M. Bartošek, J. Staudek, J. Wiedermann (Eds.), SOFSEM '95: Theory and Practice of Informatics. Proceedings, 1995. XI, 499 pages. 1995.

Vol. 1013: T.W. Ling, A.O. Mendelzon, L. Vieille (Eds.), Deductive and Object-Oriented Databases. Proceedings, 1995. XIV, 557 pages. 1995.

Vol. 1014: A.P. del Pobil, M.A. Serna, Spatial Representation and Motion Planning. XII, 242 pages. 1995.

Vol. 1015: B. Blumenthal, J. Gornostaev, C. Unger (Eds.), Human-Computer Interaction. Proceedings, 1995. VIII, 203 pages. 1995.

VOL. 1016: R. Cipolla, Active Visual Inference of Surface Shape. XII, 194 pages. 1995.

Vol. 1017: M. Nagl (Ed.), Graph-Theoretic Concepts in Computer Science. Proceedings, 1995. XI, 406 pages. 1995.

Vol. 1018: T.D.C. Little, R. Gusella (Eds.), Network and Operating Systems Support for Digital Audio and Video. Proceedings, 1995. XI, 357 pages. 1995.

Vol. 1019: E. Brinksma, W.R. Cleaveland, K.G. Larsen, T. Margaria, B. Steffen (Eds.), Tools and Algorithms for the Construction and Analysis of Systems. Selected Papers, 1995. VII, 291 pages. 1995.

Vol. 1020: I.D. Watson (Ed.), Progress in Case-Based Reasoning. Proceedings, 1995. VIII, 209 pages. 1995. (Subseries LNAI).

Vol. 1021: M.P. Papazoglou (Ed.), OOER '95: Object-Oriented and Entity-Relationship Modeling. Proceedings, 1995. XVII, 451 pages. 1995.

Vol. 1022: P.H. Hartel, R. Plasmeijer (Eds.), Functional Programming Languages in Education. Proceedings, 1995. X, 309 pages. 1995.

Vol. 1023: K. Kanchanasut, J.-J. Lévy (Eds.), Algorithms, Concurrency and Knowlwdge. Proceedings, 1995. X, 410 pages. 1995.

Vol. 1024: R.T. Chin, H.H.S. Ip, A.C. Naiman, T.-C. Pong (Eds.), Image Analysis Applications and Computer Graphics. Proceedings, 1995. XVI, 533 pages. 1995.

Vol. 1025: C. Boyd (Ed.), Cryptography and Coding. Proceedings, 1995. IX, 291 pages. 1995.

Vol. 1026: P.S. Thiagarajan (Ed.), Foundations of Software Technology and Theoretical Computer Science. Proceedings, 1995. XII, 515 pages. 1995.

Vol. 1027: F.J. Brandenburg (Ed.), Graph Drawing. Proceedings, 1995. XII, 526 pages. 1996.

Vol. 1028: N.R. Adam, Y. Yesha (Eds.), Electronic Commerce. X, 155 pages. 1996.

Vol. 1029: E. Dawson, J. Golić (Eds.), Cryptography: Policy and Algorithms. Proceedings, 1995. XI, 327 pages. 1996.

Vol. 1030: F. Pichler, R. Moreno-Díaz, R. Albrecht (Eds.), Computer Aided Systems Theory - EUROCAST '95. Proceedings, 1995. XII, 539 pages. 1996.

Vol. 1031: M. Toussaint (Ed.), Ada in Europe. Proceedings, 1995. XI, 455 pages. 1996.

Vol. 1032: P. Godefroid, Partial-Order Methods for the Verification of Concurrent Systems. IV, 143 pages. 1996.

Vol. 1033: C.-H. Huang, P. Sadayappan, U. Banerjee, D. Gelernter, A. Nicolau, D. Padua (Eds.), Languages and Compilers for Parallel Computing. Proceedings, 1995. XIII, 597 pages. 1996.

Vol. 1034: G. Kuper, M. Wallace (Eds.), Constraint Databases and Applications. Proceedings, 1995. VII, 185 pages. 1996.

Vol. 1035: S.Z. Li, D.P. Mital, E.K. Teoh, H. Wang (Eds.), Recent Developments in Computer Vision. Proceedings, 1995. XI, 604 pages. 1996.

Vol. 1036: G. Adorni, M. Zock (Eds.), Trends in Natural Language Generation - An Artificial Intelligence Perspective. Proceedings, 1993. IX, 382 pages. 1996. (Subseries LNAI).

Vol. 1037: M. Wooldridge, J.P. Müller, M. Tambe (Eds.), Intelligent Agents II. Proceedings, 1995. XVI, 437 pages. 1996. (Subseries LNAI).

Vol. 1038: W: Van de Velde, J.W. Perram (Eds.), Agents Breaking Away. Proceedings, 1996. XIV, 232 pages. 1996. (Subseries LNAI).

Vol. 1039: D. Gollmann (Ed.), Fast Software Encryption. Proceedings, 1996. X, 219 pages. 1996.

Vol. 1040: S. Wermter, E. Riloff, G. Scheler (Eds.), Connectionist, Statistical, and Symbolic Approaches to Learning for Natural Language Processing. Proceedings, 1995. IX, 468 pages. 1996. (Subseries LNAI).

Vol. 1041: J. Dongarra, K. Madsen, J. Waśniewski (Eds.), Applied Parallel Computing. Proceedings, 1995. XII, 562 pages. 1996.

Vol. 1042: G. Weiß, S. Sen (Eds.), Adaption and Learning in Multi-Agent Systems. Proceedings, 1995. X, 238 pages. 1996. (Subseries LNAI).

Vol. 1043: F. Moller, G. Birtwistle (Eds.), Logics for Concurrency. XI, 266 pages. 1996.

Vol. 1044: B. Plattner (Ed.), Broadband Communications. Proceedings, 1996. XIV, 359 pages. 1996.

Vol. 1045: B. Butscher, E. Moeller, H. Pusch (Eds.), Interactive Distributed Multimedia Systems and Services. Proceedings, 1996. XI, 333 pages. 1996.

Vol. 1046: C. Puech, R. Reischuk (Eds.), STACS 96. Proceedings, 1996. XII, 690 pages. 1996.

Vol. 1047: E. Hajnicz, Time Structures. IX, 244 pages. 1996. (Subseries LNAI).

Vol. 1048: M. Proietti (Ed.), Logic Program Syynthesis and Transformation. Proceedings, 1995. X, 267 pages. 1996.

Vol. 1049: K. Futatsugi, S. Matsuoka (Eds.), Object Technologies for Advanced Software. Proceedings, 1996. X, 309 pages. 1996.

Vol. 1050: R. Dyckhoff, H. Herre, P. Schroeder-Heister (Eds.), Extensions of Logic Programming. Proceedings, 1996. VII, 318 pages. 1996. (Subseries LNAI).

Vol. 1051: M.-C. Gaudel, J. Woodcock (Eds.), FME'96: Industrial Benefit and Advances in Formal Methods. Proceedings, 1996. XII, 704 pages. 1996.

Vol. 1052: D. Hutchison, H. Christiansen, G. Coulson, A. Danthine (Eds.), Teleservices and Multimedia Communications. Proceedings, 1995. XII, 277 pages. 1996.